Boundary Element Technology VI

SIXTH INTERNATIONAL CONFERENCE
ON
BOUNDARY ELEMENT TECHNOLOGY
BETECH/91

SCIENTIFIC COMMITTEE

The illustration on the front cover of this book can be found on page 198.

Boundary Element Technology VI

Editor: C.A. Brebbia, Wessex Institute of Technology, U.K.

Computational Mechanics Publications
Southampton Boston

CMP

Co-published with

Elsevier Applied Science
London New York

ELSEVIER

C.A.Brebbia
Computational Mechanics Institute
Wessex Institute of Technology
Ashurst Lodge
Ashurst
Southampton SO4 2AA
U.K.

Co-published by

Computational Mechanics Publications
Ashurst Lodge, Ashurst, Southampton, UK

Computational Mechanics Publications Ltd
Sole Distributor in the USA and Canada:

Computational Mechanics Inc.
25 Bridge Street, Billerica, MA 01821, USA

and

Elsevier Science Publishers Ltd
Crown House, Linton Road, Barking, Essex IG11 8JU, UK

Elsevier's Sole Distributor in the USA and Canada:

Elsevier Science Publishing Company Inc.
655 Avenue of the Americas, New York, NY 10010, USA

British Library Cataloguing in Publication Data
Boundary element technology
1. Engineering. Mathematics
I. Brebbia, C.A. (Carlos Alberto)
620.001515353

ISBN 1-85312-139-8

ISBN 1-85166-668-0 Elsevier Applied Science, London, New York
ISBN 1-85312-139-8 Computational Mechanics Publications, Southampton
ISBN 1-56252-067-9 Computational Mechanics Publications, Boston, USA

Library of Congress Catalog Card Number 91-71739

Printed and bound by Bell and Bain Ltd, Glasgow

PREFACE

This book presents a series of papers dealing mainly with the applications of the boundary element method rather than its more theoretical aspects. The contributions are all written by scientists and engineers working on the latest developments of the technique and its applications to solve engineering problems.

The volume presents the state of the art in boundary elements in fields such as fluid flow, electromagnetics and electrical engineering, hydrodynamics, heat conduction, different potential type problems, stress analysis, fracture mechanics and some of the latest developments in numerical and mathematical techniques which are essential to write efficient boundary element systems.

The book is the edited version of the Proceedings of the 6th International Boundary Element Technology Conference held at Southampton in June 1991 and organized by the Wessex Institute of Technology.

The Boundary Element Technology series of conferences (BETECH) is now well established as the pre-eminent forum for the presentation of application oriented work in Boundary Elements and associated computational techniques. From its inception in 1985 in Australia the BETECH conferences have documented the increasing application of the Method and the Proceedings have become a standard reference for the engineering community. The BETECH conferences have always had an international emphasis with venues in Adelaide, Australia 1985, M.I.T., USA 1986, Rio de Janeiro, Brazil 1987, Windsor, Canada 1989, Delaware, USA 1990, and Southampton, UK 1991.

The editor is grateful for the support received by the members of the Scientific Advisory Committee, i.e. C. Alessandri (University of Florence, Italy), M.H. Aliabadi (WIT, UK), A. Cheng (University of Delaware, USA), J.J. Connor (M.I.T., USA), S. Grilli (University of Rhode Island, USA), T. Honma (Hokkaido University, Japan), M. Ingber (University of New Mexico, USA), C. Kuhn (University of Erlangen-Nurnberg, Germany), A. Miranda (CETIM, France), R. Rangogni (ENEL, Italy), R.N.L. Smith (Royal Military College of Science, UK), G.T. Symm (National Physical Laboratory, UK), M. Tanaka (Shinshu University, Japan), J.L. Wearing (University of Sheffield, UK), L.C. Wrobel (WIT, UK), and N.G. Zamani (University of Windsor, Canada).

Carlos A Brebbia
Southampton, June 1991

CONTENTS

SECTION 1: FLUID FLOW

Numerical Investigation of Fluid Flow with Natural Convection due to Non-linear Material Behavior and Double-Diffusion Effects — 3
I. Zagar, P. Skerget, A. Alujevic

High Reynolds Number Flow about a Prolate Spheroid Moving near Ground — 13
H. Schmitt

Dual Reciprocity BIE in Lagrangian Form for Incompressible Unsteady Viscous Flow — 29
X. Jin, D.K. Brown

Forced Convection Cooling of Flat Electronic Substrates — 39
G. De Mey, S. De Smet, M. Driscart

An Analysis of the Kutta-Joukovski-Carafoli Condition at Axial Profile Cascades using the Boundary Element Method — 51
I.N. Carte

SECTION 2: ELECTRICAL AND ELECTROMAGNETIC PROBLEMS

Boundary Element Analysis of Electromagnetic Fields in Helically Symmetric System — 65
H. Igarashi, T. Honma

A Subdomain Method for FEM-BEM Coupling in Electromagnetism — 77
F. Delincé, P. Dular, A. Genon, H. Hedia, W. Legros, A. Nicolet

Modelization of Movement in Electromagnetic Devices with the Boundary Element Method — 89
N. Bamps, F. Delincé, P. Dular, A. Genon, W. Legros, A. Nicolet

Comparison of Two Different BEM Formulations for Nonlinear Eddy Currents — 101
A. Kost

Procat System: Recent Advances and Future Developments in Numerical Simulation of Cathodic Protection Problems 117
W.J. Mansur, J.C.F. Telles, J.A.F. Santiago

SECTION 3: POTENTIAL TYPE PROBLEMS

Transient Saltwater Intrusion in Confined Aquifers using a B-Spline Boundary Element Formulation 135
J.J.S.P. Cabral, L.C. Wrobel

Sound Radiation of a Cantilever Plate - Comparisons of BEM Results with Experimental Data 147
O. von Estorff, A. Homm, F. Bartels

SECTION 4: STRESS ANALYSIS

Anisotropic Effective Stress Soil Behaviour by BEM 161
F. Lin, L.A. Wood

The Use of a Foundation Element in a Three-Dimensional Structural Analysis Code 173
M.A. Ramalho, W.S. Venturini

On the Use of Geometrical Symmetry in the Boundary Element Methods for 3D Elasticity 185
M. Bonnet

The Multiple Reciprocity Method Applied to Thermoelastic Problems with Concentrated and Distributed Heat Sources 201
A.C. Neves, C.A. Brebbia

Solving Plate Bending Problems using Boundary Elements Method 215
L.M. Castro, J. Pinho, P. Parreira

Kernel Functions of Mindlin's Solution for Evaluation of Internal Stresses 227
Y.S. Jiang

Boundary Element Formulation for Non-linear Problems using the Reissner's Theory 239
G.O. Ribeiro, W.S. Venturini

Contact Analysis using BEASY: Theory and Application 253
S.M. Niku, R.A. Adey, J. Baynham

Transient Dynamic Elastoplastic Analysis by the Boundary Element Method 265
J.A.M. Carrer, J.C.F. Telles

A General Adjoint Approach to Shape Design Sensitivity Analysis 279
J. Unzueta, E. Schaeidt, A. Longo, J.J. Anza

SECTION 5: NUMERICAL AND MATHEMATICAL TECHNIQUES

A new B.E. Formulation and its Applications in Engineering 295
C.A. Brebbia, T. DeFigueiredo

Boundary Elements for Square-Root Singularities on Curves 309
J.C. Mason

Improving the Accuracy of BEM in the use of Non-equidistant Elements 321
M. Karamanoglu, G.E. Beswick

High Precision Numerical Integration Method for 3-D BEM and its Error Analysis using Complex Function Theory 335
K. Hayami

Penalty-Combined Boundary Methods and Applications in Solving Elliptic Problems with Singularities 339
Z.C. Li, T.D. Bui

SECTION 6: FRACTURE MECHANICS

Fracture Diagrams for Short Cracks at a Stress Concentration 357
R.D. Stark, D.J. Cartwright

Cracks Grow in Cement Structures using Laplace Transform and Boundary Element Method 371
M. Kassem

Dual Boundary Element Analysis of Pin-Loaded Lugs 381
A. Portela, M.H. Aliabadi, D.P. Rooke

Three-Dimensional Weight Functions for Circular Cracks in Cylindrical Bars 393
R. Bains, M.H. Aliabadi, D.P. Rooke

Stress Analysis of Cyclic Symmetric Structures by the New Type Boundary Integral Equation 407
J.F. Xie, Y.L. Wu

SECTION 1: FLUID FLOW

Numerical Investigation of Fluid Flow with Natural Convection due to Non-linear Material Behavior and Double-Diffusion Effects

I. Zagar, P. Skerget, A. Alujevic

Faculty of Engineering, University of Maribor, Slovenia

ABSTRACT

The paper reports numerical simulation results of natural convection when the flow is driven by density variations due to combined heat and mass transfer. The Boussinesq approximation is used to take into account the temperature and concentration dependent density. Non-steady state solutions are performed for various combinations of boundary conditions and for different values of the thermal and mass transfer Grasshoff numbers. The influence of the fluid density inversion on the flow structure is also studied. Boundary-domain integral method has been applied.

INTRODUCTION

The transport phenomenon of natural convection in general fluid flow is included into governing equations through the Boussinesq approximation. In this paper two different problems of natural convection in fluids are studied.

In the first problem, natural convection motion due to simultaneous existence of temperature and concentration gradients in fluids is considered. A practical example of such complex phenomenon is a storage tank of liquified natural gas. Fluid properties of liquified natural gas ($Pr = 2.2$; $Sc = 130$; $Gr_t = 2500$; $Gr_m = -1000$) are taken into account in the numerical example of a closed cavity subject to different kinds of boundary condition.

In the second problem, the nonlinear Boussinesq approximation was considered. Neglecting the nonlinear term in the Boussinesq approximation for the same fluids may lead to a completely wrong picture of the real situation. Example of such fluid is water in the range of 0^oC to 8^oC. Density of water is increasing to 4^oC and

then decreasing. Effect of such anomaly of water density is evident from the development of the temperature and velocity field.
The boundary-domain integral method is used for the numerical simulation of the time-dependent set of governing equations in velocity-vorticity formulations. The numerical results agree with those obtained from experiments [1] and calculations of other authors[2],[5].

GOVERNING EQUATIONS

The partial differential equations set, governing the motion of viscous incompressible fluid is known as nonlinear Navier-Stokes equations expressing the basic conservation balances of mass, momentum, energy and mass concentration. Introducing vorticity ω and stream function ψ of the solenoidal velocity field, the computation of the flow is divided into kinematics given by the Poisson's elliptic equation, written in plane $x-y$ by

$$\frac{\partial^2\psi}{\partial x^2}+\frac{\partial^2\psi}{\partial y^2}+\omega=0 \tag{1}$$

and into the kinetics described by the vorticity equation

$$\frac{\partial\omega}{\partial t}+v_x\frac{\partial\omega}{\partial x}+v_y\frac{\partial\omega}{\partial y}=\nu\left(\frac{\partial^2\omega}{\partial x^2}+\frac{\partial^2\omega}{\partial y^2}\right)+g_y\frac{\partial F}{\partial x}-g_x\frac{\partial F}{\partial y} \tag{2}$$

The buoyancy effect is included by energy and molar fraction equations

$$\frac{\partial T}{\partial t}+v_x\frac{\partial T}{\partial x}+v_y\frac{\partial T}{\partial y}=a\left(\frac{\partial^2 T}{\partial x^2}+\frac{\partial^2 T}{\partial y^2}\right) \tag{3}$$

$$\frac{\partial C}{\partial t}+v_x\frac{\partial C}{\partial x}+v_y\frac{\partial C}{\partial y}=D\left(\frac{\partial^2 C}{\partial x^2}+\frac{\partial^2 C}{\partial y^2}\right) \tag{4}$$

and Boussinesq approximation, given for linear normalised difference of density

$$\frac{\rho-\rho_o}{\rho_o}=F=-\beta_t\,(T-T_o)-\beta_m\,(C-C_o) \tag{5}$$

For water in the region of anomaly from $0^o\ C$ to $8^o\ C$ the normalized difference of density is given [5]

$$\frac{\rho-\rho_o}{\rho_o}=F=(0.066576\,T-0.008322\,T^2)/\rho_o \tag{6}$$

BOUDARY-DOMAIN INTEGRAL EQUATIONS

The boundary domain integral statement for the flow kinematics can be derived from the vector elliptic equation for vector potential ψ_i [3] applying Green's theorem for the vector functions and the elliptic fundamental solution u^*, resulting in the following statement

$$c(\xi)v_x(\xi) + \int_\Gamma v_x \frac{\partial u^*}{\partial n}\, d\Gamma = \int_\Gamma v_y \frac{\partial u^*}{\partial t}\, d\Gamma - \int_\Omega \omega \frac{\partial u^*}{\partial y} d\Omega \tag{7}$$

$$c(\xi)v_y(\xi) + \int_\Gamma v_y \frac{\partial u^*}{\partial n}\, d\Gamma = -\int_\Gamma v_x \frac{\partial u^*}{\partial t}\, d\Gamma + \int_\Omega \omega \frac{\partial u^*}{\partial x} d\Omega \tag{8}$$

Describing the laminar transport of the vorticity, temperature and molar fraction in the integral statement, one has to take into account that the vorticity, temperature and mass concentration obey a nonhomogeneous parabolic equation[3],[6]. So the following boundary-domain integral formulations can be derived for the plane

$$c(\xi)\omega(\xi, t_F) + \nu \int_\Gamma \int_{t_{F-1}}^{t_F} \omega \frac{\partial u^*}{\partial n}\, dt\, d\Gamma = \nu \int_\Gamma \int_{t_{F-1}}^{t_F} \frac{\partial \omega}{\partial n} u^*\, dt\, d\Gamma$$

$$- \int_\Gamma \int_{t_{F-1}}^{t_F} (\omega v_n + g_t\, F) u^*\, dt d\Gamma$$

$$+ \int_\Omega \int_{t_{F-1}}^{t_F} (\omega\, v_x + g_y\, F,\ \omega\, v_y - g_x\, F)\, (\frac{\partial u^*}{\partial x}, \frac{\partial u^*}{\partial y})\, dt d\Omega + \int_\Omega \omega_{F-1}\, u^*{}_{F-1}\, d\Omega \tag{9}$$

$$c(\xi)T(\xi, t_F) + a \int_\Gamma \int_{t_{F-1}}^{t_F} T \frac{\partial u^*}{\partial n}\, dt\, d\Gamma = a \int_\Gamma \int_{t_{F-1}}^{t_F} \frac{\partial T}{\partial n} u^*\, dt\, d\Gamma$$

$$- \int_\Gamma \int_{t_{F-1}}^{t_F} T\, v_n u^*\, dt\, d\Gamma$$

$$- \int_\Omega \int_{t_{F-1}}^{t_F} (T\, v_x,\ T\, v_y)\, (\frac{\partial u^*}{\partial x}, \frac{\partial u^*}{\partial y})\, dt\, d\Omega + \int_\Omega T_{F-1}\, u^*{}_{F-1}\, d\Omega \tag{10}$$

$$c(\xi)C(\xi, t_F) + D \int_\Gamma \int_{t_{F-1}}^{t_F} C \frac{\partial u^*}{\partial n}\, dt\, d\Gamma = D \int_\Gamma \int_{t_{F-1}}^{t_F} \frac{\partial C}{\partial n} u^*\, dt\, d\Gamma$$

$$- \int_\Gamma \int_{t_{F-1}}^{t_F} C\, v_n u^*\, dt\, d\Gamma$$

$$- \int_\Omega \int_{t_{F-1}}^{t_F} (C\, v_x,\ C\, v_y)\, (\frac{\partial u^*}{\partial x}, \frac{\partial u^*}{\partial y})\, dt\, d\Omega + \int_\Omega C_{F-1}\, u^*{}_{F-1}\, d\Omega \tag{11}$$

NUMERICAL RESULTS

Water anomaly in a closed cavity

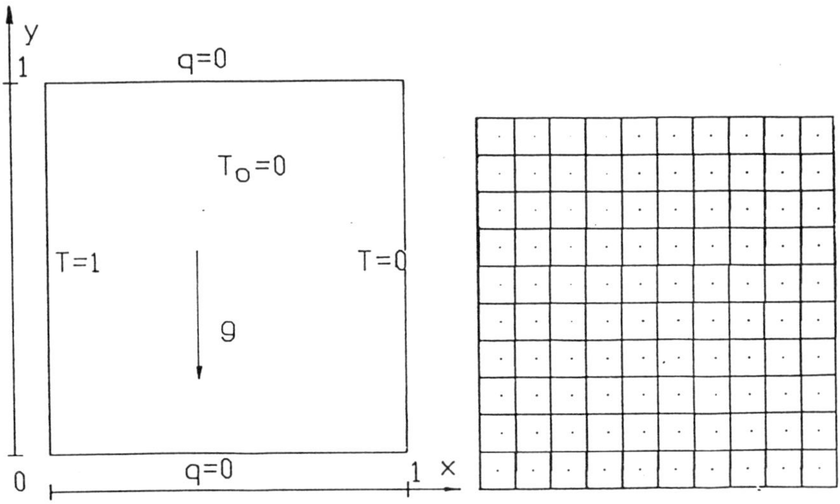

Fig. 1: Problem geometry and boundary conditions

Geometry and boundary condition for buoyancy driven laminar flow in a closed square cavity, filled by water exhibiting an anomaly of density is shown on Fig.1. A mesh of 40 elements (80 nodes) and 100 internal cells has been used.A model was taken as in [4] where linear Boussinesq approximation for air has been used. [1],[5].

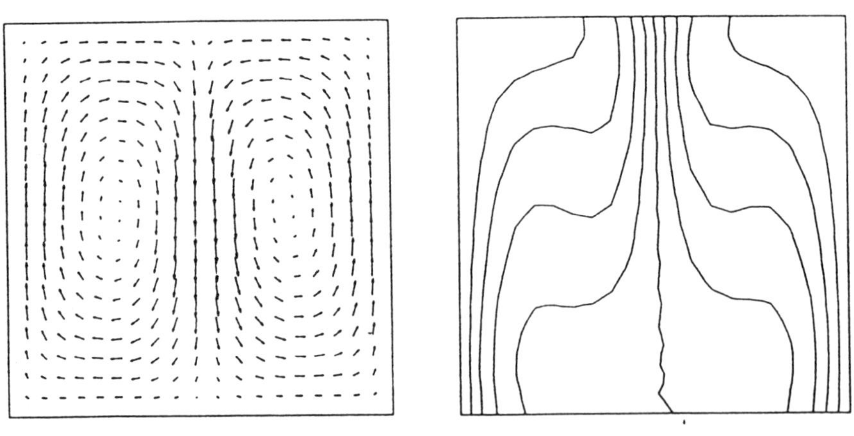

Fig. 2: Steady state results $Ra = 10^5$

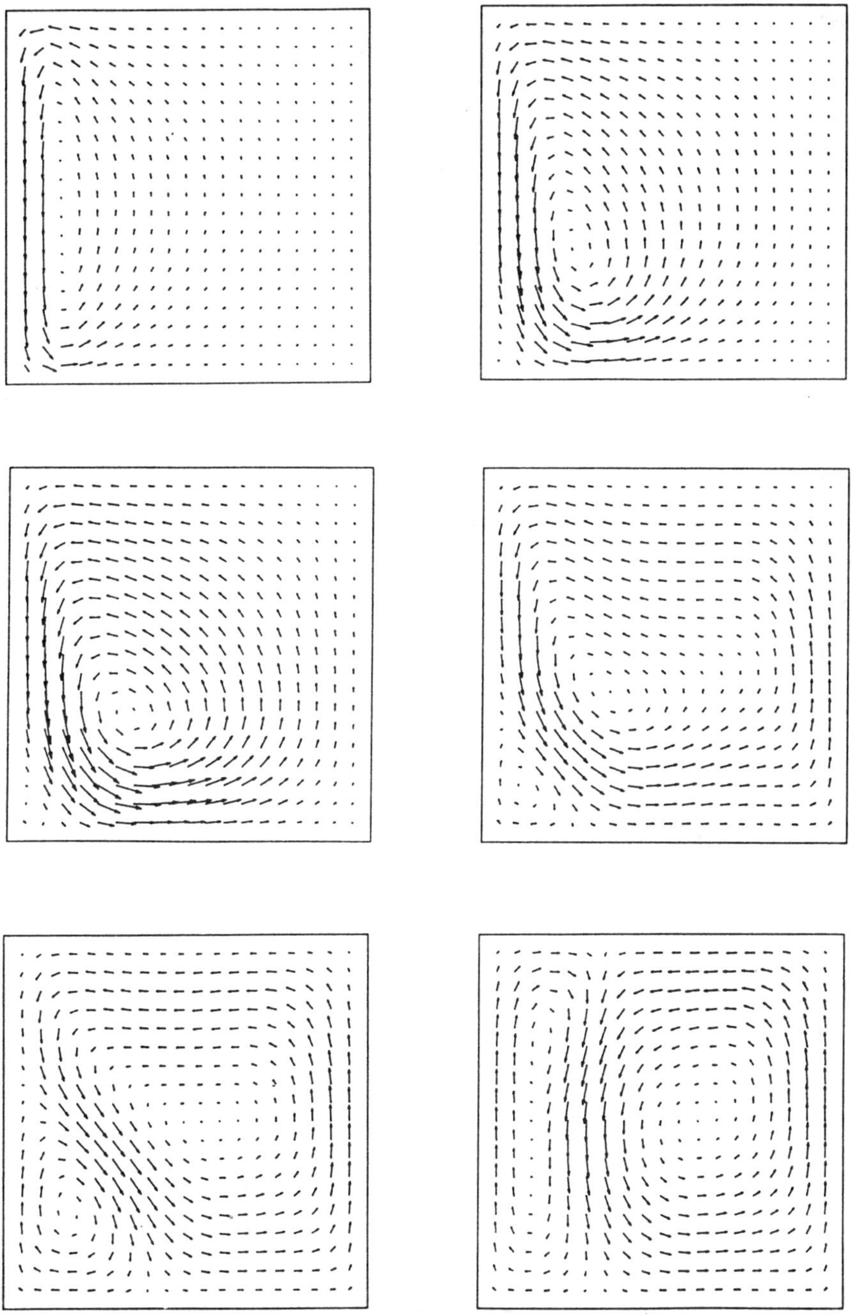

Fig. 3: Velocity field development, $Ra = 10^5$, $t = 0.2$, 0.8, 1.2, 3.2, 6.4 and $40s$

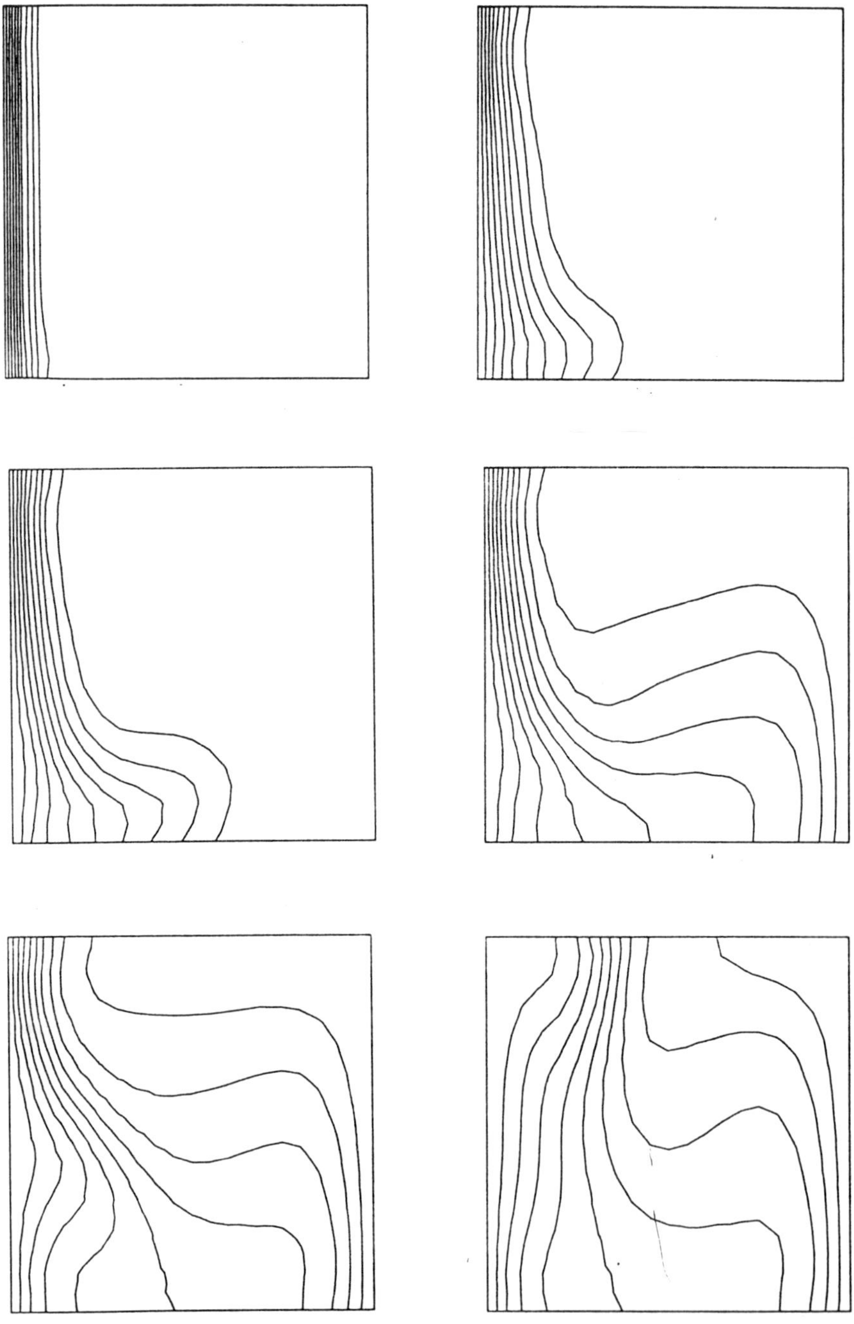

Fig. 4: Temperature field development $Ra = 10^5$, $t = 0.2, 0.8, 1.2, 3.2, 6.4$ and $40s$

Comparison of velocity (Fig. 3) and temperature field (Fig. 4) shows evident difference in formation and development of water transients whith those obtained for air [4] in same range. Figure 4 gives the final steady state of both velocity and temperature distributions. There are two separated circulation zones observed, while in the middle a symmetry separation line developed.

Double-diffusion in closed cavity

On the same mesh as in previously case and similar boundary conditions exept for vertical walls, where linear variation of temperature and molar fraction is imposed

$$X = 0,1 \;\; T = 1 - Y \quad C = 1 - Y$$

problem of double-diffusion was studied. The linear variation of molar fraction can be interpreted as diffusion of the component through the wall. Fluid was considered for which body forces due to heat and mass transfer are opposed, $Ra_t = 5.5\,10^3$ ($Gr_t = 2500$, $Pr = 2.2$) and $Ra_m = -1.3\,10^5$ ($Gr_m = -1000$, $Sc = 130$) [2]. Figure 5,6,7 show development of velocity, temperature and molar fraction.

REFERENCES

[1] Inaba,H., Fukuda,T.:*Natural Convection in an Inclined Square Cavity in Regions of Density Inversion of Water.* J. Fluid Mech. 142, pp.363-381, 1984.

[2] Seveleder,V., Petit,J.P.:Flow Structures Induced by Opposing in Double-Diffusive Natural Convection in a Cavity. Numerical Heat Transfer, Part A, Vol.15, pp.431-444, 1989.

[3] Skerget,P., Alujevic,A., Zagar,I., Brebbia,C.A., Kuhn,G.:*Time Dependent Three Dimensional Laminar Isochoric Viscous Fluid Flow by BEM*, 10th Int. Conf. on BEM, Southampton, Springer-Verlag, Berlin, 1988.

[4] Skerget,P., Alujevic,A., Brebbia,C.A., Kuhn,G.:*Natural and Forced Convection Simulation Using the Velocity-Vorticity Approach*, Topics in Boundary Element Research (Ed. by Brebbia C.A.), Vol.5, Ch.4, Springer-Verlag , Berlin, 1989.

[5] Vasič,S., Hanjalič,K.:*Free Convection in Fluid with Density Anomaly* (in srbocroate) YU TERM 90, pp.219-225, Neum, 1990.

[6] Zagar,I., Skerget,P., Alujevic, A.:*Boundary-Domain Integral Method for the Space Time Dependent Viscous Incompressible Flow.* Symposium of the International association for Boundary Element Method. IABEM-90, Roma, Italy, 1990.

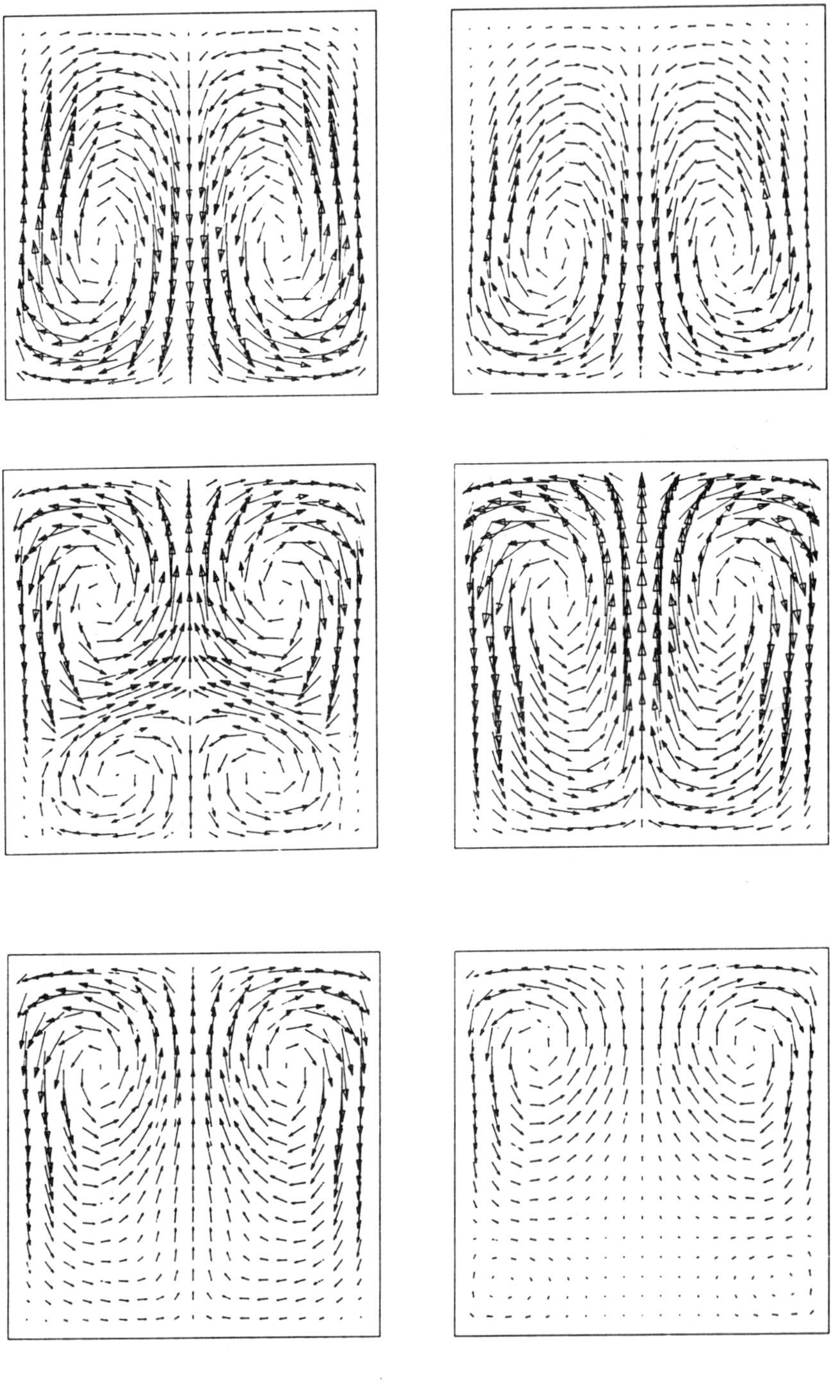

Fig. 5: Velocity field development, $t = 4, 8$, 19.2, 57.6, 124.8., 240 and 480s

Fig. 6: Temperature field development, $t = 4, 8$, 19.2, 57.6, 124.8., 240 and 480s

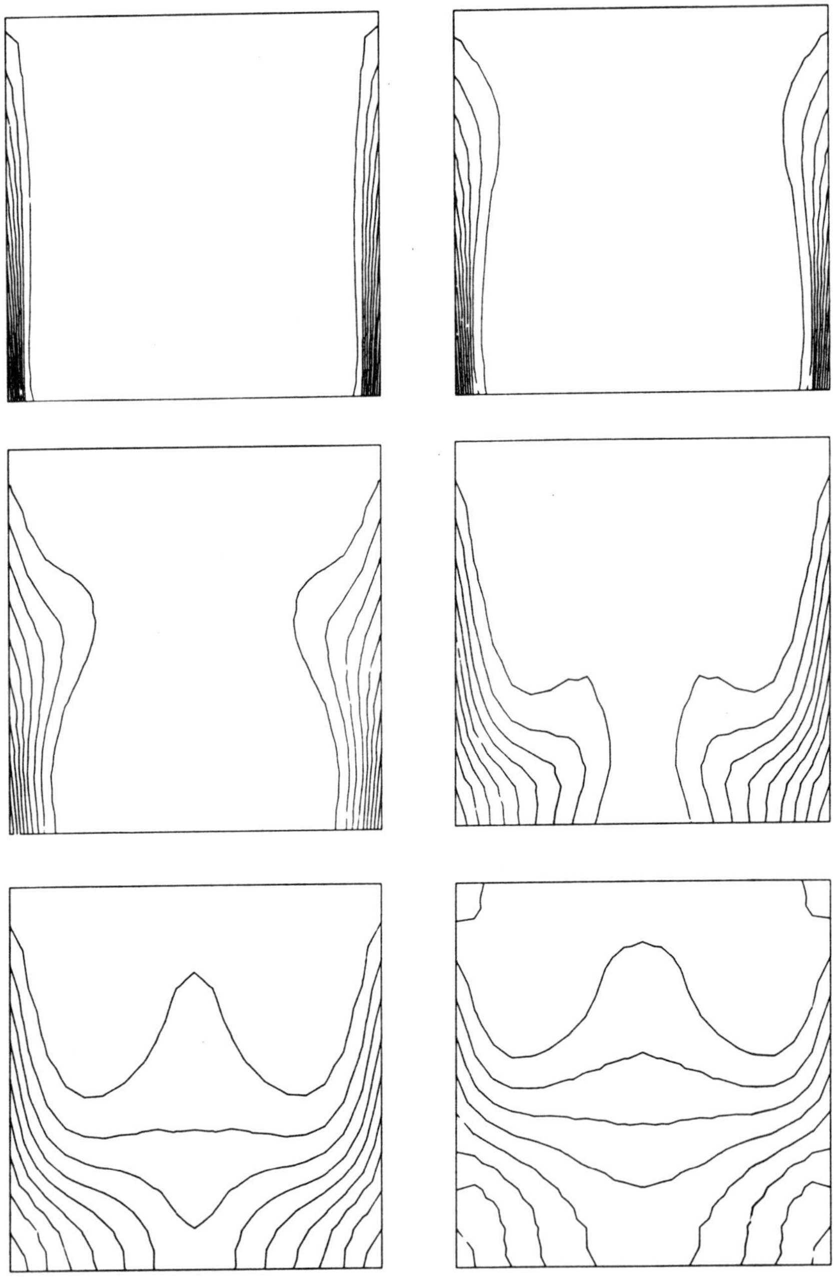

Fig. 7: Molar fraction field development, $t = 4, 8$, 19.2, 57.6, 124.8., 240 and 480s

High Reynolds Number Flow about a Prolate Spheroid Moving near Ground

H. Schmitt

Institute of Theoretical Fluid Mechanics, Deutsche Forschungsanstalt für Luft- und Raumfahrt (DLR), D-3400 Göttingen, Germany

ABSTRACT

The three-dimensional steady flow of an incompressible fluid about a prolate spheroid moving very near a plane wall is studied. The Reynolds number is assumed to be high enough to allow a flow with an inviscid core between the spheroid and the wall. The accomplishment of this condition is verified by an estimation of the thickness of the boundary layer. The potential flow is obtained by means of superposition of potentials of sources distributed on the surface of the spheroid and of source doublets situated on the vortex sheet originating from the spheroid. The distance between the spheroid and the wall as well as the angle which it forms with the wall are varied in order to study their effect on the pressure distribution. Further, the resulting pressure distribution is compared with experimental values of other authors.

1. INTRODUCTION

A prolate spheroid causes a displacement of the upstream uniform flow. On the assumption that this is the only effect, the inviscid flow around the spheroid is described by the analytic solution of Heine (see Lamb [8]). This can be approximated with the boundary element method of Hess and Smith [5,6]. The quality of the approximation rises with the number of boundary elements (see Schmitt [12]). If the spheroid is inclined with an angle of incidence against the upstream flow, a vortex sheet separates from it, which causes a lift. It has been shown (see Schmitt [12]), that the regard of this vortex sheet improves substantially the agreement with the pressure distribution on the surface measured by Meier und Kreplin [9].

For a body moving near a plane wall, the usual method of potential theory is to consider also the image of the body with respect to this wall. This has been performed in a previous paper (see Schmitt and Schneider [15]) for one arrangement of the spheroid and without regarding the vortex sheet. Its influence will be incorporated in the following calculations.

If gradually smaller distances between the spheroid and the wall are considered, the relation of the thickness of the boundary layers on their surfaces to the width of the inviscid flow region in the slot grows. Therefore, the assumption of such a region must be justified. This will be done with use of the estimation of the boundary layers given in the third section.

2. POTENTIAL FLOW

The three-dimensional steady flow of an incompressible fluid about a prolate spheroid moving near an infinite plane wall is studied. The geometry of the flow is shown in figure 1. The axis x of revolution forms the angle α of incidence with the wall. At infinity, the uniform flow velocity V_∞ is parallel to the wall and parallel to the projection of the axis of revolution on the wall. A system of Cartesian coordinates x, y, z with the origin in the center of the spheroid is used. The distance h between the spheroid and the wall is defined as the distance between the center of the spheroid and the wall. All lengths are related to the major axis a of the spheroid. Therefore, this axis has the length 1. The length of the minor axis is given as 1/6, because only this ratio of the two axes shall be studied in the following examples.

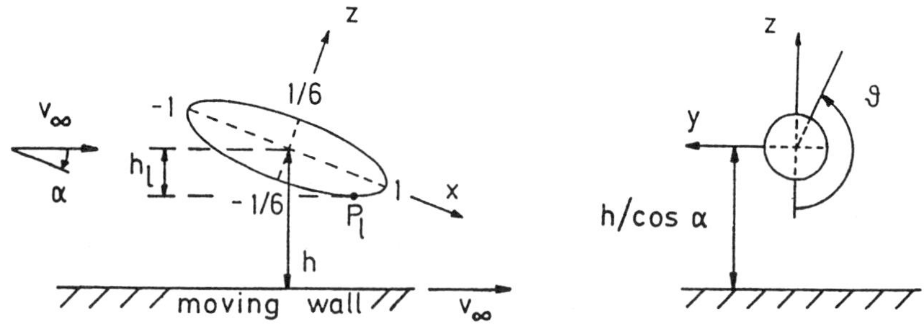

Fig. 1. Geometry of the flow

In the following, only flows with large Reynolds numbers $Re = aV_\infty/\nu$ are considered; here, ν is the kinematic viscosity; then mathematically, the problem is a singular perturbation problem. The zeroth order outer solution is that of the inviscid flow; the first order inner solution is given by Prandtl's boundary layer equations.

For the inviscid flow, when the velocity is irrotational, the disturbance velocity potential φ can be introduced:

$$\mathbf{v} = \mathbf{i} + grad\ \varphi \ . \qquad (1)$$

Here, the quantity $\mathbf{v}$ is the relation of the velocity to V_∞. $\mathbf{i}$ is a unity vector in the direction of the flow at infinity. With this, the mass-conservation equation becomes (see for example [14]) :

$$\Delta\ \varphi = 0\ . \qquad (2)$$

In the limit $Re \to \infty$, the wake region of the body degenerates into a vortex sheet originating from the body; this sheet is a stream surface, across which the pressure is continuous. Therefore, the boundary conditions are (see [14]):

$$\mathbf{v} \cdot \mathbf{n} = 0 \qquad \textit{on the surfaces of the spheroid and of the wall and on the vortex sheet}\ , \qquad (3)$$

$$p_+ = p_- \qquad \textit{across the vortex sheet} \ , \tag{4}$$

$$\lim_{|\mathbf{x}| \to \infty} \ |grad\ \varphi| = 0 \ . \tag{5}$$

Here, **n** is the unity vector in the direction of the outward normal, and the sides of the vortex sheet are denoted by + and -, respectively. p is the deviation of the pressure from the value p_∞ at infinity related to ρV_∞^2, where ρ is the density. **x** is the nondimensional position vector.

The potential flow can be calculated by means of superposition of elementary solutions of Laplace's equation (2). In the following, the potentials of sources on the surface S of the spheroid are used, which describe the displacement effect of the spheroid on the uniform flow. Further, the potentials of source doublets on the vortex sheet S_V are used, which is produced by the spheroid; these solutions represent the modification of the flow by the lift acting on the spheroid:

$$\varphi(\mathbf{x}) = -\int_S \frac{\sigma(\mathbf{q})}{r(\mathbf{x},\mathbf{q})}\, dS + \int_{S_V} \mu(\mathbf{q})\, \frac{\partial}{\partial n_V} \left(\frac{1}{r(\mathbf{x},\mathbf{q})} \right) dS \ . \tag{6}$$

In this equation, the strength σ of the sources and the strength μ of the source doublets depend on the position **q** on the surface. $r(\mathbf{x},\mathbf{q})$ is the distance between the reference point **x** and the source point **q** . $\partial/\partial n_V$ is the derivative in the direction of the source doublet, which is directed perpendicular to the vortex sheet.

The strengths of the sources and the source doublets are determined by putting (6) into the boundary condition (3). In the special case, in which no source doublets are needed, the following Fredholm integral equation of the second kind results for a point **p** on the surface of the spheroid:

$$2\pi\ \sigma(\mathbf{p}) - \int_S \left(\frac{\partial}{\partial n_p}\ \frac{1}{r(\mathbf{p},\mathbf{q})} \right) \sigma(\mathbf{q})\ dS = -\ \mathbf{n}_p \cdot \mathbf{i} \ . \tag{7}$$

Here, $\mathbf{n}_p$ is the outward normal of the surface in the point **p** .

In order to solve (7) numerically, one half of the surface ($y \leq 0$) of the spheroid is approximated by a large number of boundary elements, the panels (540 in the following examples). On every boundary element, the source density is assumed to be constant. The details of the method are given in [12,14], which are based on the works of Hess [4] and Hess and Smith [5,6].

In the case, in which source doublets are used in addition to the sources in equation (6), also the vortex sheet S_V is approximated by boundary elements, and the source doublet is assumed to be constant on every element. Again, the details of the distribution of the boundary elements and the approximation are given in [12,14]. In the following examples, 9 longitudinal stripes of elements are used on the half sheet $y \leq 0$. In this case, the numerical method is based on the studies of Kraus and Sacher [7], who calculated the flow around an airfoil.

The plane $z = 0$ is taken as an approximate position of the vortex sheet. This position is suggested by experimental results for a spheroid, which has the thickness, i.e. the ratio of the two axes of the spheroid, of the small value 1/6. Further, a theory, in which a linearisation in this parameter is introduced, leads to the position $z = 0$.

The boundary condition of vanishing normal velocity at the plane wall is fullfilled by treating the flow also about the images of the spheroid and of the vortex sheet with respect to this wall. The surfaces of these images are approximated by boundary elements in the same way as the originals.

3. ESTIMATE OF THE THICKNESS OF THE BOUNDARY LAYER

For finite Reynolds numbers, the viscous boundary layers that exist on the different solid surfaces in the configuration must be taken into account. On the spheroid, the boundary layer begins in the front stagnation point of the spheroid (see Schlichting [11] or [14]).

Considering the boundary layer at the plane wall, it must be taken into account, that in the system of coordinates x, y, z fixed at the spheroid, the wall moves with the upstream velocity. Therefore, at infinity no boundary layer does exist at the wall. In the neighborhood of the spheroid, the velocity is different from that of the upstream flow, and thus a continuously increasing boundary layer forms at the wall; the beginning of this layer is not known a priori.

The boundary layers can be calculated with finite difference methods (see for instance [14]). Instead of this, an estimate of their thickness shall be given in the following. If the distance between the spheroid and the wall is large enough that the thickness of the potential flow in the slot is large in comparison to the thickness of each of the two boundary layers, a confinement to the calculation of the potential flow following the last section is justified. Thus, only the orders of magnitude of the thicknesses of the boundary layers are of interest. These are estimated using the results for a flat plate given in the following.

If the flow is laminar, the nondimensional thickness δ of the boundary layer along a flat plate is related to the nondimensional distance s from its beginning by (see for example Schlichting [11, p.140]):

$$\delta = \frac{5}{\sqrt{Re}} \sqrt{s} \ . \tag{8}$$

For high Reynolds numbers depending on initial and boundary conditions, the flow can become turbulent. The thickness of the turbulent boundary layer along a smooth flat plate is specified by Schlichting [11, p.638] as:

$$\delta = \frac{0.37}{Re^{\frac{1}{5}}} \, s^{\frac{4}{5}} \ . \tag{9}$$

Rotta [10, p.188] gives a more complex relationship including also the laminar beginning. From his graph, at $s = 1$ the displacement thickness $\delta_1 = 2.6 \cdot 10^3 / Re$ can be inferred for $Re = 1.2 \cdot 10^6$, and therefore, $\delta_1 = 0.00217$. If the simple 1/9-th-power law of the velocity distribution is assumed for $Re = 1.2 \cdot 10^6$, $\delta_1 / \delta = 1/10$ (see Schlichting [11, p.638]) and thus $\delta = 0.0217$ at $s = 1$. Since equation (9) gives $\delta = 0.0225$ at this distance, the relationship of Rotta gives the same order of magnitude.

These results for the thickness of the boundary layer adjacent to a flat plate are used as an approximation for the thickness of that adjacent to the spheroid. Here, the current distance s begins in the front stagnation point, which lies closer to the ground than the nose. The relation of δ to the thickness of the potential flow in the slot between the spheroid and the ground has its highest value at the lowest point P_l with the coordinates

$$x_l = \frac{6\, tg\alpha}{\sqrt{1+36\, tg^2\alpha}}\,, \quad y_l = 0, \quad z_l = -\frac{1}{6}\,\frac{1}{\sqrt{1+36\, tg^2\alpha}}\,, \tag{10}$$

and P_l has the distance $h - h_l$ from the ground, where

$$h_l = \frac{\cos\alpha}{6}\sqrt{1+36\, tg^2\alpha}\,. \tag{11}$$

The boundary layer thickness δ_l at this point can be estimated using $s_l = 1$, which leads to:

$$\delta_l \approx \begin{cases} 5\,/\,\sqrt{Re} & \textit{for laminar flow,} \\ 0.37\,/\,Re^{\frac{1}{5}} & \textit{for turbulent flow.} \end{cases} \tag{12}$$

The quality of this estimation can be tested by comparison of the results of some full laminar boundary layer calculations for $h \to \infty$, that is, without wall, which are given by Schmitt and Schneider [14].

First, for the case $\alpha = 0°$ treated in the next section, for $Re = 7.6 \cdot 10^5$ equation $(12)_1$ leads to $\delta_l = 0.00573$, and in [14] the value $\delta_l = 0.00474$ has been found at the largest cross-section of the spheroid. Though it must be taken into account, that for smaller distance h the velocity between the body and the wall is greater and therefore δ is smaller, it can be inferred, that equation $(12)_1$ gives a good estimation in this case.

For the case $\alpha = 10°$, which also will be treated in the following, and for the same Reynolds number, in [14] the maximal value $\delta = 0.01$ has been found in the leeward plane of symmetry, before separation occurs. From (10) it follows, that $x_l = 0.725$ and with use of $s = 1 + x_l$ equation (8) leads to $\delta = 0.00753$. Thus the estimated value of the boundary layer thickness at the windward side is smaller than the calculated value at the leeward side, as would be expected. Therefore, the estimation is good enough for the present intention.

The maximal thickness δ_w of the boundary layer adjacent to the plane wall is estimated by the same value as that of the boundary layer adjacent to the spheroid:

$$\delta_w \approx \delta_l\,. \tag{13}$$

The minimal width h_P remaining for the potential flow between the spheroid and the wall is:

$$h_P = h - h_l - \delta_l - \delta_w\,. \tag{14}$$

Now the condition explained in the beginning of this section can be expressed in the form, that the ratio h_P/δ_l has to be large compared with 1, which shall be met by the assumption:

$$\frac{h_P}{\delta_l} \geq 2\,. \tag{15}$$

From the equations (12) to (15), a condition follows for the relation between the distance $h - h_l$ and the Reynolds number:

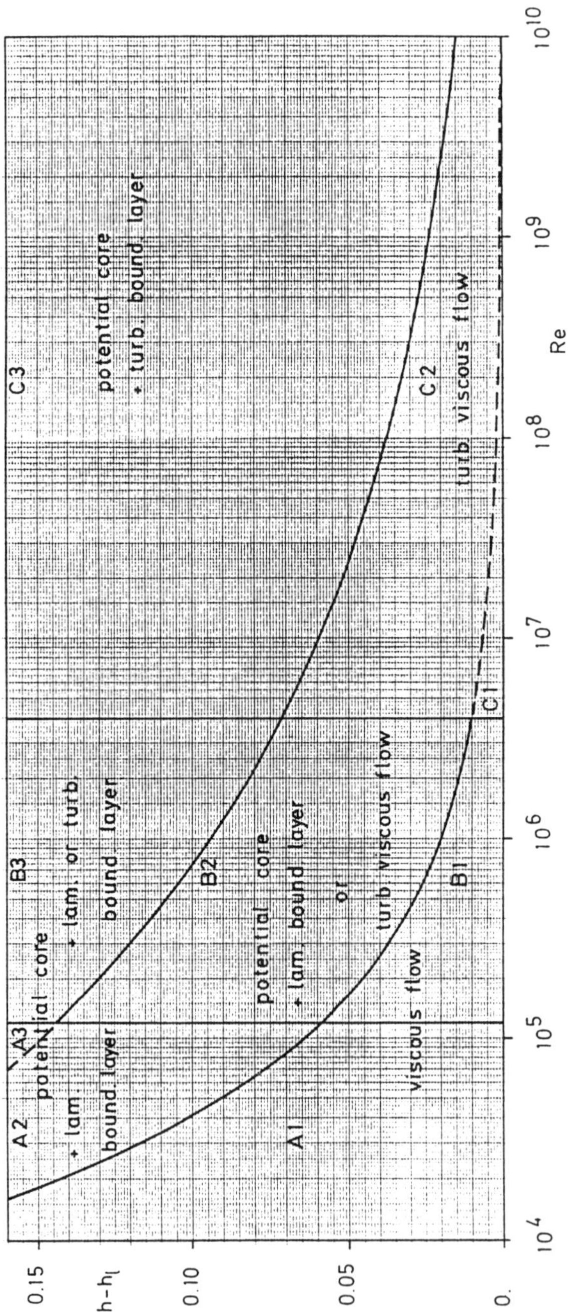

Fig. 2. Existence of a potential flow regime

$$h - h_l \geq \begin{cases} 20 / \sqrt{Re} & \textit{for laminar flow,} \\ 1.48 / Re^{\frac{1}{5}} & \textit{for turbulent flow.} \end{cases} \qquad (16)$$

In the laminar case, Beese [2] has studied the two-dimensional flow around an airfoil near ground using matched asymptotic expansions. He has found an estimate analogous to $(16)_1$ with the constant value $10 \cdot \sqrt{6}$ instead of 20.

The results given by (16) are shown in figure 2. The lower curve is the limit line of the estimate $(16)_1$; if the flow is characterized by a point in the region 1 below this line, the calculation of a potential flow in the slot on the assumption of small boundary layers is not justified. In such a case, the full Navier-Stokes equation must be used for the evaluation of the entire flow field. The upper curve is the limit line of the estimate $(16)_2$; if the flow is characterized by a point in the region 3 above this line, it has a potential core in the slot and turbulent or laminar boundary layers along the spheroid and the wall. For points in the region 2 between the two lines, the flow has a potential core in the slot only if it is laminar. Turbulent flows corresponding to points in this region can only be calculated with use of the Reynolds equations holding for such flows.

According to Batchelor [1, p.313], the flow of air or water at normal temperature along a flat plate is laminar for $Re < 1.2 \cdot 10^5$ regardless of any disturbing condition. On the other side, the flow is turbulent for $Re > 4 \cdot 10^6$, even if a very smooth stream is obtained with modern wind tunnel techniques. If these boundaries are taken as estimates for the present case, the flow is laminar in region A, laminar or turbulent in region B, and turbulent in region C.

At last, for the laminar flow the assumption of incompressibility shall be examined. This assumption is justified for $M_\infty \leq 0.2$, where $M_\infty = V_\infty / \sqrt{\gamma R T}$ is the Mach number at infinity and γ, R and T are the ratio of the specific heats, the gas constant, and the temperature, respectively. With help of the relation $\nu = C\sqrt{T} / [\rho(1 + D/T)]$ (see for instance Vincenti and Kruger [16, p.23]) with constants C and D depending on the special gas, the Reynolds number can be written in the form:

$$Re = \frac{1}{C} \sqrt{\frac{\gamma}{R}} \, \frac{a \tilde{p} M_\infty}{T} \left(1 + \frac{D}{T}\right), \qquad (17)$$

where $\tilde{p}$ is the dimensional pressure. With the values $\gamma = 1.4$, $R = 2.97 \cdot 10^6\ cm^2/s^2K$, $C = 1.4 \cdot 10^{-5}\ g/cm\,s\,K^{1/2}$ and $D = 103\ K$ for nitrogen and $a = 1.2\ m$, $\tilde{p} = 1\ at$, $T = 300\ K$, the condition $Re \leq 5.3 \cdot 10^6$ results, which is met for laminar flows according to figure 2.

4. EXAMPLES OF POTENTIAL FLOWS ABOUT SPHEROIDS WITHOUT INCIDENCE

At first, the potential flow about the spheroid has been studied for the case $\alpha = 0°$, i. e. without incidence. In this case for $h \to \infty$, the flow is axisymmetric and the vortex sheet does not play any role; therefore for finite values of h, the influence of the vortex sheet can be neglected in a good approximation and the flow can be calculated with the help of equation (7).

4.1 Distance between the spheroid and the wall $h = 0.25$

For $h = 0.25$, equation (11) gives $h_l = 1/6$ for $\alpha = 0°$, and therefore, $h - h_l = 0.0833$. From figure 2 it can be deduced, that for $Re > 5.8 \cdot 10^4$ a flow with a potential core is possible. The boundary layers can be turbulent for $Re > 1.8 \cdot 10^6$.

Figures 3a-d show the pressure coefficient $c_p(x) = 2p = 1 - \mathbf{v}^2$. At first, the properties of the flow without wall shall be recalled shortly, which already have been studied in [14]. Since in this case the flow is axisymmetric, the stagnation point is situated at $x = -1$. Here $\mathbf{v} = 0$ and hence $c_p = 1$. With increasing x, the cross-sectional area of the flow decreases. Hence, as a consequence of the equation of continuity, the velocity

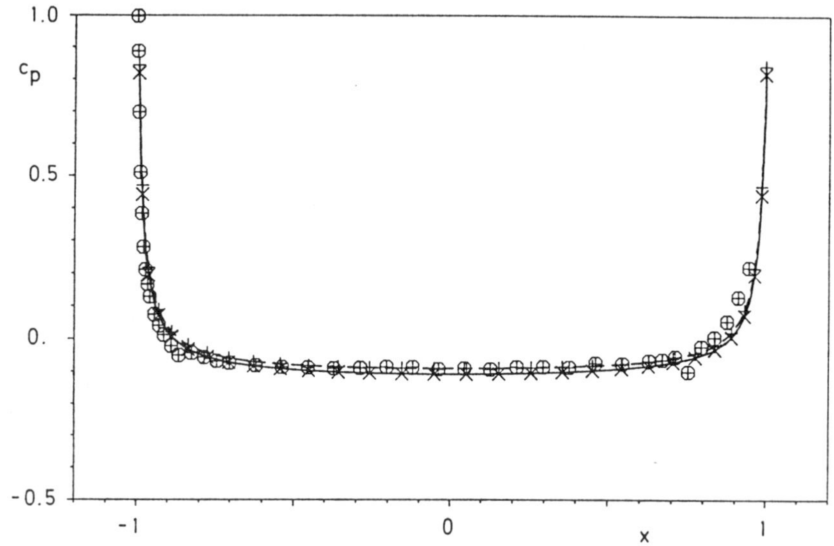

Fig. 3. Pressure coefficient on the surface along a meridian for $\alpha = 0°$ and $h = 0.25$. —×— Theory with wall, + – + theory without wall, ⊕⊕⊕ experiment Böddener [3]. (a) $\vartheta = 180°$

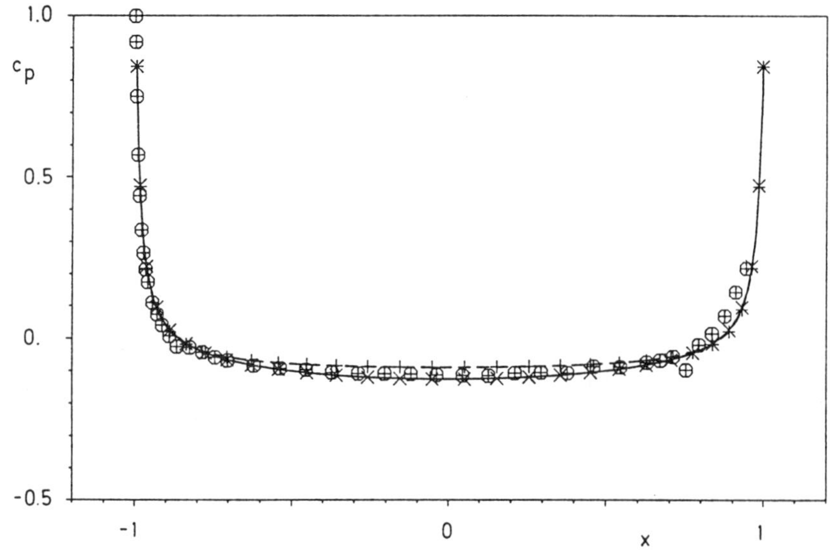

Fig. 3b. $\vartheta = 90°$

increases. Due to Bernoulli's equation, the pressure decreases up to the largest cross-section at $x = 0$. Behind this section, the pressure increases again in a symmetric manner.

The presence of the wall prevents the displacement of the flow due to the spheroid; this leads to a stronger increase of the velocity in the neighborhood of the body, and therefore to a smaller pressure. This effect is small for $\vartheta = 180°$ that is, on the upper

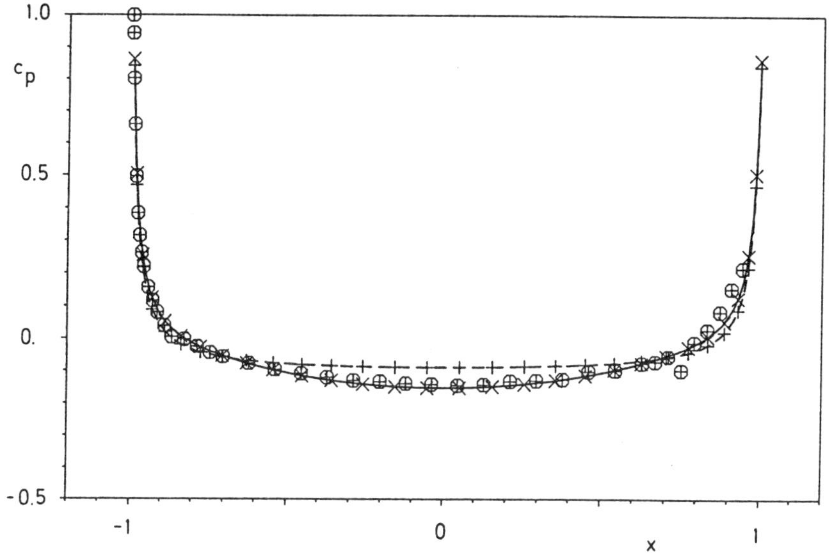

Fig. 3c. $\vartheta = 45°$

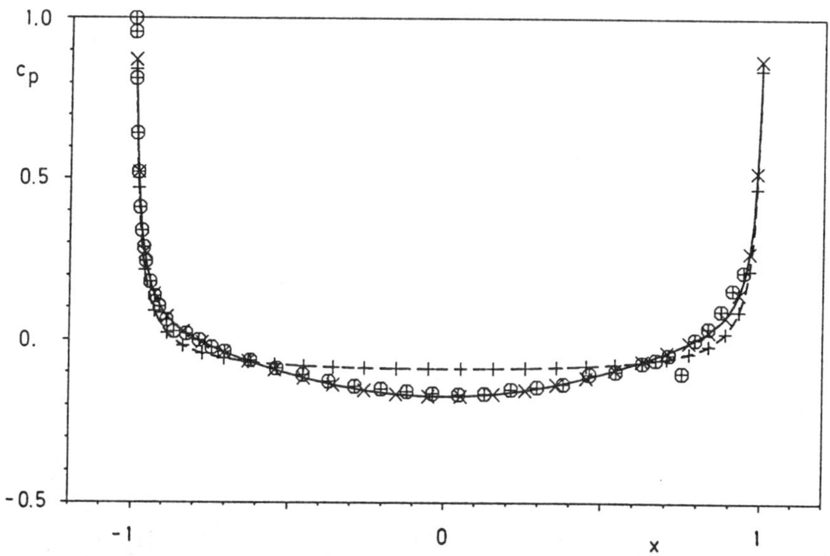

Fig. 3d. $\vartheta = 0.2°$

side (figure 3a), and increases with decreasing ϑ, that is approaching the lower side (figures 3b-d). On the meridians $\vartheta = 45°$ and $\vartheta = 0.2°$ lying close to the wall the opposite effect appears in the neighborhood of the nose and the rear of the spheroid. This can be understood, since here the flow avoids the wall by going to the other side of the spheroid. A comparison of the calculated results with experimental results of Böddener [3] obtained at a Reynolds number $Re = 3.6 \cdot 10^6$ for turbulent boundary layer flow shows very good agreement.

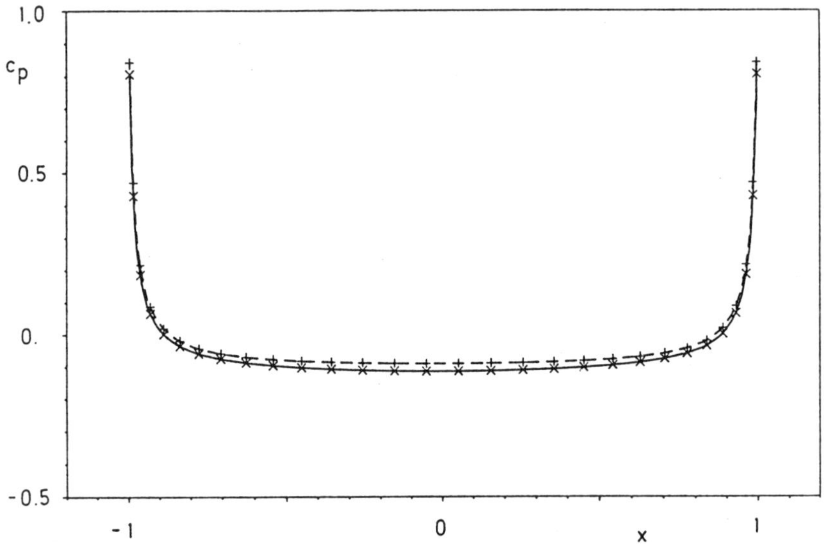

Fig. 4. Pressure coefficient on the surface along a meridian for $\alpha = 0°$ and $h = 0.19$. —×— Theory with wall, + – + theory without wall. (a) $\vartheta = 180°$

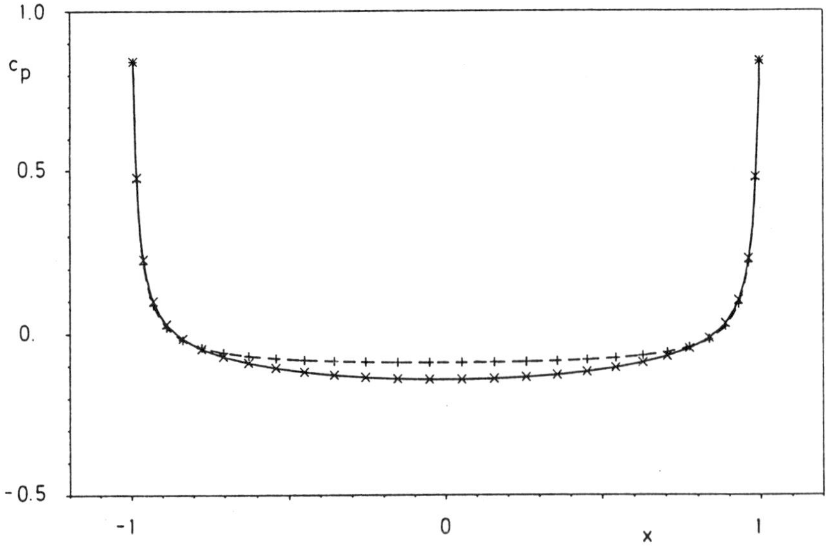

Fig. 4b. $\vartheta = 90°$

In [13] the flow around a prolate spheroid moving near a lateral plane wall has been studied. Without incidence, the geometry coincides with that of the present case for $\alpha = 0°$. Comparison shows, that for the greater distance $h = 1/3$, which only has been considered in that paper, the influence of the wall is smaller.

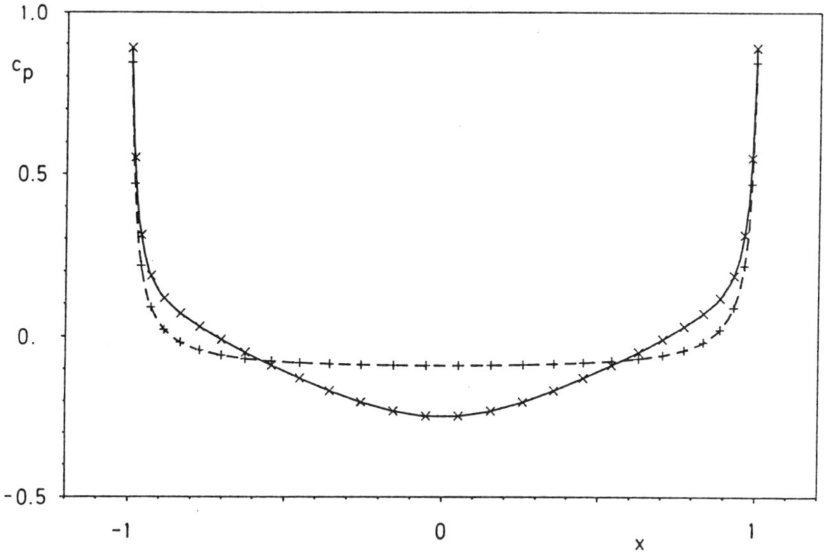

Fig. 4c. $\vartheta = 0°$

4.2 Distance between the spheroid and the wall $h = 0.19$

For $h = 0.19$ it follows from equation (11), that $h - h_l = 0.0233$. Figure 2 shows, that in this case for $7.5 \cdot 10^5 < Re < 4 \cdot 10^6$ a potential core together with laminar boundary layers are possible. For $Re > 10^9$ a potential core together with turbulent boundary layers are possible.

The results shown in figure 4 agree qualitatively with those in figure 3, but the influence of the wall is stronger. Even along the lateral meridian (figure 4b), the pressure in the neighborhood of the nose and the rear of the spheroid is higher than in the case without wall.

5. EXAMPLE OF POTENTIAL FLOW ABOUT A SPHEROID WITH INCIDENCE

It is known from the results for the case without wall, that for a finite angle of incidence the vortex sheet, which describes the lateral separation from the spheroid, plays an important part in the potential flow (see [12,14]).

For $\alpha = 10°$ and $h = 0.25$, equation (11) leads to $h - h_l = 0.0111$. Figure 2 shows that a potential core together with laminar boundary layers is possible for $3.4 \cdot 10^6 < Re < 4 \cdot 10^6$. A potential core together with turbulent boundary layers is possible for $Re > 4.22 \cdot 10^{10}$.

Figure 5 demonstrates that as a consequence of the incidence, the symmetry to $x = 0$ is lost. In the case without wall, a further consequence is that the minimal value of c_p is diminished; for, the spheroid at incidence represents a body of larger cross-section blocking the flow than the spheroid without incidence.

The presence of the wall causes an increase of the velocity and therefore a decrease of the pressure as in the case $\alpha = 0°$. Since this effect is higher on the lower side of the

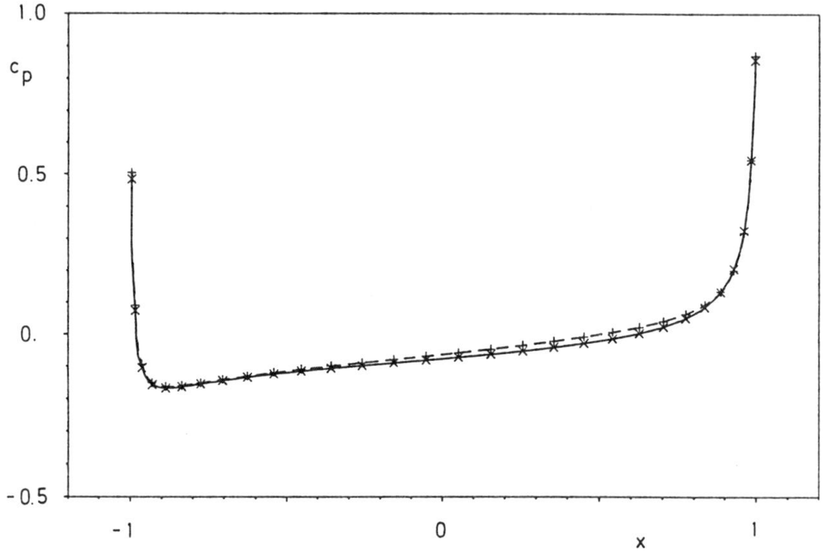

Fig. 5. Pressure coefficient on the surface along a meridian for $\alpha = 10°$ and $h = 0.25$. —×— Theory with wall, + – + theory without wall. (a) $\vartheta = 180°$

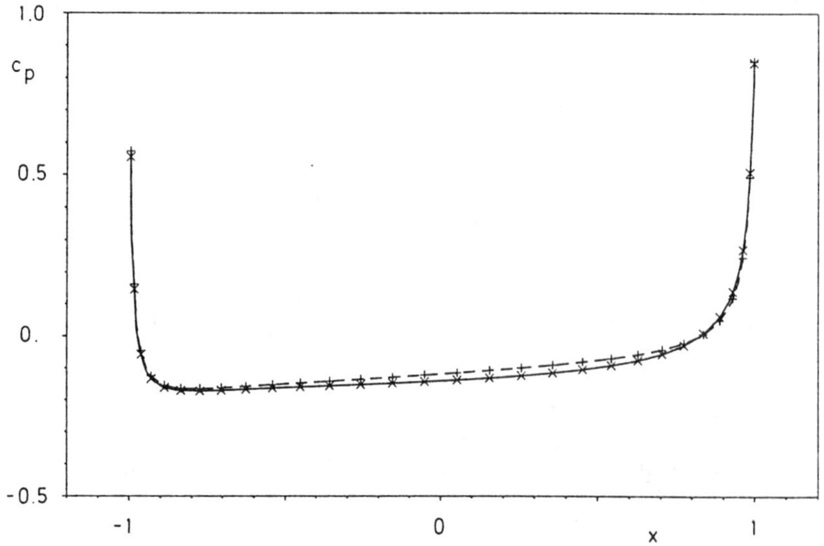

Fig. 5b. $\vartheta = 135°$

spheroid, it can be stated, that the wall exerts an additional force attracting the spheroid. The results along the lateral meridian $\vartheta = 90°$ (figure 5c) are not as good as along the other meridians, because the model for the shape of the vortex sheet used for the calculations differs from the real vortex sheet mainly in this region. For the case without wall, a very good agreement of the calculated data with experimental ones of other authors, except for the lateral meridian, has been demonstrated in [12].

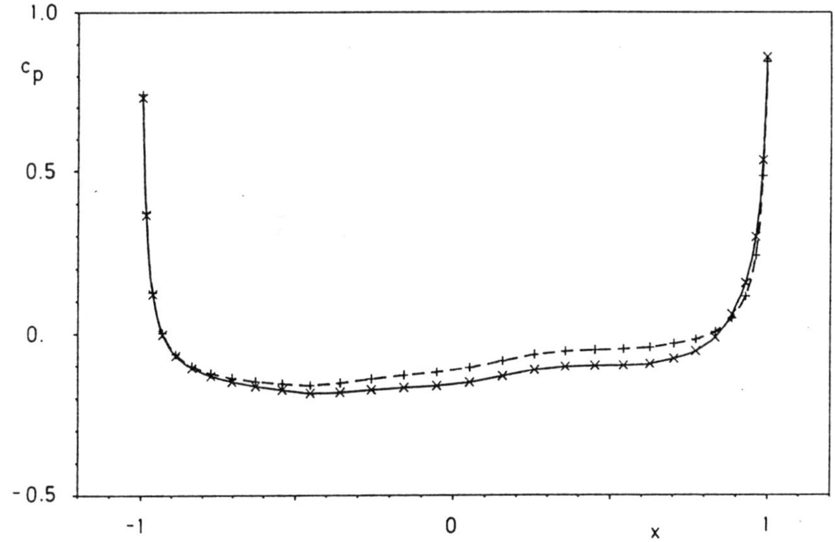

Fig. 5c. $\vartheta = 90°$

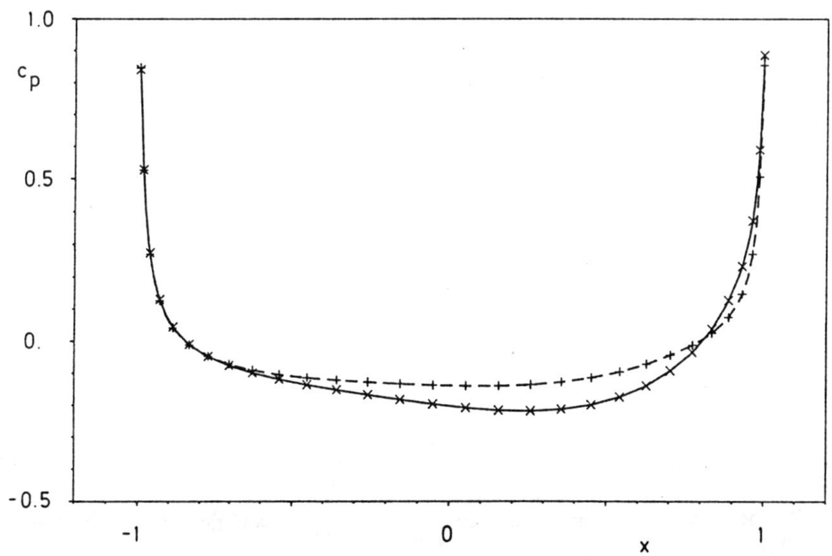

Fig. 5d. $\vartheta = 65°$

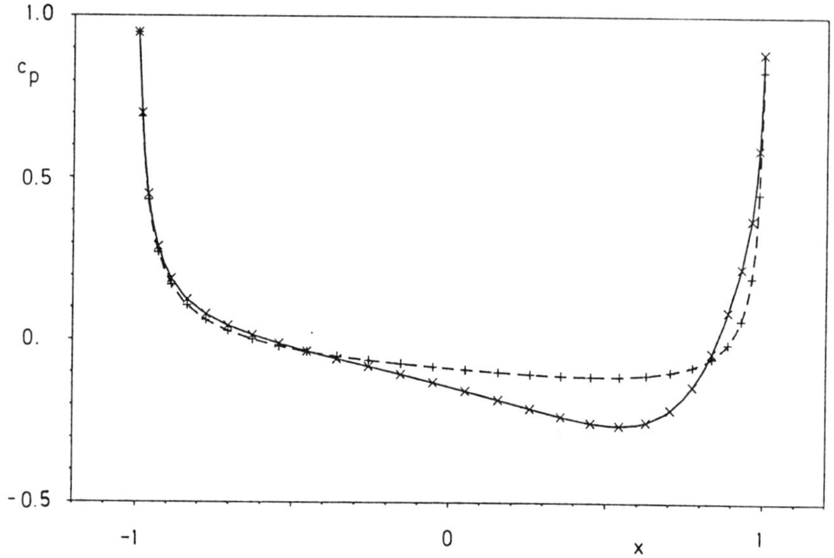

Fig. 5e. $\vartheta = 35°$

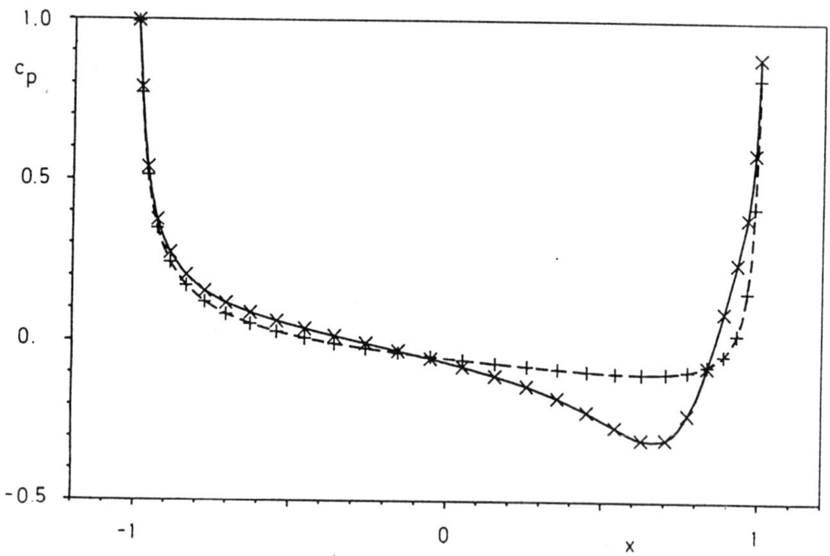

Fig. 5f. $\vartheta = 0°$

REFERENCES

1. Batchelor, G.K. An Introduction to Fluid Dynamics, Cambridge University Press, 1967.

2. Beese, E. Untersuchungen zum Einfluß der Reynoldszahl auf die aerodynamischen Beiwerte von Tragflügelprofilen in Bodennähe, Thesis, Ruhr-University, Bochum, 1982.
3. Böddener, W. Measurements of the Pressure Distribution on a Prolate Spheroid in the Neighborhood of a Plane Wall, Private Communication, 1987.
4. Hess, J. L. Review of Integral-Equation Techniques for Solving Potential-Flow Problems with Emphasis on the Surface-Source Method, Comp. Meth. Appl. Mech. Eng., Vol.5, pp 145-196, 1975.
5. Hess, J. L. and Smith, A. M. O. Calculation of Nonlifting Potential Flow about Arbitrary Three-Dimensional Bodies, J. Ship Research, Vol.8, pp. 22-44, 1964.
6. Hess, J. L. and Smith, A. M. O. Calculation of Potential Flow about Arbitrary Bodies. Progress in Aeronautical Sciences (Ed. Küchemann, D.), Vol.8, pp. 1-138, Pergamon, London, 1967.
7. Kraus, W. and Sacher, P. Das Panelverfahren zur Berechnung der Druckverteilung von Flugkörpern im Unterschallbereich, Zeitschrift für Flugwissenschaften, Vol.21, pp. 301-311, 1973.
8. Lamb, H. Hydrodynamics, 6th Edition, Dover Publications, New York, 1932.
9. Meier, H.U. and Kreplin, H.-P. Experimental Study of Boundary Layer Velocity Profiles on a Prolate Spheroid at Low Incidence in the Cross Section $x_0/L = 0.64$, in: Viscous and Interacting Flow Field Effects (Ed. Fiore, A.W.), pp. 169-189, Proc. 5th U.S. Air Force and F.R.G. Data Exchange Agreement Meeting, Techn. Rep. AFFDL-TR-80-3088, Wright-Patterson Air Force Base, Ohio, 1980.
10. Rotta, J.C. Turbulent Boundary Layers in Incompressible Flow. Progress in Aeronautical Sciences (Ed. Ferri, A., Küchemann, D. and Sterne, L.H.G.), Vol.2, pp. 1-219, Pergamon Press, Oxford and New York, 1962.
11. Schlichting, H. Boundary Layer Theory, 7th Edition, McGraw-Hill Book Company, New York, 1979.
12. Schmitt, H. Druckverteilung an einem schräg angeströmten Rotationsellipsoid , Ingenieur -Archiv, Vol.53, pp. 337-344, 1983.
13. Schmitt, H. Strömung eines inkompressiblen Fluids um ein Rotationsellipsoid in der Nähe einer ebenen Wand, ZAMM, Vol.67, pp. T326-T327, 1987.
14. Schmitt, H. and Schneider, G.R. Calculation of the Potential Flow with Consideration of the Boundary Layer. Chapter 6, Topics in Boundary Element Research (Ed. Brebbia, C.A.), Vol.5, pp. 110-133, Springer-Verlag, Berlin and New York, 1989.
15. Schmitt, H. and Schneider, G.R. Einfluß des Bodens auf die Druckverteilung an einem umströmten Rotationsellipsoid , ZAMM, Vol.70, pp. T404-T406, 1990.
16. Vincenti, W.G. and Kruger, C.H. Introduction to Physical Gas Dynamics, John Wiley and Sons, New York and London, 1965.

Dual Reciprocity BIE in Lagrangian Form for Incompressible Unsteady Viscous Flow

X. Jin, D.K. Brown

Department of Mechanical Engineering, University of Glasgow, Glasgow, G12 8QQ, U.K.

ABSTRACT

Boundary Integral Equations for incompressible unsteady viscous flow in primitive variables are presented. They are based on a group of governing equations in Lagrangian form locally in time, and time independent fundamental solutions. In order to preclude the need for meshing over the medium, the Lagrangian time derivatives are treated as a series of coordinate functions multiplied by unknown time dependent functions, which gives rise to the dual reciprocity method in fluid dynamics. The method makes use of the advantages of the Lagrangian approach which can drop the convection terms, and the dual reciprocity boundary element method to avoid the traditional mesh distortion problem which blocks the development of the Lagrangian approach. A matrix boundary element equation is then obtained, and the equation needs only the boundary discretization and a few internal points fixed in the medium to increase the accuracy; all the moving boundary and internal points concerned would be traced in each time step. The pressure, however, can be solved explicitly by substituting the velocity and traction components obtained on the boundary into the decoupled pressure equation using a different fundamental solution. The method can be used for solving finite fluid flow with free surface or interface. If all the boundaries are fixed, the method would be very efficient because most of the matrix coefficients would be calculated only once at the beginning.

1.INTRODUCTION

In recent years, the boundary element methods have been rapidly developed in fluid mechanics. Many efforts have been made by Wu[1] on viscous flow by using the integral representations of both velocity and vorticity components as the basic unknowns; Coleman[2,3] on Newtonian and non-Newtonian flows by using the stream and stress functions; Bush and Tanner[4], Tosaka and Onisaka[5,6] by using the primitive variables; Kitagawa and Brebbia[7,8] by using penalty function method, etc. Newly published books [9,10] have introduced those and some other significant developments in more recent years on BIEM for both steady and unsteady viscous flow. Almost all of the formulations were described in Eulerian coordinate system. But for some problems in fluid dynamics, the finite fluid medium concerned is better described in a Lagrangian system. The numerical techniques in Lagrangian form, however, have been developed for fluid dynamics both in finite difference method and in finite element method for many years. Since the coordinate system follows the fluid motion, they have significant advantages of sharp resolution of material interface and no convection across the boundary of each element.The Lagrangian techniques have, however, a major disadvantage of mesh tangling at severe distortions, and it has been thought to block their developments. Some efforts have been made to overcome the mesh distortion problems, such as various methods of remeshing, finite volume method and other Eulerian-Lagrangian combination methods. The boundary element method has its well-known advantage that it can transform the unknowns for solving into the boundary, so it naturally has the possibility to avoid the problem mentioned. For general fluid dynamics the problem is that the domain integral terms are often unavoidable. To avoid the convection terms, some developments have been done recently by Hebeker [11] on the boundary element technique in Lagrangian system.

In the light of Nardini and Brebbia's work[12] and Wrobel's work[13], the Dual Reciprocity method can be used for precluding the need for domain integration for both dynamic problems and diffusion problems. It can be applied to fluid dynamics problems as well. In this paper, we wish to use the Lagrangian approximate equations to develop the boundary integral equations, and suggest a group of boundary-only equations for the unsteady viscous flows.

2. LAGRANGIAN APPROXIMATE GOVERNING EQUATIONS

The governing equations for incompressible viscous unsteady fluid are given by the continuity equation

$$u_{i,i} = 0 \tag{1}$$

and the Navier-Stokes equation

$$\rho \frac{Du_i}{Dt} = T_{ij,j} + \rho f_i \tag{2}$$

in which the stress tensor is given by

$$T_{ij} = -p\delta_{ij} + \mu(u_{i,j} + u_{j,i}) \tag{3}$$

where D/Dt is the Lagrangian time derivative, u_i, p velocity and pressure, ρ, μ density and viscosity. The initial and boundary conditions about the fluid medium are given as follows

$$\left.\begin{array}{ll} u_i = u_i^o(\vec{x}), & T_{ij} = T_{ij}^o(\vec{x}) \\ u_i = \bar{u}_i(\vec{x}, t), & \tau_i = n_j \bar{T}_{ij}(\vec{x}, t) \end{array}\right\} \tag{4}$$

Equations (1)-(4) provide a complete system for solving the unsteady viscous flow problems. If we assume the solution of the system exists, then the motion of any medium point can be traced by a tracer as the time increases. The curve formed by the tracer is the Lagrangian coordinates

$$X_i = X_i(\vec{x}, t); \qquad \vec{x} = (x_1, x_2, x_3) \tag{5}$$

in which x_k is the Eulerian coordinate of the tracer at the initial state.

Now suppose that the time t is the sum of discretized time increments Δt :

$$t_m = t_{m-1} + \Delta t \qquad (m = 1, 2, 3, \cdots) \tag{6}$$

and the solution of the system (1)-(4) can be obtained on each time interval Δt, then we can say the state of the medium concerned at the beginning of each time interval is the initial state of the system solved. If Δt is small enough, the coordinates of any tracer can then be approximately expressed by

$$x_i^{m+1} = x_i^m + u_i^m \Delta t \tag{7}$$

in which the velocity of any tracer in mth time interval can be solved through the following Lagrangian governing equation (the body force is neglected)

$$\rho \dot{u}_i^m = - p_{,i}^m + \mu(u_{i,kk}^m + u_{k,ik}^m) \tag{8}$$

together with equation (1) for the same interval. The time derivative in the left hand side can be approximately expressed as

$$\dot{u}_i^m = \frac{u_i(\vec{x}^m, t_m) - u_i(\vec{x}^{m-1}, t_{m-1})}{\Delta t}; \quad \vec{x} = (x_1, x_2, x_3) \tag{9}$$

3. INTEGRAL REPRESENTATIONS

In order to obtain the integral representations for the unknown velocity components and pressure, the weighted residual function method is used. A group of tensor weight functions u_{ik} are used for equation (8) because it is a vector equation, and a group of vector weight functions p_k for equation (1)

$$\int_\Omega [- p_{,i}^m + \mu(u_{i,kk}^m + u_{k,ik}^m)] u_{ij}^* d\Omega = \int_\Omega \rho \dot{u}_i^m u_{ij}^* d\Omega \tag{10}$$

$$\int_\Omega u_{i,i}^m p_j^* d\Omega = 0 \tag{11}$$

The divergence theorem can be applied in the left hand sides of both equations, and after some mathematical manipulations, both equations can be rewritten as the following integral equations

$$\int_\Gamma (\tau_i^m u_{ij}^* - u_i^m \tau_{ij}^*) d\Gamma + \int_\Omega [p u_{j,i}^* + u_i^m(\mu u_{ij,kk}^* + \mu u_{kj,ik}^* - p_{j,k}^* \delta_{ki})] d\Omega$$

$$= \int_{\Omega} \rho \dot{u}_i^m u_{ij}^* d\Omega \qquad (12)$$

where Γ is the boundary which may move as the medium deforms.

The requirement for the weight functions u_{ij} and p_j is to eliminate the body integrals in the left hand side of the equation; both

$$u_{ij,i}^* = 0; \qquad \mu u_{ij,kk}^* - p_{j,k}^* \delta_{ki} = \delta_{ij} \delta(\vec{\xi} - \vec{x}) \qquad (13)$$

and

$$u_{ij,i}^* = \delta_{ij} \delta(\vec{\xi} - \vec{x}); \qquad \mu(u_{ij,kk}^* + u_{kj,ik}^*) - p_{j,k}^* \delta_{ki} = 0 \qquad (14)$$

can satisfy it. The requirement for pressure can further reduce the functions in equation (14) as follows

$$u_{i,i}^{**} = \delta(\vec{\xi} - \vec{x}); \qquad \mu(u_{i,kk}^{**} + u_{k,ik}^{**}) - p_{,k}^{**} \delta_{ki} = 0 \qquad (14')$$

The solutions of equations (13) and (14') in two dimensional case are given in the Appendix. By applying those weight functions solved from equations (13) and (14'), we can split equation (12) to obtain the integral representations both for the velocity components and the pressure

$$C\delta_{ij} u_i^m - \int_{\Gamma} u_i^m \tau_{ij}^* d\Gamma + \int_{\Gamma} \tau_i^m u_{ij}^* d\Gamma = \int_{\Omega} \rho \dot{u}_i^m u_{ij}^* d\Omega \qquad (15)$$

$$Cp^m - \int_{\Gamma} u_i^m \tau_i^{**} d\Gamma + \int_{\Gamma} \tau_i^m u_i^{**} d\Gamma = \int_{\Omega} \rho \dot{u}_i^m u_i^{**} d\Omega \qquad (16)$$

where C is 1 for interior point and 1/2 for a point on a smooth boundary.

4. DUAL RECIPROCITY

Equations (15) and (16) supply a group of integral equations for the unknown velocity and pressure. But the right hand sides of the equations, though they can be carried out by using element integrations after body discretization, are the main problem of mesh distortion and all the internal points would be involved in the equations. In order to transform these integrals into the boundary to simplify the problem, an approximating approach called the Dual Reciprocity method can be used. The idea was used by Nardini and Brebbia for dynamic analysis[12] and was introduced by Wrobel for diffusion problems[13]. In the light of this idea, we can assume that the Lagrangian time derivative can be approximately expressed by the sum of N coordinate functions f^j multiplied by the unknown time dependent functions a^j_i:

$$\dot{u}_i^m(\vec{x}, t) = \sum_{j=1}^{N} \alpha_i^j(t) f^j(\vec{x}) \qquad (17)$$

Substituting this expression into the right hand side of equation (15), we have

$$\int_{\Omega} \rho \dot{u}_i^m u_{ik}^* d\Omega = \rho \sum_{j=1}^{N} \alpha_1^j(t) \int_{\Omega} \delta_{li} f^j(\vec{x}) d\Omega \tag{18}$$

The resulting body integral can now be transformed into the boundary integrals as we can always assume a fluid field corresponding to the jth coordinate function as follows

$$\mu\left(U^j_{l,ss} + U^j_{sl,ls}\right) - \delta_{ls} P^j_{i,s} = \delta_{il} f^j \tag{19}$$

whose velocity solution can be used to give the following boundary expression

$$\int_{\Omega} \delta_{li} f^j u_{ik}^* d\Omega = C\delta_{ik} U^j_{il} - \int_{\Gamma} U^j_l \tau^*_{ik} d\Gamma + \int_{\Gamma} \sigma^j_{li} u^*_{ik} d\Gamma \tag{20}$$

where σ^j_{ii} are the corresponding tractions on the boundary.

Substituting this result into equation (15) and after rearrangement it gives

$$C\delta_{ik} u_i^m - \int_{\Gamma} u_i^m \tau^*_{ik} d\Gamma + \int_{\Gamma} \tau_i^m u^*_{ik} d\Gamma$$

$$= \rho \sum_{j=1}^{N} \left(C\delta_{ik} U^j_{il} - \int_{\Gamma} U^j_l \tau^*_{ik} d\Gamma + \int_{\Gamma} \beta^j_{li} u^*_{ik} d\Gamma \right) \alpha^j_1 \tag{21}$$

where Γ is the medium's boundary at the mth time step. In the same way the integral equation for pressure can be obtained

$$Cp^m - \int_{\Gamma} u_i^m \tau_i^{**} d\Gamma + \int_{\Gamma} \tau_i^m u_i^{**} d\Gamma$$

$$= \rho \sum_{j=1}^{N} \left(C\delta_{il} U^j_i - \int_{\Gamma} U^j_i \tau_i^{**} d\Gamma + \int \beta^j_i u_i^{**} d\Gamma \right) \alpha^j_1 \tag{22}$$

Equations (21) and (22) give a group of integral equations from which a numerical solution procedure can be derived.

5. THE NUMERICAL APPROACH

In order to solve equations (21) and (22), the boundary element method can be employed. From equation (17) one can see that a linear relation can be set up for obtaining the time dependent functions α^j_i by evaluating the equation at all nodes concerned

$$\{\dot{u}\} = [F]\{\alpha\} \tag{23}$$

The choice of the coordinate functions should be linearly independent and the number of them should be equal to the number of the nodes to form a square matrix [F] from whose inversion the relation for the vector $\{\alpha\}$ is produced

$$\{\alpha\} = [F]^{-1}\{\dot{u}\} \tag{24}$$

Since all the integrals can be evaluated in each boundary element by a suitable numerical integration method (in most of the cases the standard Gaussian formulae are adopted) to leave the unknowns as the vectors, an assembled matrix system can be set up

$$[H]\{u\} - [G]\{\tau\} = [C]\{\dot{u}\} \tag{25}$$

where $[C]=\rho[HU\text{-}GT][F]^{-1}$.

Now we employ a two-level time integration scheme by introducing a linear approximation $\{u\}$ and $\{\tau\}$

$$\left.\begin{aligned}\{u\} &= (1-\lambda)\{u\}^{m} + \lambda\{u\}^{m+1}\\ \{\tau\} &= (1-\lambda)\{\tau\}^{m} + \lambda\{\tau\}^{m+1}\end{aligned}\right\} \tag{26}$$

where λ $(0\leq\lambda\leq1)$ is a parameter which positions the middle values of $\{u\}$ and $\{\tau\}$ in each time step, and the Lagrangian time derivative, from equation (9), can be written as follows

$$\{\dot{u}\} = \frac{1}{\Delta t}\left(\{u\}^{m+1} - \{u\}^{m}\right) \tag{27}$$

Substituting these three relations into equation (25) we have

$$\begin{aligned}[\lambda\Delta tH - C]^{m+1}\{u\}^{m+1} &- \lambda\Delta t[G]^{m+1}\{\tau\}^{m+1}\\ &= [(\lambda-1)\Delta tH - C]^{m}\{u\}^{m} + (1-\lambda)[G]^{m}\{\tau\}^{m}\end{aligned} \tag{28}$$

in which the superscripts refer to the mth time step and the left hand side is the beginning of the time step and the right hand side is the end of the time step. The pressure equation (22) can be set up in the same matrix form and the unknown pressure at each boundary node and internal point can be directly obtained through the equation from the velocity and traction component results at all boundary nodes solved in equation (28). After obtaining all the boundary unknowns, the values of any internal point in the medium can be worked out. One must notice that the positions of all the points concerned in the calculation are updated by equation (7) in each step since those points are regarded as Lagrangian tracers fixed in the medium.

From this equation we can see that the coefficients of matrices [H], [G] and [C] have to be updated for each time step as the boundary of the medium deforms and moves. If all the boundaries are fixed or moving in the same direction at the same speed, matrices [H] and [G] will not change. For matrix[C], because some internal points must be needed, all of the coefficients need to be evaluated for each time step. Therefore the efficiency of the method would strongly depend on the choice of the coordinate functions. One of the choices can be the conical function, as [11] suggested, whose corresponding solutions are given in the Appendix. The choice of the coordinate functions for the internal points can also be different from those for the boundary points. From this point of view, we can suggest such a way to choose the coordinate functions that the efficiency can be achieved significantly to form matrix [C]. As can be seen, matrix [C] consists of two matrices, say $[D][F]^{-1}$, which can be divided into the following four parts

$$[D] = \begin{bmatrix} D_{BB} & D_{BI} \\ D_{IB} & D_{II} \end{bmatrix}; \quad [F] = \begin{bmatrix} F_{BB} & F_{BI} \\ F_{IB} & F_{II} \end{bmatrix} \tag{29}$$

where the subscripts BB and II stand for boundary points related and internal points related coefficients, respectively; and BI and IB are the interrelated coefficients. If all the boundaries of a problem are fixed or moving in the same direction at the same speed, the boundary related coefficients will not change at all. The other three parts depend on the moving internal points. If we choose one of the interrelated coefficient parts in [F] matrix as zero, for example, $F_{BI}=0$, the inversion of it is then reduced to

$$[F]^{-1} = \begin{bmatrix} F_{BB}^{-1} & 0 \\ -F_{II}^{-1}F_{IB}F_{BB}^{-1} & F_{II}^{-1} \end{bmatrix} \tag{30}$$

in which only the internal points related part has to be inverted anew.

The most direct application of this method is probably to the problems of an object moving in an infinite incompressible fluid field if a group of proper coordinate functions can be chosen. A basic condition of the choice of the coordinate functions is that the influence on the infinite boundary points are zero.

6. CONCLUSION

Boundary integral equations in Lagrangian form for the components of velocity, traction and the pressure of unsteady incompressible viscous fluid flow are presented. Although the method can be quite time-consuming for free surface problems because the coefficients of the matrices [H], [G] and [C] have to be updated for each time step, it is an efficient method for the problems of a moving body in an infinite incompressible fluid field. The advantages of Lagrangian form include the linearized governing equations, sharp resolution of material interface or free surface. The method is based on the time-independent fundamental solutions; and the body integral of the inertial term is transformed into the boundary by using the Dual Reciprocity method. The formulations can be solved numerically by the boundary element method and no domain mesh is needed.

REFERENCES

1.Wu, J.C., Problems of General Viscous Flow, Developments in Boundary Element Methods (Ed. Banerjee,P.K., Shaw,R.P.), Vol. 2, pp. 69-109, Applied Science Publishes, London, 1982.

2.Coleman, C.J., A Contour Integral Formulation of Plane Creeping Newtonian Flow, Q. J. Mech. appl. Math.,Vol. XXXIV, Pt. 4, pp. 453-464, 1981.

3.Coleman, C.J., On the Use of Boundary Integral Methods in the Analysis of Non-Newtonian Fluid Flow, J. of Non-Newtonian Fluid Mechanics, Vol.16, pp. 347-355, 1984.

4.Bush, M.B. and Tanner, R.I., Numerical Solution of Viscous Flows Using Integral Equation Methods, Int. J. Numerical Methods in Fluids, Vol.3, pp. 71-92, 1983.

5.Tosaka, N. and Onishi, K., Boundary Integral Equation Formulations for Steady Navier-Stokes Equations Using the Stokes Fundamental Solutions, Engineering Analysis, Vol..2, pp.128-132, 1985.

6.Tosaka, N. and Onishi, K., Boundary Integral Equation Formulations for Unsteady Incompressible Viscous Fluid Flow by Time-differencing, Engineering Analysis, Vol.3, pp. 101-104, 1986.

7.Kitagawa, K., Brebbia, C.A., Wrobel, L.C. and Tanaka, M., Boundary Element Analysis of Viscous Flow by Penalty Function Formulation, Engineering Analysis, Vol.3, pp. 194-200, 1986.

8.Kitagawa, K., Brebbia,C.A. and Tanaka, M. A Boundary Element Analysis of Natural Convection Problems by Penalty Function Formulation, Boundary Elements X, (Ed. Brebbia, C. A.), Vol.2, pp. 323-341, Springer-Verlag, Berlin and New York, 1988.

9.Brebbia,C.A.(Ed.), Viscous Flow Applications, Topics in Boundary Element Research Vol.5,Springer-Verlag, Berlin, Heidelberg, 1989.

10.Banerjee, P.K., Morino, L.(Ed.), Boundary Element Methods in Nonlinear Fluid Dynamics, Developments in Boundary Element Methods Vol.6, Elsevier Applied Science, London, New York, 1990.

11.Hebeker, F.K., On Lagrangian and Boundary Element Methods for Some Unsteady Isothermal Navier Stokes Flows, Boundary Elements X, (Ed. Brebbia, C.A.), pp. 373-380, 1988.

12.Nardini, D. and Brebbia,C.A., Boundary Element Integral Formulation of Mass Matrices for Dynamic Analysis. Chapter 7, Topics in Boundary Element Research, (Ed. Brebbia, C.A.), Vol.2 pp. 192-207, Springer-Verlag, Berlin and New York, 1985.

13.Wrobel, L.C., On the use of Time-Dependent Fundamental Solutions for the Boundary Element Formulation of Diffusion Problems, Boundary Elements X, Vol.2, (Ed. Brebbia, C.A.), pp. 1-13, 1988.

APPENDIX

The fundamental solutions of equation (13) are

$$u^*_{ij} = \frac{1}{8\pi\mu}\left[(2 \log r + 3)\delta_{ij} - 2r_{,i}r_{,j}\right] \tag{A. 1}$$

$$\tau^*_{ij} = \frac{1}{\pi r} r_{,i} r_{,j} \frac{\partial r}{\partial n} \tag{A. 2}$$

The fundamental solutions of equation (14') are

$$u^{**}_i = -\frac{r_{,i}}{2\pi r} \tag{A. 3}$$

$$\tau^{**}_i = \frac{\mu\left(2r_{,i}r_{,j} - \delta_{ij}\right)n_j}{\pi r^2} \tag{A. 4}$$

The conical function is the simplest choice of the coordinate functions. In this case, f^j is assumed to be the Euclidean distance between a pole A_j and any point x

$$f^j = r\left(A_j, x\right) \tag{A. 5}$$

The corresponding solutions to equation (19) are

$$U_{ik} = \frac{r^3}{45\mu}\left(4\delta_{ik} - 3r_i r_k\right) \tag{A. 6}$$

$$\sigma_{ikj} = \frac{r^2}{15}\left[3\left(\delta_{ik}r_j + \delta_{ij}r_k + \delta_{kj}r_i\right) - 2r_i r_k r_j\right] \tag{A. 7}$$

$$\sigma_{ij} = \frac{r^2}{15}\left[3\left(\delta_{ik}r_j n_k - \delta_{ij}\frac{\partial r}{\partial n} + \delta_{kj}r_i n_k\right) + 2r_i r_j \frac{\partial r}{\partial n}\right] \tag{A. 8}$$

Forced Convection Cooling of Flat Electronic Substrates

G. De Mey, S. De Smet, M. Driscart
Laboratory of Electronics, University of Ghent, Sint Pietersnieuwstraat 41, 9000 Ghent, Belgium

ABSTRACT

If flat electronic substrates are subjected to a forced convection cooling in a parallel direction, the fluid velocity can be considered as being a known value. For the temperature distribution only the linear diffusion equation extended with a drift term remains to be solved. This equation, having a known fundamental solution, can be easily solved numerically using the boundary element method.

1 INTRODUCTION

Cooling of electronic equipment has been the subject of a growing interest during the past years [1] [2] [3] [4]. Due to the higher densities in present day microelectronics, the power densities have reached such a high level that electronic designs cannot be carried out without a thermal analysis. Thermal problems in electronics have been the subject of several papers [5] [6] [7] [8] [9] [10] [11] [12] [13]. Only a few contributions also deal with the boundary element method [14] [15] [16] [17] [18] [19].

In the literature some papers treat the problem of convection cooling in electronic applications [20] [21] [22] [23]. In this paper a particular example of forced convection is considered. Using fan power, the cooling fluid is blown over the substrate with a constant and uniform velocity. It is assumed that the substrate is flat and sufficiently thin, so that the fluid velocity will not be disturbed. This assumption is reasonable due to the small dimensions of electronic substrates. In this case the mathematical approach can be largely simplified. The Navier Stokes equations can be disregarded, the fluid velocity being a known quantity . Only the equation for the fluid temperature remains to be solved. Fortunately, this equation turns out to be linear with a known fundamental solution. The introduction of a boundary element approach becomes then obvious. An analogue approach to a related problem has been published recently [24].

2 BASIC EQUATIONS

Let us consider a flat substrate in the (x, y) plane (fig.1), immersed in a

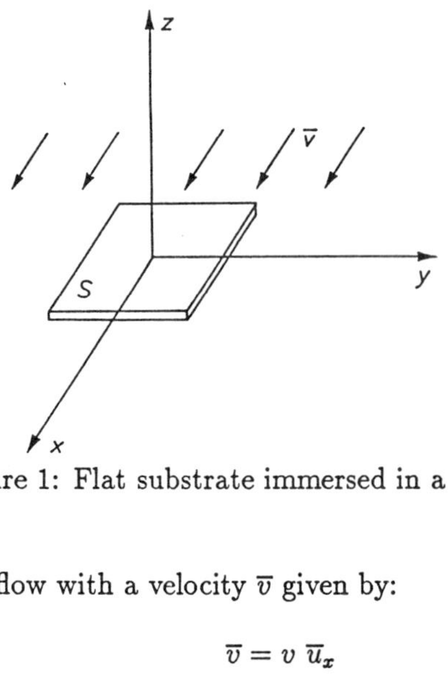

Figure 1: Flat substrate immersed in a uniform flow

uniform fluid flow with a velocity $\overline{v}$ given by:

$$\overline{v} = v\ \overline{u}_x \tag{1}$$

The cooling fluid can be air or a liquid. If heat is dissipated in the electronic components mounted on the substrate, a temperature rise will result. Normally substrates are made from ceramic materials, which are relatively good thermal conductors. The heat generated in the individual components is spread over the entire substrate [8] [9] [10] [11] [12], which behaves like an isothermal surface.

The diffusion convection equation for the fluid temperature $T(x, y, z)$ is given by:

$$\lambda \nabla^2 T = C_v \overline{v} \cdot \nabla T \tag{2}$$

where λ is the thermal conductivity of the fluid (in $W/m\ °C$) and C_v the volumetric thermal capacity expressed in $J/m^3\ °C$. Using (1), the equation (2) can be rewritten as:

$$\lambda \nabla^2 T = C_v v \frac{\partial T}{\partial x} \tag{3}$$

The boundary condition is quite simple: on the substrate the fluid temperature should be equal to the known substrate temperature T_s:

$$T(x, y, 0) = T_s \quad if \quad (x, y) \in S \tag{4}$$

where S denotes the substrate area. Note however that the temperature field T is built up on both sides of the substrate. Due to the symmetry of the problem only the upper half space $z > 0$ will be considered.

For technical applications one is primarily interested in the heat transfer from the substrate to the ambient fluid rather then a temperature distribution. The heat flux at the upper surface of the substrate is given by:

$$q = -\lambda (\frac{\partial T}{\partial z})_{z=0} \quad in \ W/m^2 \tag{5}$$

Note that the same heat flux is removed along the bottom surface. Integrating (5) over the entire substrate gives the total dissipated power P (W).

The equation (3) can be easily transformed to a more flexible form by introducing a new unknown function ψ defined by:

$$T = \psi exp(\frac{C_v v}{2\lambda} x) \tag{6}$$

Inserting (6) in the equation (3) gives the following equation for ψ:

$$\nabla^2 \psi - (\frac{C_v v}{2\lambda})^2 \psi = \nabla^2 \psi - \frac{\psi}{L^2} = 0 \tag{7}$$

The new equation (7) has the advantage that first derivatives no longer appear. The equation (7) allows radially symmetric solutions, which is not possible for (3). The new constant L appearing in (7) has the dimensions of a length and is called characteristic length:

$$L = \frac{2\lambda}{C_v v} \tag{8}$$

For air cooling with $\lambda = 0.02 W/m\ °C$, $C_v = 1.17 kJ/m^3\ °C$ and a velocity $v = 1 m/s$, the characteristic length $L = 34 \mu m$, which is very small even compared with substrate dimensions used in microelectronics.

3 FUNDAMENTAL SOLUTION

The fundamental solution or the Green's function of (3) is easily found from (7), having a solution of the form $e^{-r/L}/r$ with $r = \sqrt{x^2+y^2+z^2}$. The fundamental solution $G(\bar{r}|\bar{r}')$ is defined as the solution of (3) provided a Dirac source is applied in the point $\bar{r}'$:

$$\nabla^2 G(\bar{r}|\bar{r}') - \frac{2}{L}\frac{\partial G(\bar{r}|\bar{r}')}{\partial x} = \delta(\bar{r}-\bar{r}') \tag{9}$$

$G(\bar{r}|\bar{r}')$ is easily found to be:

$$G(\bar{r}|\bar{r}') = \frac{e^{(x-x')/L} \; e^{-|\bar{r}-\bar{r}'|/L}}{4\pi|\bar{r}-\bar{r}'|} \tag{10}$$

It is useful to get some physical understanding of (10). If the source point

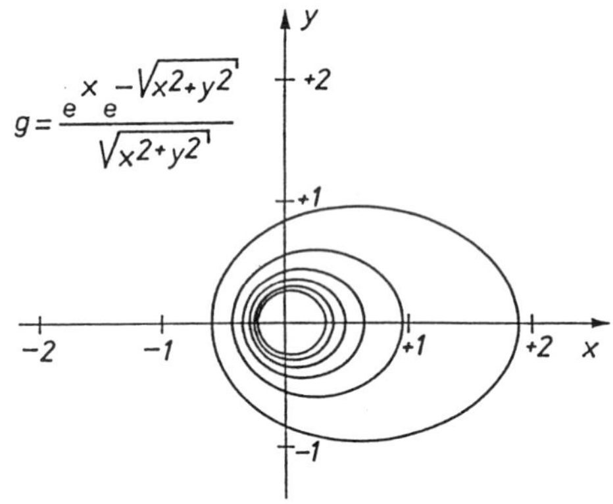

Figure 2: Isovalued lines of the fundamental solution

$\bar{r}'$ is put in the origin, fig.2 displays some isovalue lines of G. The shape is quite similar to other well known phenomena such as the distribution of smoke particles from a chimney scattered by the wind. In all directions the function (10) drops exponentially with a length constant comparable to L. In the downstream direction (= x-axis), (10) behaves like $1/x$, a much less decreasing behaviour.

4 BOUNDARY INTEGRAL EQUATION

Starting from the equations:

$$\nabla^2 T - \frac{2}{L}\frac{\partial T}{\partial x} = 0 \tag{11}$$

$$\nabla^2 G - \frac{2}{L}\frac{\partial G}{\partial x} = \delta(\overline{r} - \overline{r}') \tag{12}$$

One finds after applying Green's theorem:

$$T(\overline{r}') = \int\int_S [T(\overline{r})\frac{\partial G(\overline{r}|\overline{r}')}{\partial n} - G(\overline{r}|\overline{r}')\frac{\partial T(\overline{r})}{\partial n}]dS \tag{13}$$

where n is the outward directed normal unity vector or $n = -z$ for the flat substrate problem considered here. As already mentioned only the upper half space is considered here. The integration (13) is limited to the upper surface S of the substrate. Note that due to symmetry $\partial T/\partial n = 0$ outside the substrate in the (x, y)-plane.

Inserting the boundary condition (4) in (10) and taking into account that $\partial G(\overline{r}|\overline{r}')/\partial n = 0$ if $\overline{r}' \in S$ one obtains finally:

$$T_S = -\int\int_S G(\overline{r}|\overline{r}')\frac{\partial T(\overline{r})}{\partial n}dS \qquad \overline{r}' \in S \tag{14}$$

(14) is an integral equation in the unknown function $\partial T/\partial n$. The solution yields directly the heat flux distribution (5) we are interested in.

For the numerical solution, the substrate is divided into elements. In each of them the unknown function is replaced by an unknown constant. In order to get the coefficient matrix, the Green's function has to be integrated over the individual elements. This must be done carefully because the element dimensions can be much greater than the characteristic length L. If these integrations are done by taking only a limited number of points per element, meaningless results would be obtained.

Once the heat flux $\partial T/\partial n$ has been calculated, the temperature can be obtained in any point $\overline{r}'$ by using (13).

5 RESULTS

Fig.3 displays the temperature distribution in the (x, y)-plane. Calculations have been carried out for a square substrate with a side length of $50mm$. Air was taken as the cooling medium with a velocity of $1m/s$. For the numerical solution the substrate has been divided into $25 \times 25 = 625$ uniform elements. The substrate temperature T_S was normalised to unity, due to the linearity of the problem. The results of fig.3 clearly indicate the heating of the air in the dowmstream direction. Note that these results have been obtained with a mesh size of $2mm$ which is much larger than the characteristic length (8). Nevertheless quite accurate results are found, which have been verified by solving the same problem with different mesh sizes.

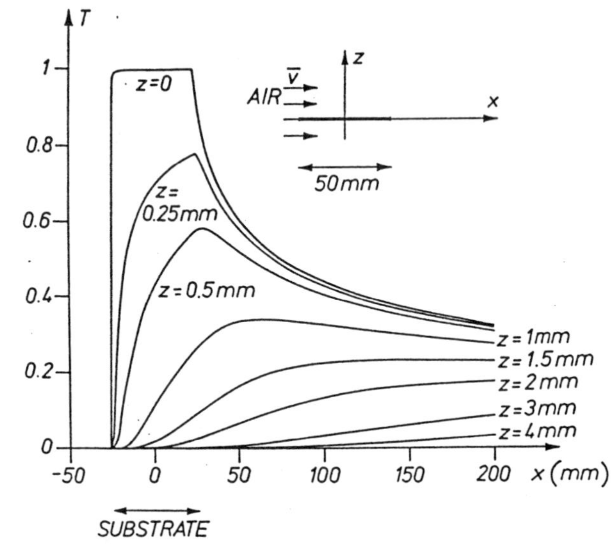

Figure 3: Air temperature above the substrate

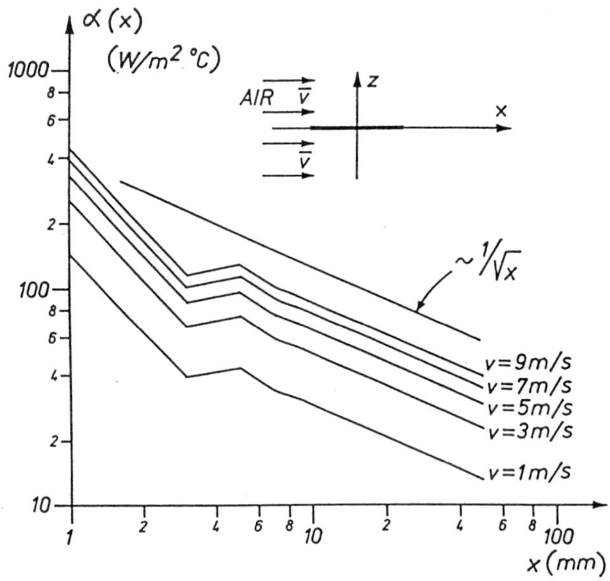

Figure 4: Convection coefficient along the x-axis

Fig.4 shows the local convection coefficient $\alpha(x)$ along the x-axis. $\alpha(x)$ is defined as the ratio of the heat flux q given by (5) and the temperature drop T_S:

$$\alpha(x) = \frac{q(x)}{T_S} = -\frac{\lambda}{T_S}\left(\frac{\partial T}{\partial z}\right)_{z=0} \tag{15}$$

Calculations have been done for various values of the air speed. As expected, the heat transfer coefficient increases with v. The results of fig.4 are also found to be in good agreement with the relation:

$$\alpha(x) \sim \frac{1}{\sqrt{x}} \tag{16}$$

A proof of (16) can be found in several textbooks on basic heat transfer [2] [25]. The decrease of the convection coefficient with x can be easily understood physically. At the upstream side, the air has not been warmed up yet, yielding a maximum cooling efficiency. At the downstream side the air has been heated by flowing over the substrate. Consequently less heat can then be removed. In electronic design, temperature sensitive elements should be placed in upstream positions for better heat removal. The irregularities in the curves of fig.4 are due to numerical noise.Remember that the unknown function was represented by a piece-

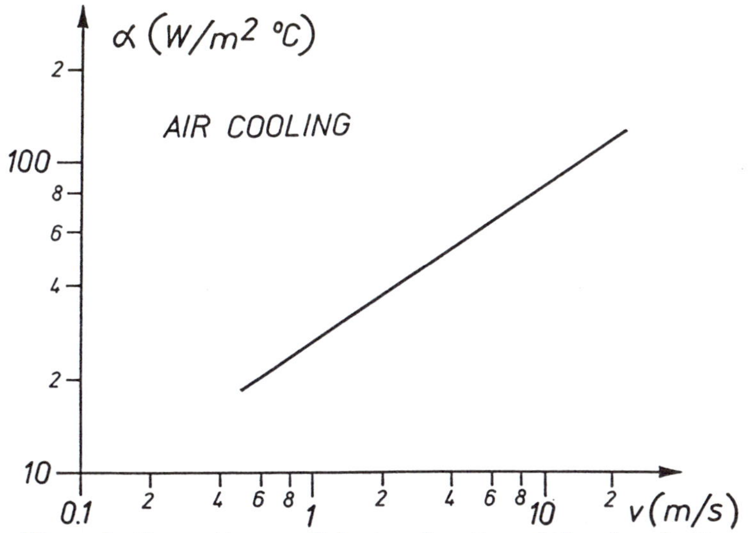

Figure 5: Convection coefficient as function of the air velocity

wise constant function which is a rather crude approximation. Note also that the logarithmic scale expands the leftmost elements.

The global heat transfer coefficient can be expressed by the average convection coefficient α which can be obtained by integrating (15) over the entire substrate. The variation of α as a function of the air speed v is plotted on fig.5. Note that the numerical results coincide perfectly with the relationsship $\alpha \sim \sqrt{v}$, generally adopted in forced convection cooling [2] [25]. The numerical values for α are all in the range found for forced air cooling [1] [2].

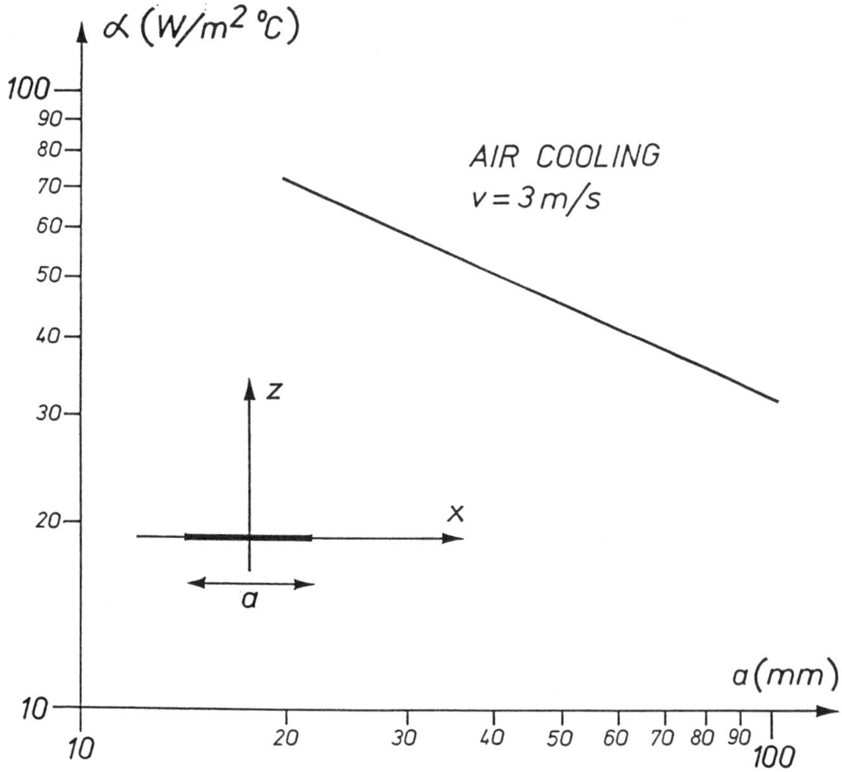

Figure 6: Convection coefficient as a function of the substrate dimension

Fig.6 gives the average convection coefficient as a function of the side length a of the substrate. The results coincide with the theoretical relation $\alpha \sim 1/\sqrt{a}$, which can be obtained by integrating (16). The larger the substrate dimensions the less efficient the cooling per unit area will be.

6 CONCLUSION

The forced convection cooling of flat substrates has been treated using a boundary element approach. The technique has proved to be very adequate and the results were found to be in perfect agreement with the basic data published in most textbooks on heat transfer.

7 ACKNOWLEGDEMENTS

The authors wich the thank the IWONL (Belgium) for their financial support.

8 REFERENCES

1. D. J. DEAN:
"Thermal design of electronic circuit boards and packages"
Electrochemical Publications, Ayr, Scotland, 1985.

2. G. ELLISON:
"Thermal computations for electronic equipment"
Van Nostrand, New York, 1984.

3. R. T. HOWARD (ed.):
"Thermal management concepts in microelectronic packaging"
ISHM Technical Monograph, 6984-003, Silver Spring, 1984.

4. A. BAR-COHEN and A. D. KRAUS (eds.):
"Advances in thermal modelling of electronic components and systems"
Hemisphere Publications, New York, 1988.

5. G. ELLISON:
"The thermal design of an LSI single chip package"
IEEE Transactions on Parts, Hybrids and Packaging, 1973, vol. PHP-12, p. 371-378.

6. M. MALALINGAM:
"Thermal management in semiconductor device packaging"
Proceedings IEEE, 1985, vol.73, p.1396-1404.

7. R. F. DAVID:
"Computerised thermal analysis of hybrid circuits"

IEEE Transactions on Parts, Hybrids and Packaging, 1977, vol. PHP-13, p.283-290.

8. E. WEHRHAHN:
"Temperature distribution on substrates of hybrid integrated circuits"
Proceedings of the 5th European Hybrid Microelectronics Conference, Stresa, 1985, p.288-295.

9. G. CASSELMAN and G. DE MEY:
"A thermal model for hybrid circuits"
Hybrid Circuits, 1986, vol.10, p.9-13.

10. G. DE MEY and L. VANSCHOOR:
"Thermal analysis of hybrid circuits with mounted components"
Hybrid circuits, 1988, vol.15, p.28-32.

11. E. WEHRHAHN:
"Temperature distribution on multilayer substrates"
Proceedings 6th European Microelectronics Conference, 1987, Bournemouth, p.264-280.

12. G. DE MEY, L. ROTTIERS, M. DRISCART and L. VANSCHOOR:
"Thermal studies on hybrid circuits"
Hybrid Circuits, 1988, vol.17, p.8-11.

13. G. ELLISON:
"TAMS: A thermal analyser for multilayer structures"
Electrosoft, 1990, vol.1, p. 85-97.

14. A. L. PALISOC, J. MIN and C. C. LEE:
"The effect of die bond voids on the thermal performance degradation of solid state devices"
Proceedings BEM10 Conference, 1988, Soutampton, vol.2, p. 543-554.

15. C. C. LEE and A. L. PALISOC:
"Thermal analysis of semiconductor devices"
Topics in Boundary Element Research, C. A. Brebbia (ed.), vol.7, 1990, p. 12-33.

16. M. DRISCART and G. DE MEY:
"A boundary element approach for thermal modelling of microelectonic components and packages"
Proceedings EUROBEM Conference, 1990, Nice, p. 393-404.

17. E. BOONE, G. DE MEY and L. ROTTIERS:
"Temperature distribution on ceramic substrates"
Electronics Letters, 1986, vol.22, p. 442-443.

18. L. ROTTIERS and G. DE MEY:
"Simulation of the thermal behaviour in hybrid circuits"
Proceedings of the Intersociety Conference on Thermal Phenomena in Electronic Components, Los Angeles, 1988, p.15-20.

19. J. C. LIU, D. E. HUNTER and R.L. KOZAREK:
"Electronic package thermal design using boundary elements"
Proceedigs SEMI-THerm Symposium, Phoenix, 1990, p. 65-69.

20. A. BAR-COHEN:
"Bounding relations for natural convection heat transfer from vertical printed circuit boards"
Proceedings IEEE, 1985, vol.73, p.1388-1395.

21. W. A. CAMPO, G. KERJILIAN and H. SHAUKATULLAH:
"Prediction of component temperatures on circuit cards, cooled by natural convection"
IEEE Transactions on Components, Hybrids and Manufacturing Technology, 1982, vol. CHMT-5, p.499-501.

22. A. VIAULT, V. PIMONT, A. LATROBE et M. CADRE:
"THEBES, un logiciel de calcul des temperatures d'air dans les equipements electroniques"
Proceedings 3rd Conference on Heat Transfer, Energy and Environment applied to Communication and Computer fields, Perros-Tregastel, France, 21-23/06/89.

23. G. FUCHS:
"Rack thermal measurements and a new rack simulation model"
Proceedings 3rd Conference on Heat Transfer, Energy and Environment applied to Communication and Computer fields, Perros-Tregastel, France, 21-23/06/89.

24. D. B. FIGUEIREDO and L. C. WROBEL:
"A boundary element analysis of convective heat diffusion problems"
Proceedings Advanced Computational Methods in Heat Transfer, Portsmouth 1990, p. 230-238.

25. F. P. INCROPERA and D. P. DE WITT:
"Introduction to heat transfer"
John Wiley, New York, 1985, p. 292.

An Analysis of the Kutta-Joukovski-Carafoli Condition at Axial Profile Cascades using the Boundary Element Method

I.N. Carte

Polytechnical Institute "Traian Vuia" Timişoara România

ABSTRACT

The circulation determination in the case of the ideal incompressible fluid motion round isolated profiles and round axial profile cascades is made by imposing an exterior condition, the Kutta-Joukovski-Carafoli condition, it means null velocity in the trailing edge of the profile. An analysis of this condition by imposing null velocity in different points from the neighbouring of the trailing edge and the influence upon the velocity and pressure field, locally and globally, on the profile boundary, is presented in the paper.
For this, the boundary element method was used for solving a boundary-value problem for the Laplace equation in the stream function ψ. The analysis was done upon a pump axial cascades.

INTRODUCTION

The solving of the direct problem of the axial profile cascades means the motion determination with circulation, round these cascades, of the incompressible and ideal fluid, the cascade geometry and cinematic conditions in a section, accepted as entrance being given.
The determination of these problems requires a supplementary boundary condition - the Kutta-Joukovski-Carafoli condition.

THE TYPICAL ANALYSIS DOMAIN. THE BOUNDARY-VALUE PROBLEM FORMULATION

The analysis domain is defined by the boundaries of two neighbouring profiles disposed at step t, and the boundary conditions on the influx respectively flux boundary will be put at the t/2 from the loading edges front, respectively trailing edges at the profiles (fig.1) [1].

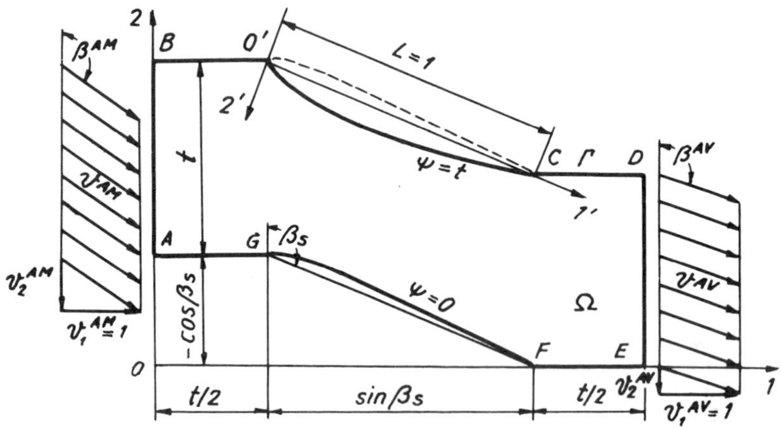

Fig.1. The analysis domain for a pumps axial cascades

If O12 is the fundamental plan at the motion, the potential flow at the ideal incompressible fluid leads to the Laplace equation for the stream function

$$\psi_{/ii} = 0 \tag{1}$$

and the velocity fluid is given by

$$v_i = \varepsilon_{ij}\,\psi_{/j} \tag{2}$$

The velocity field periodicity imposes for ψ function

$$\begin{aligned} \psi(x_1, x_2 + kt) &= \psi(x_1, x_2) + kt \qquad k \in \mathbb{Z} \\ \psi_{/i}(x_1, x_2 + kt) &= \psi_{/i}(x_1, x_2) \end{aligned} \tag{3}$$

Thus we have to solve a mixed boundary value problems with periodicity, for equation (1) with the following boundary conditions on the Γ boundary at the Ω domain.

$$\begin{cases} \psi = 0 & \text{on GF} \\ \psi = t & \text{on O'C} \\ \psi_{/n} = -\psi_{/1} = ctg\,\beta^{AM} & \text{on AB} \\ \psi_{/n} = \psi_{/1} = a & \text{on DE} \\ \psi \text{ and } \psi_{/n} & \text{periodical on AG and BO', respectively FE and CD} \\ \psi_{/n} = 0 & \text{in the stagnation posterior point} \end{cases} \quad (4)$$

accepting on the AB and DE

$$v_1^{AM} = v_1^{AV} = 1 \quad (5)$$

and the a constant, once determined, will define the exit angle β^{AV} through

$$\beta^{AV} = arcctg\,(-a) \quad (6)$$

THE UTILIZATION OF THE BOUNDARY ELEMENT METHOD

In the plane potential motion hypothesis, thus governed by equation (1), the connection between unknown values of the function ψ and normal derivative $\psi_{/n}$ on the boundary is given by the integral equation on the boundary [2].

$$c(\xi)\psi(\xi) + \int_\Gamma \psi(x)[\psi^*(\xi,x)]_{/n} d\Gamma(x) = \int_\Gamma [\psi(x)]_{/n} \psi^*(\xi,x) d\Gamma(x) \quad (7)$$

where $\psi^*(\xi,x)$ is the fundamental solution of Laplace equation. If we make a discretization of the boundary in N elements Γ_j, the discretized form of the equation (7) leads to the algebric system

$$\sum_{j=1}^{N} H_{ij}\psi_j = \sum_{j=1}^{N} G_{ij}[\psi_{/n}]_j \quad (8)$$

in which, for linear elements

$$H_{ij} = h^2_{ij-1} + h^1_{ij} \qquad i \neq j \quad (9)$$

$$H_{ii} = -\sum_{\substack{j=1 \\ j\neq i}}^{N} H_{ij} \quad (10)$$

$$G_{ij} = g^2_{ij-1} + g^1_{ij} \qquad i \neq j \tag{11}$$

$$G_{ii} = \frac{l_{i-1}}{2}\left[\frac{3}{2} - \ln(l_{i-1})\right] + \frac{l_i}{2}\left[\frac{3}{2} - \ln(l_i)\right] \tag{12}$$

with

$$h^{\alpha}_{ij} = \int_{\Gamma_j} \varphi^{\alpha} \psi^{*}_{/n} \, d\Gamma \qquad \alpha = 1,2 \tag{13}$$

$$g^{\alpha}_{ij} = \int_{\Gamma_j} \varphi_{\alpha} \psi^{*} \, d\Gamma \qquad \alpha = 1,2 \tag{14}$$

in which

$$\varphi_1 = \frac{l}{2}(1-\eta) \; ; \; \varphi_2 = \frac{l}{2}(1+\eta) \tag{15}$$

are the linear interpolation functions, in η - natural coordinates.
Let m respectively n the number of intervals on BO' and AG respectively CD and FE. The number of uniknowns is N+2m+2n+4. N equations of type (8) can be written, from the periodicity conditions (3) we have m+n-2 relations for ψ function values

$$\psi^{u}_{i} = \psi^{l}_{i'} + t \qquad \text{i and i' have the same abscissa } x_1 \tag{16}$$

m+n-2 relations for the normal derivatives $\psi_{/n}$

$$\left[\psi_{/n}\right]^{u}_{i} = \left[\psi_{/n}\right]^{l}_{i'} \qquad \text{i and i' have the same abscissa } x_1 \tag{17}$$

on parts AG, BO' and FE, CD and 8 linear relations between the lateral normal derivatives in corner nodes (the direction of the velocity vector is known).
Therefore we have N+2m+2n+4 equations.

THE VELOCITY AND PRESSURE FIELD ON THE PROFILE BOUNDARY

The velocity on the profile boundary is given by the normal derivative

$$\psi_{/n} = \psi_{/i} \, n_i = v_{tan} \tag{18}$$

or reported to the entrance velocity v^{AM}

$$\bar{v} = v \left(v^{AM}\right)^{-1} \tag{19}$$

The pressure field is obtained writing a Bernoulli theorem between entrance and curent point

$$\bar{p} = 1 - \bar{v}^2 \qquad (20)$$

NUMERICAL REZULTS

The method was applied to a pump axial cascade formed from NACA 8410 profiles with parameters t/L = 0,75; $\beta^{AM} = 130^\circ$; $\beta_s = 120^\circ$; $\beta^{AV} = 110^\circ$.
In fig.2 the domain of analysis considerind the profile cord equal to the unity and the discretization of the boundary in linear elements are given.

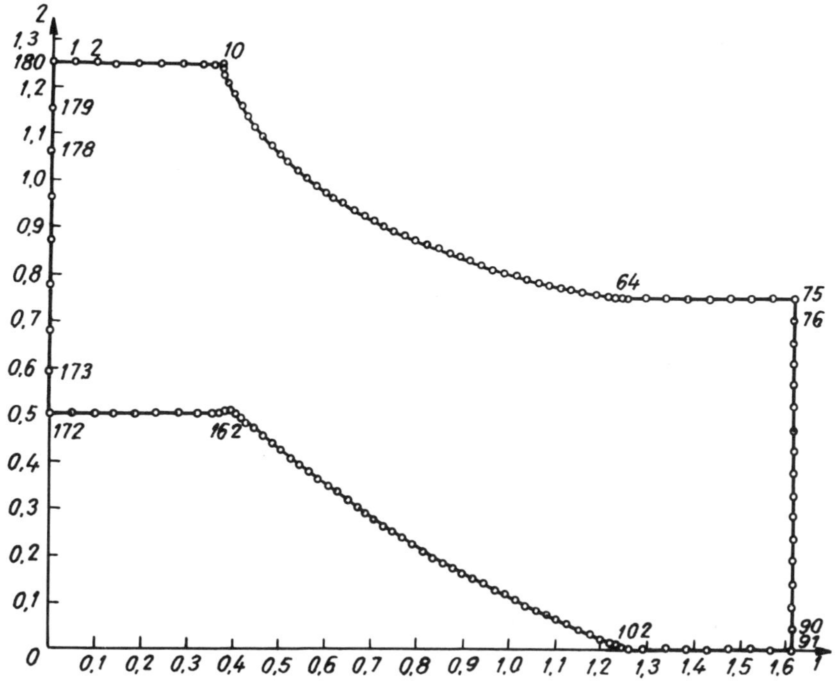

Fig.2. The analysis domain and the discretization of the boundary

In fig.3 NACA 8410 profile, the discretization of his boundary in linear elements and the correspondence between the nodes on his boundary and the nodes on the analysis domain boundary are given and in fig.4 the boundary discretization from the trailing edge zone of the profile is given in detail.

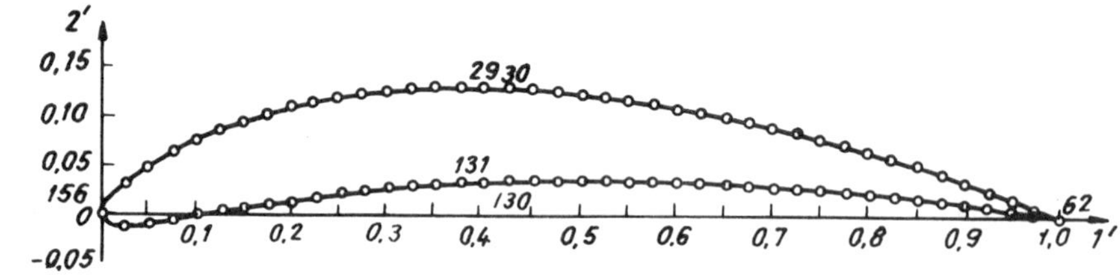

Fig. 3. NACA 8410 profile. His boundary discretization and the correspondence between the nodes on the profile boundary and the nodes on the analysis domain boundary

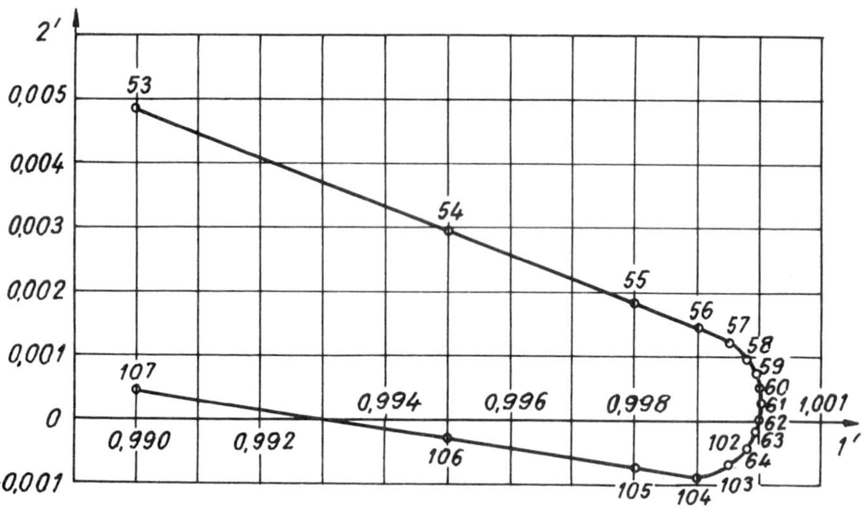

Fig.4. The boundary discretization from the trailing edge zone of the profile

In the following an analysis is presented concerning the imposing of the Carafoli condition in different points on the profile boundary from the trailing edge zone and its influence on the velocity and pressure field on the profile boundary. Thus in fig.5, the velocity field $\overline{v}$ on the profile boundary is given for three points in which the Carafoli condition was imposed: node 62 (the trailing edge), node 59 (suction side) and node 103 (pressure side) (fig.4), and in fig.6 the pressure field p on the profile boundary is given in the same conditions

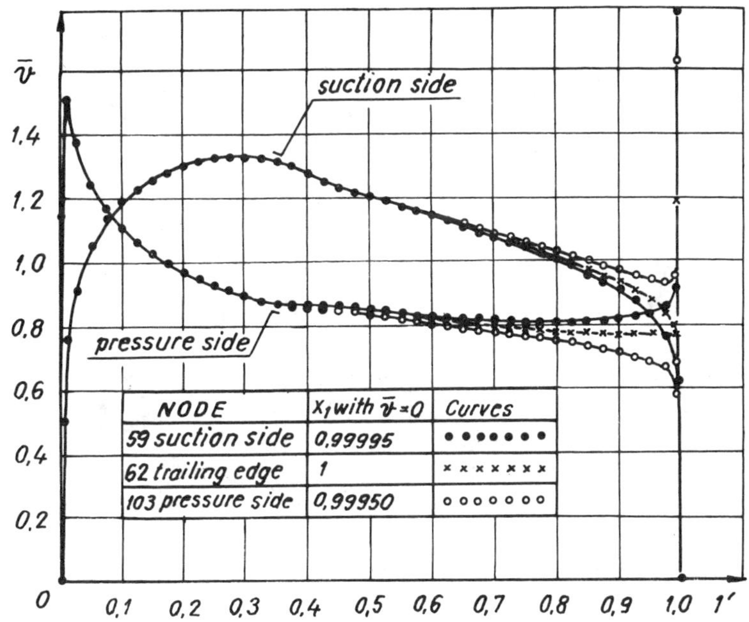

Fig.5. The velocity field $\overline{v}$ on the profile boundary for different points in which the Carafoli condition was imposed

In fig.7 the velocity field $\overline{v}$ in the trailing edge zone of the profile is given in detail for seven points in which the Carafoli condition was imposed: node 62 (trailing edge), nodes 59, 60 and 61 on suction side and nodes 63, 64 (102) and 103 on pressure_side (fig.4), and in fig.8 the pressure field $\overline{p}$ in the trailing edge zone of the profile is given in detail in the same conditions

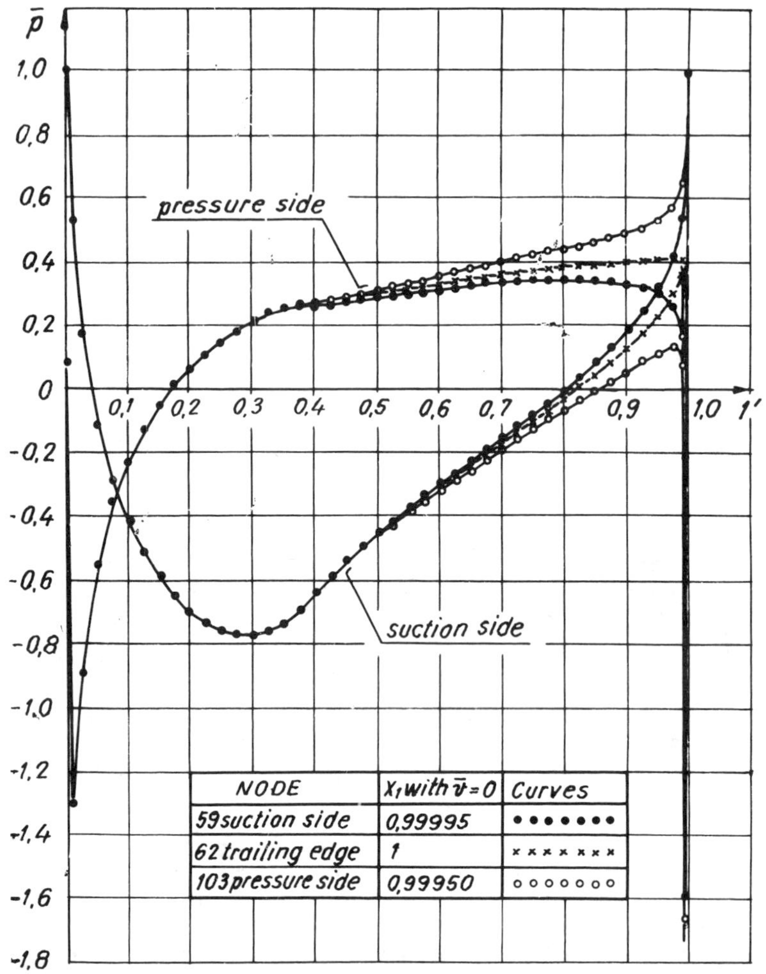

NODE	X_1 with $\bar{v}=0$	Curves
59 suction side	0,99995	● ● ● ● ● ● ●
62 trailing edge	1	× × × × × × ×
103 pressure side	0,99950	○ ○ ○ ○ ○ ○ ○

Fig.6. The pressure field $\bar{p}$ on the profile boundary for different points in which the Carafoli condition was imposed

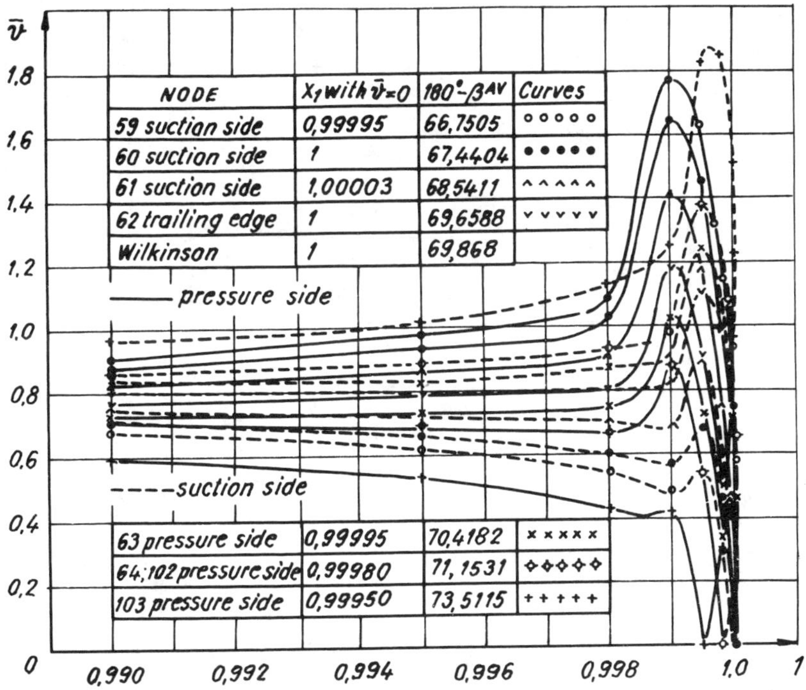

Fig.7. The velocity field $\overline{v}$ on the profile boundary from the trailing edge zone for different points in which the Carafoli condition was imposed

CONCLUSIONS

The use of the boundary element method in the study of the plane potential motion around profile cascades proved to be efficient and operative. The used of the linear element for the boundary discretization permited the determination with a great acuracy, of the velocity an pressure field on the profile boundary from the trailing edge as well as the imposing of the Carafoli condition.

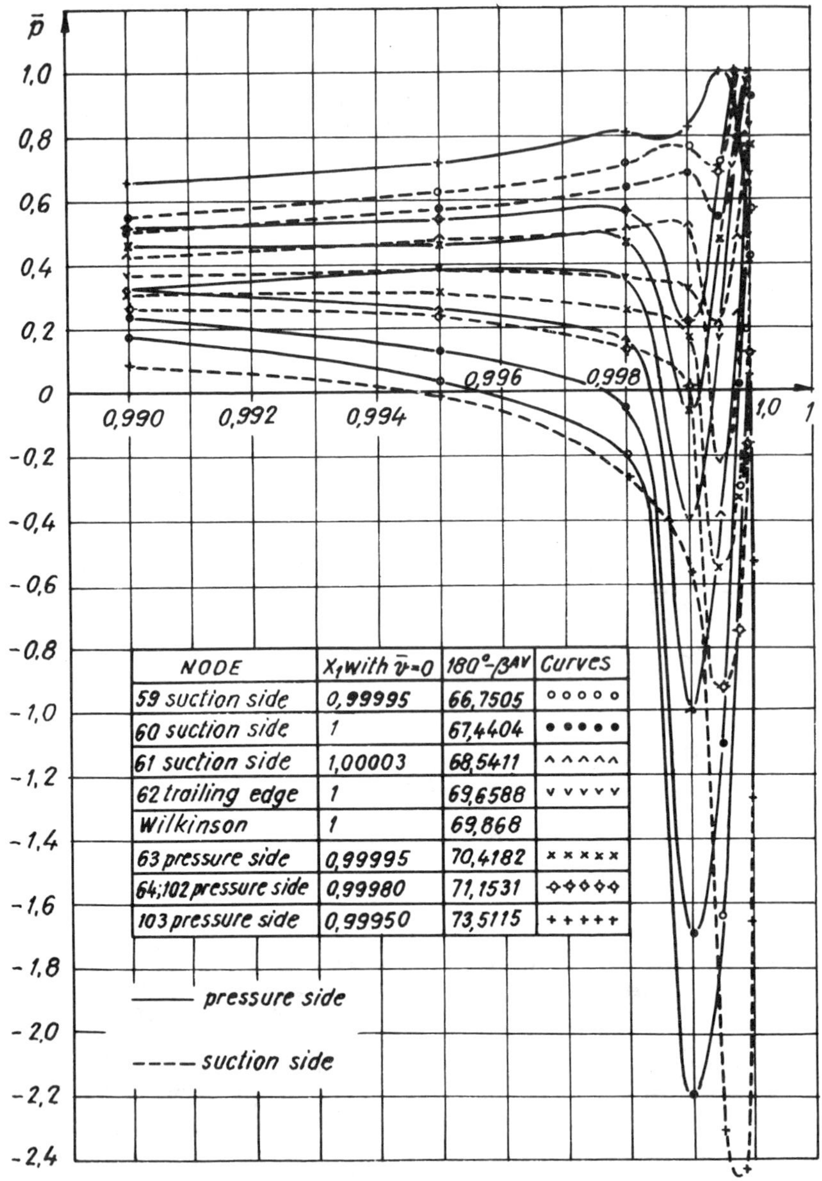

Fig. 8. The pressure field $\bar{p}$ on the profile boundary from the trailing edge zone for different points in which the Carafoli condition was imposed

REFERENCES

1. Anton, I. and Carte, I.N. The Finite Element Method and the Boundary Element Method in the Hydrodynamics of Axial Profile Cascades in Hydroturbo, Vol.1, pp AI1 - AI9, Proceeding of the International Conference Hydroturbo 89 on Hydrodynamic Machines in Power Engineering, Brno, Czechoslovakia, 1989.
2. Brebbia, C.A., Telles, J.C.F. and Wrobel, L.C. Boundary Element Techniques. Chapter 2, Potentials Problems, pp 47-107. Springer-Verlag, Berlin, Heidelberg, New-York and Tokyo, 1984.

SECTION 2: ELECTRICAL AND ELECTROMAGNETIC PROBLEMS

Boundary Element Analysis of Electromagnetic Fields in Helically Symmetric System

H. Igarashi, T. Honma

Department of Electrical Engineering, Faculty of Engineering, Hokkaido University, Kita 13, Nishi 8, Kita-ku, Sapporo, 060 Japan

ABSTRACT

This paper presents a boundary element analysis of electromagnetic fields in helical systems. A two dimensional integral equation is introduced from the scalar Helmholtz-type equation in the general curvilinear coordinates. An integral equation for helical systems is derived from the above equation using the twisted coordinates. The boundary element method based on the present formulation is applied to the analyses of potential fields in a helical column, electromagnetic fields in a twisted waveguide and magnetohydrodynamic equilibria in a helical vessel.

INTRODUCTION

In computational analyses of field problems, systems can often be described in two dimension. For this reason, a number of numerical schemes have been studied especially for translationally symmetric and axisymmetric problems. On the other hand, we find helical structures in many engineering devices such as stellarator-type nuclear fusion machines [1], twisted waveguides [2], optical fibers [3] and helical antennas [4]. It is convenient to use the twisted coordinates (X, Y, Z) in which the axes X and Y rotate with the rotation of cross section for the analyses of the electromagnetic fields in these devices [2]. Figure 1 illustrates a straight helical system of a helical pitch h and the twisted coordinates (X, Y, Z). In this coordinates, the fields can be reduced to be two dimensional because we can assume some dependence of physical quantities on the axis Z,

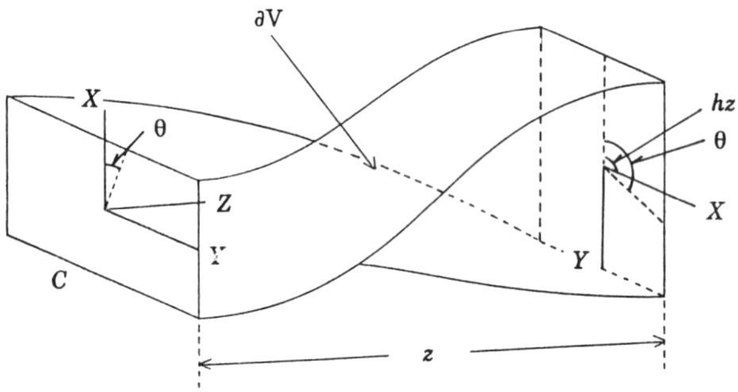

Fig. 1 Straight helical system

usually *exp(-jβZ)*. (When $\beta=0$, the system is called helically symmetric system.)

The above helical systems are often analyzed by the perturbation method [2-4] with the assumption of a small helical pitch. When analyzing the problems by the perturbed technique, we can understand the physical properties such as field configurations and dispersion relations etc. more systematically than numerical approaches because we can obtain those properties in closed forms. However, the dependence of the accuracy of the perturbed solutions on the helical pitch has not investigated because we do not have any standard numerical schemes in which the accuracy of the solutions are guaranteed for arbitrary helical pitch values. In practice, there are few studies [5] on the numerical schemes for the helical systems in contrast with translationally symmetric and axisymmetric systems. In this paper, we present boundary element analyses of electromagnetic field problems in straight helical structures. The main purpose of this study is to show the validity and usefulness of the boundary element analysis of helical electromagnetic field problems.

The remainder of this paper is organized as follows. Firstly, we derive a two dimensional boundary integral equation from a Helmholz-type differential equation in the general curvilinear coordinates. Next, from the above equation, we introduce a boundary integral equation described in the twisted coordinates for the analysis of helical electromagnetic problems. Finally, the present method is applied to the analyses of potential fields in a helical column, electromagnetic fields in a twisted waveguide and magnetohydrodynamic (MHD) equilibria in a helical vessel.

FORMULATION

Two dimensional boundary integral equation in the general curvilinear coordinates

Before we carry out the boundary element formulation for helical systems, we summarize below the relations associated with the general curvilinear coordinates (u^1, u^2, u^3).

$$\mathbf{e}^i \equiv \nabla u^i \ , \tag{1}$$

$$\mathbf{e}_1 \equiv \sqrt{g}(\mathbf{e}^2 \times \mathbf{e}^3), \mathbf{e}_2 \equiv \sqrt{g}(\mathbf{e}^3 \times \mathbf{e}^1), \mathbf{e}_3 \equiv \sqrt{g}(\mathbf{e}^1 \times \mathbf{e}^2) \ , \tag{2}$$

$$\mathbf{A} = A^i \mathbf{e}_i = A_i \mathbf{e}^i \ , \tag{3}$$

$$A^i = \mathbf{A} \cdot \mathbf{e}^i \ , \ A_i = \mathbf{A} \cdot \mathbf{e}_i \ , \tag{4}$$

$$\mathbf{e}^i = g^{ij} \mathbf{e}_j \ , \ A^i = g^{ij} A_j \ , \tag{5}$$

$$g^{ij} \equiv \mathbf{e}^i \cdot \mathbf{e}^j \ , \ g \equiv 1 / det[g^{ij}] \ , \tag{6}$$

$$\nabla^2 = \frac{1}{\sqrt{g}} \frac{\partial}{\partial u^i} \left(g^{ij} \sqrt{g} \frac{\partial}{\partial u^j} \right) \tag{7}$$

where $\mathbf{e}^i$ and $\mathbf{e}_i$ are the covariant and contravariant basis vectors, respectively, g^{ij} is the metric tensor, A_i and A^i are the covariant and contravariant components of vector $\mathbf{A}$ and we use Einstein's convention in eqs. (3), (5) and (7).

Let us consider the Helmholtz-type equation

$$\frac{1}{\sqrt{g}} \frac{\partial}{\partial u^i} \left(g^{ij} \sqrt{g} K \frac{\partial \Psi}{\partial u^j} \right) + k^2 \Psi$$

$$\equiv L \Psi + k^2 \Psi = F \ , \tag{8}$$

in the coordinates (u^1, u^2, u^3), where $K = K(u^1, u^2)$ and $F = f(u^1, u^2, \psi)$ $exp(-j\beta u^3)$. Note that, in eq. (8), the operator L corresponds to Laplacian ∇^2 when the function K is 1. To solve eq. (8) in the system in which the function Ψ takes the form $\psi(u^1, u^2)\, exp(-j\beta u^3)$ by the use of boundary element method, it is convenient to derive a two dimensional boundary integral equation on the surface of u^3=constant. For this purpose, multiplying eq. (8) by the fundamental solution $\Psi^* = \psi^*(u^1, u^2)\, exp(j\beta u^3)$, which satisfies

$$\frac{1}{\sqrt{g}} \sum_{i=1}^{2} \sum_{j=1}^{2} \frac{\partial}{\partial u^i} \left(g^{ij} \sqrt{g} K \frac{\partial \psi^*}{\partial u^j} \right) + \frac{4\pi}{\sqrt{g}} \delta(u^1 - u^1{}_0)\, \delta(u^2 - u^2{}_0)$$

$$\equiv L_t \psi^* + \frac{4\pi}{\sqrt{g}} \delta(u^1 - u^1{}_0)\, \delta(u^2 - u^2{}_0) = 0 \ , \tag{9}$$

and integrating over a region V enclosed by a boundary ∂V yield

$$C_0\psi_0\int du^3 = \int_{\partial V} K\Psi^* \frac{\partial \Psi}{\partial n} dS - \int_{\partial V} K\Psi \frac{\partial \Psi^*}{\partial n} dS + \int_V [\Psi(L - L_t + k^2) - F]\Psi^* dV, \tag{10}$$

where $\partial / \partial n = \mathbf{n} \cdot \nabla$, $\mathbf{n}$ denotes the unit vector in the normal direction of the surface ∂V and C_0 is the geometric factor related to the Cauchy principal value of the boundary integrals and if the boundary ∂V is smooth, then $C_0 = 2\pi$. Moreover, the surface element dS and operator $\partial / \partial n$ in eq. (10) can be expressed as

$$dS = |\mathbf{N}|\, ds\, du^3, \tag{11}$$

$$\frac{\partial}{\partial n} = \frac{\mathbf{N}}{|\mathbf{N}|} \cdot \mathbf{e}^i \frac{\partial}{\partial u^i}, \tag{12}$$

where

$$\mathbf{N} \equiv \left(\frac{du^1}{ds}\mathbf{e}_1 + \frac{du^2}{ds}\mathbf{e}_2 \right) \times \mathbf{e}_3, \tag{13}$$

and ds denotes the differential line element on the intersection C between surface of $u^3 = \text{constant}$ and ∂V. Figure 2 shows the definition of the vector $\mathbf{N}$.

From eqs. (10) - (13), we can derive the two dimensional integral equation as follows :

$$C_0\psi_0 = \oint_C K\psi^* q P\, ds - \oint_C K\psi Q(\psi^*)\, ds + \int_\Omega [\psi R(\psi^*) - f\psi^*\sqrt{g}]\, du^1 du^2, \tag{14}$$

where

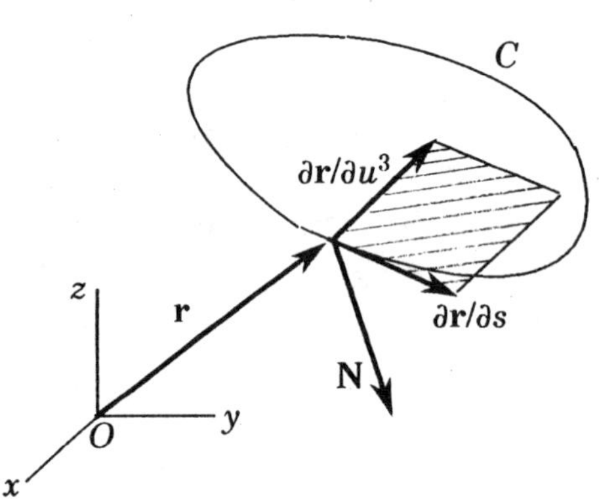

Fig. 2 Definition of the vector $\mathbf{N}$. In this figure, $\partial\mathbf{r}/\partial u^3$ and $\partial\mathbf{r}/\partial s$ correspond to $\mathbf{e}_3$ and $du^1/ds\,\mathbf{e}_1 + du^2/ds\,\mathbf{e}_2$, respectively.

$$P \equiv \sqrt{g}\left[g^{22}\left(\frac{du^1}{ds}\right)^2 - 2g^{12}\frac{du^1}{ds}\frac{du^2}{ds} + g^{11}\left(\frac{du^2}{ds}\right)^2\right]^{\frac{1}{2}},$$

$$Q \equiv \sqrt{g}\left[-\frac{du^1}{ds}\left(g^{12}\frac{\partial}{\partial u^1} + g^{22}\frac{\partial}{\partial u^2} + j\beta g^{23}\right) + \frac{du^2}{ds}\left(g^{11}\frac{\partial}{\partial u^1} + g^{12}\frac{\partial}{\partial u^2} + j\beta g^{13}\right)\right],$$

$$R \equiv j\beta\left[\sqrt{g}K\left(g^{13}\frac{\partial}{\partial u^1} + g^{23}\frac{\partial}{\partial u^2}\right) + \left(\frac{\partial}{\partial u^1}g^{13} + \frac{\partial}{\partial u^2}g^{23}\right)\sqrt{g}K\right] + \sqrt{g}(k^2 - \beta^2 g^{33}K), \tag{15}$$

and $q \equiv \partial\Psi/\partial n\ exp(j\beta u^3)$. When we specify the system, *i.e.*, determine the metric tensor g^{ij} , we can easily obtain the two dimensional integral equation for the system from eq. (14).

Boundary integral equation for straight helical systems

For the analysis of straight helical systems, we introduce the twisted coordinates (X, Y, Z) [2]

$$\begin{aligned} X &= x\cos(hz) + y\sin(hz), \\ Y &= y\cos(hz) - x\sin(hz), \\ Z &= z, \end{aligned} \tag{16}$$

where (x, y, z) are the Cartesian coordinates. Figure 3 illustrates the relation between the twisted and Cartesian coordinates. In straight helical systems, as shown in Fig. 1, physical quantities may be dependent on Z in the form $exp(-j\beta Z)$. Hence we can apply the formulation in the previous section to these systems. (Here note that we can also choose the axes r and $\zeta \equiv \theta - hz$ instead of X, Y, which are often used in the plasma physics [1], to describe the straight helical systems)

The substitution of eqs. (1) and (16) into eq. (6) yields the metric tensor for the twisted coordinates g^{ij} as follows :

$$\left[g^{ij}\right] = \begin{bmatrix} 1+h^2Y^2 & -h^2XY & hY \\ -h^2XY & 1+h^2X^2 & -hX \\ hY & -hX & 1 \end{bmatrix}, \tag{17}$$

and it can be easily shown that $g = 1$.

Substituting eq.(17) into eqs. (14) and (15), it can be shown that the Helmholtz-type differential equation of our interest takes the following form in the twisted coordinates

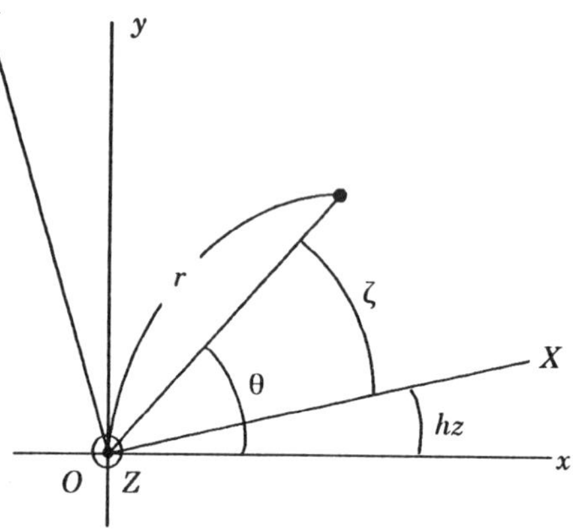

Fig. 3. Twisted Coordinates (X, Y, Z)

$$(1+h^2Y^2)K\frac{\partial^2\psi}{\partial X^2}-2h^2XYK\frac{\partial^2\psi}{\partial X\partial Y}+(1+h^2X^2)K\frac{\partial^2\psi}{\partial Y^2}$$
$$-\left[K(2j\beta hY+h^2X)-(1+h^2Y^2-j\beta hY)\frac{\partial K}{\partial X}+h^2XY\frac{\partial K}{\partial Y}\right]\frac{\partial\psi}{\partial X}$$
$$+\left[K(2j\beta hX-h^2Y)+(1+h^2X^2+j\beta hX)\frac{\partial K}{\partial Y}-h^2XY\frac{\partial K}{\partial X}\right]\frac{\partial\psi}{\partial Y}+(k^2-K\beta^2)\psi=f. \tag{18}$$

Moreover, the boundary integral equation to be numerically solved can be derived from eqs. (14), (15) and (17) as follows :

$$C_0\psi_0=\oint_C K\psi^* q\,[1+h^2(XX'+YY')^2]^{1/2}\,ds$$
$$-\oint_C K\psi\left[X'\left\{h^2XY\frac{\partial}{\partial X}-(1+h^2X^2)\frac{\partial}{\partial Y}+j\beta hX\right\}\right.$$
$$\left.+Y'\left\{-h^2XY\frac{\partial}{\partial Y}+(1+h^2Y^2)\frac{\partial}{\partial X}+j\beta hY\right\}\right]\psi^*\,ds$$
$$+\int_\Omega\left[j\beta h\psi\left\{2K\left(Y\frac{\partial}{\partial X}-X\frac{\partial}{\partial Y}\right)+Y\frac{\partial K}{\partial X}-X\frac{\partial K}{\partial Y}\right\}+\psi(k^2-\beta^2K)-f\right]\psi^*\,dXdY. \tag{19}$$

NUMERICAL RESULTS

Helically symmetric potential problem ($K=1$, $k=\beta=f=0$)

We solve the helically symmetric potential problem in the helical column with a rectangular cross section shown in Fig.4. In this case, since the equation to be solved is the Laplace equation, the functions K and f, and constant k in eq. (19) are taken to be 1.0, 0.0 (constant)

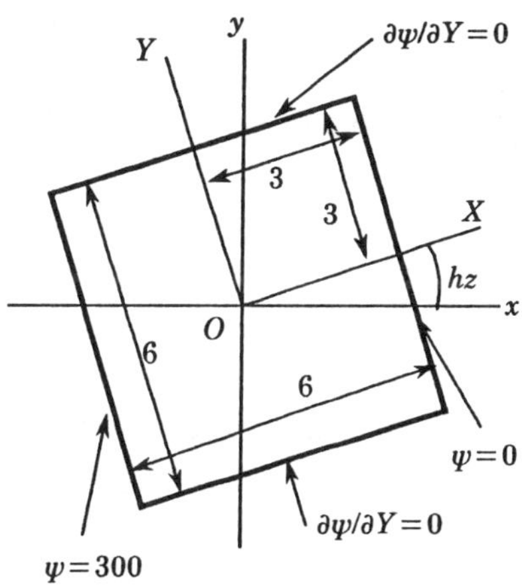

Fig. 4. Potential problem

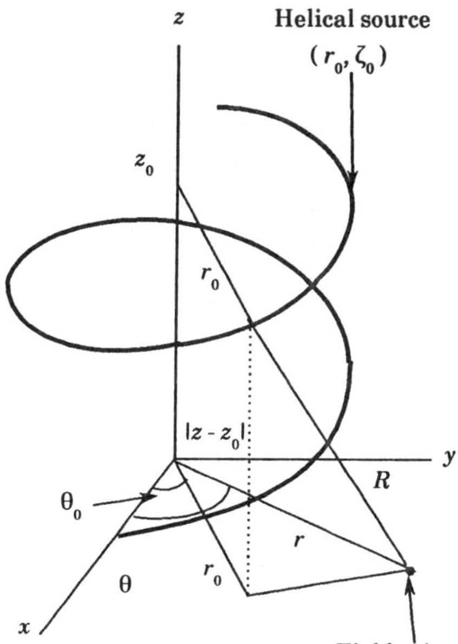

Fig. 5. Helical source

and 0.0, respectively. Moreover, the constant β is set to be 0.0 because of the helical symmetry.

The fundamental solution ψ^* for this case is defined in terms of eqs. (9) and (17) and can be expressed in the form

$$\psi^* = -2 \log r_> + 4 \sum_{m=1}^{\infty} I_m(mhr_<) K_m(mhr_>) \cos[\, m(\zeta - \zeta_0)\,], \tag{20}$$

where I_m and K_m are the modified Bessel functions of the 1st and 2nd kind and $r_> \equiv max(r, r_0)$ and $r_< \equiv min(r, r_0)$. It can be shown that the representation (20) is equivalent to the integral form [1, 6]

$$\psi^* = \int_{-\infty}^{\infty} \left[\frac{1}{R} - \frac{1}{\{a^2 + (z - z_0)^2\}^{1/2}} \right] dz_0, \tag{21}$$

where the integration is performed along the source on the helix $r = r_0$, $\zeta = \zeta_0$ and the distance R between the source and a field point (r, θ, z) is defined as $R \equiv \{ r^2 + r_0^{\,2} - 2rr_0 \cos(\theta - \theta_0) + (z - z_0)^2 \}^{1/2}$. Here, note that θ_0 depends on z_0, *i.e.*, $\theta_0 = \zeta_0 + hz_0$. In addition, the second term of the integrand in eq. (21) is the additional term which makes the integration converge and a is an arbitrary constant. A helical source and field point are illustrated in Fig. 5. Although we can clearly understand the physical meaning of representation (21), it is expensive to numerically calculate the integration. Namely, we have to subdivide the integral domain into very small elements especially when field points are near the source because the function $1/R$ vibrates with a large amplitude in such cases. Hence, we calculate the value of ψ^* by representation (20) in this paper. An effective scheme for evaluation of the infinite series in eq. (20) is discussed in detail in the references [7, 8].

To numerically solve the integral equation (19), we subdivide the contour in Fig. 4 into 80 linear portions and use linear and constant interpolations for the potential ψ and flux q, respectively, in those portions.

The solution by boundary element method for $h=0.3$, 0.5 are summarized with the 2nd-order perturbed solutions and finite element solutions [8] in Table 1. From these results, it can be seen that the boundary element solution is in good agreement with the 2nd-order perturbed solution and the finite element solution for $h=0.3$. However, there are some discrepancies between the two numerical solutions and the 2nd-order perturbed solution for $h=0.5$. This reason may be that the perturbed solution does not converge in good accuracy for $h=0.5$.

Electromagnetic fields in a twisted waveguide ($K=1, f=0$)

We apply the present scheme to the analysis of electromagnetic fields in the twisted waveguide with a rectangular cross section shown in Fig. 6. The governing equation of electromagnetic fields in

Table 1. Solutions to the potential problem (values of ψ for $Y=0$)

Helical pitch h	X	Zero-pitch solution	2nd-order perturbed solution	FE solution	BE solution
0.3	- 1.5	225.00	218.21	218.11	218.39
0.3	0.0	150.00	150.00	150.00	150.00
0.3	1.5	75.00	81.79	81.89	81.61
0.5	- 1.5	225.00	210.11	208.40	208.63
0.5	0.0	150.00	150.00	150.00	150.00
0.5	1.5	75.00	89.89	91.60	91.37

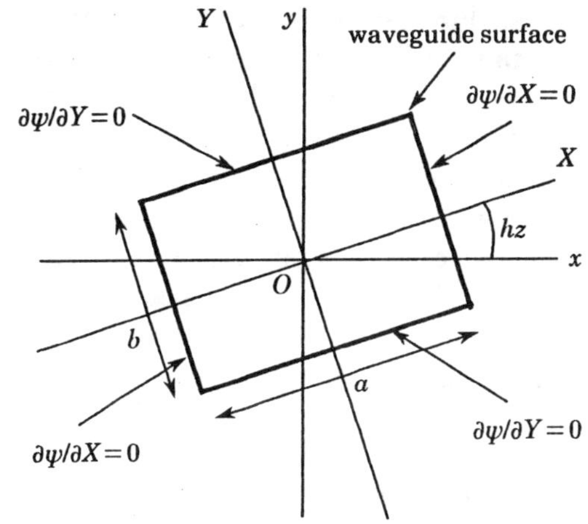

Fig. 6 Cross section of a twisted waveguide (X, Y, Z)

this waveguide is approximately expressed in terms of the scalar Helmholtz equation [2]. Hence, in this case, the functions K and f are taken to be 1.0 and 0.0, and the constants k and β correspond to the wave number and phase constant, respectively. Moreover, the function ψ corresponds to the amplitude of the magnetic Hertzian vector $\psi \mathbf{e}^Z$ which satisfies

$$E^X = \frac{\partial \psi}{\partial Y}, \quad E^Y = -\frac{\partial \psi}{\partial X}, \quad E^Z = 0, \tag{22}$$

$$\mathbf{H} = \frac{j}{\omega\mu}\left[k^2 \psi \mathbf{e}^Z + \nabla\left(\frac{\partial \psi}{\partial z}\right)\right], \tag{23}$$

where ω is the angular frequency, μ is the permeability, E^X, E^Y and E^Z are the contravariant components of the electric field, H is the magnetic field, and $\mathbf{e}^Z$ is the third covariant basis vector of the twisted coordinates. Moreover, the boundary conditions for the rectangular guide of width a and height b are

$$E^X = \frac{\partial \psi}{\partial Y} = 0, \qquad \text{at } Y = \pm\, b/2,$$

$$E^Y = -\frac{\partial \psi}{\partial X} = 0, \qquad \text{at } X = \pm\, a/2. \tag{24}$$

The dispersion relation for this waveguide can be obtained by solving eigenvalue problem (18) for given phase constants. In this paper, the phase constant β is set to be zero for simplicity (the eigenvalues k_i correspond to cut-off wave numbers for i th mode).

On the other hand, we have two ways to solve the scalar Helmholtz equation by BEM ; the method with the fundamental solution to (a) the scalar Helmholtz equation, and (b) the Laplace equation. If we use method (a), we have to calculate the eigenvalues by the determinant search method and, in general, it is expensive. In method (b), which is called the hybrid BEM [9], the term $k^2\psi$ is treated as an inhomogeneous term and the unknown values ψ_i are set at nodes in a domain and on boundaries. When using method (b), we can effectively find eigenvalues because eq. (18) reduces to a linear proper equation. For this reason, we choose the method (b) to solve eq. (18), *i.e.*, eq. (19) with $K=1, f=\beta=0$.

The resultant eigenvalues k for the lowest mode are summarized in table 2. Solutions by the perturbation method [2] and FEM [10] are also shown in this table. It can be seen that the discrepancies between the wave numbers by BEM and those by FEM are negligible. On the other hand, there are considerable differences between the wave number by perturbation method and that by other two methods for $h=0.9$ and $b=0.8$.

MHD equilibria in a straight helical vessel ($k=\beta=0$)

The MHD equilibria are described by

$$\mathbf{J}\times\mathbf{B} = \nabla p, \quad \nabla\times\mathbf{B} = \mu_0\mathbf{J}, \quad \nabla\cdot\mathbf{B} = 0, \tag{25}$$

where $\mathbf{J}$ is the current density, $\mathbf{B}$ is the magnetic field, p is the plasma pressure and μ_0 is the permeability of the free space. If an equilibrium configuration has a helical symmetry ($\beta=0$), eq. (25) can be written in terms of the stream function ψ and current potential $g(\psi)$ in the following form [1, 11],

Table 2. Cut-off wave numbers for a twisted waveguide ($a=1.0$)

b	Helical pitch h	Perturbation method	FEM	BEM
0.3	0.1	3.143	3.146	3.148
0.3	0.5	3.164	3.168	3.169
0.3	0.9	3.214	3.219	3.220
0.8	0.1	3.143	3.146	3.150
0.8	0.5	3.178	3.184	3.188
0.8	0.9	3.235	3.268	3.275

$$\nabla\cdot(K\nabla\psi) = \frac{-2K^2}{h}g(\psi) - Kg(\psi)g'(\psi) - \mu_0 p'(\psi), \tag{26}$$

where $K \equiv h^2/(1+h^2r^2)$ and the functions ψ and $g(\psi)$ are defined by

$$B_r = -\frac{h}{r}\frac{\partial\psi}{\partial\zeta}, \quad B_\theta - hrB_z = h\frac{\partial\psi}{\partial r}, \quad hrB_\theta + B_z = hg(\psi). \tag{27}$$

The MHD equilibrium can be obtained by solving (26) for given functions $p(\psi)$ and $g(\psi)$. In this case, the principal solution ψ^* is given by [1]

$$\psi^* = -\frac{2}{h^2}\left(\log r_> + \frac{1}{2}h^2r_>^2\right)$$

$$-4rr_0\sum_{m=1}^{\infty} I_m'(mhr_<)K_m'(mhr_>)\cos[m(\zeta-\zeta_0)] \tag{28}$$

The integral representation of eq. (26) can be easily obtained by setting the constants k and β to be zero in eq. (19). We can solve eq. (19) without iterations when the inhomogeneous terms in eq. (26) are independent from the function ψ, *i.e.*, $f=f(X, Y)$ (vacuum fields). Figure 7 shows the magnetic surfaces (contours of ψ) calculated by BEM [5] for vacuum fields in the conducting vessel with a circular cross section with a shift S from the origin. When the inhomogeneous terms in eq. (26) are the functions of ψ, integral equation (19) can be solved using appropriate iterative techniques [12, 13].

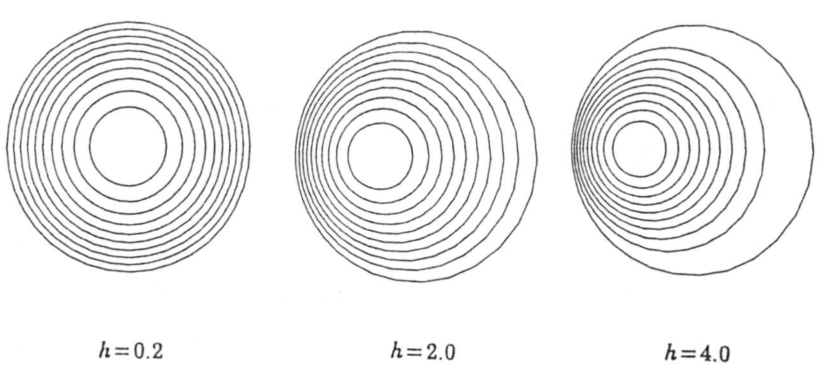

Fig. 7 MHD equilibrium configuration ($S=0.5$)

SUMMARY

In this paper, the two dimensional integral equations have been introduced from the scalar Helmholtz-type equation in the general curvilinear coordinates and the twisted coordinates. The boundary element method based on the present integral equation has been applied to the analyses of potential fields in a helical column, electromagnetic fields in a twisted waveguide and MHD equilibria in a helical vessel. As a result, the present formulation is proved to be valid. We conclude that BEM is not only useful to analyze electromagnetic fields in translationally symmetric and axisymmetric systems but also in helical systems.

REFERENCES

1. Gardner, H. J., Dewar, R. L. and Sy, W. N-C. The Free Boundary Equilibrium Problem for Helically Symmetric Plasmas, *J. Comp. Phys.*, **74**, 477-487, 1988.
2. Lewin, L. *Theory of Waveguides*, Chapter 4, 91-118, Newnes-Butherworths, London, 1975.
3. Fang, X-S. and Lin, Z-Q. Field in Single-Mode Helically-Wound Optical Fibers, *IEEE Trans. Microwave Theory Tech.*, **MTT-33**, 1150-1154, 1985.
4. Wlodarczyk, M. T. and Seshadri, S. R. Excitation and Scattering of Guided Modes on a Helically Corrugated Dielectric Cylinder, *IEEE Trans. Microwave Theory Tech.*, **MTT-34**, 8-18, 1986.
5. Igarashi, H. and Honma, T. An Equilibrium Analysis of Helically Symmetric Plasmas using Boundary Element Method, *IEEE Trans. Mag.*, **MAG-26**, 1774-1776, 1990.
6. Merkel, Z. A Green's Function Method for the Vacuum Contribution to the MHD Stability of Helically Symmetric Equilibria, *Z. Naturforsch*, **37a**, 859-866, 1982.
7. Barnes, D. C. and Cary, J. R. Free-Boundary Stability of Straight Stellarators, *Phys. Fluids*, **27**, 2522-2534, 1984.
8. Igarashi, H. and Honma, T. submitted to Eng. Analysis with Boundary Elements.
9. Tosaka, N. and Kakuda, K. An Integral Equation Method for Non-self-adjoint Eigenvalue Problems, Int. J. Num. Meth. Eng., **20**, 131, 1984.
10. Igarashi, H. and Honma, T. A Finite Element Analysis of TE Modes in Twisted Waveguides, Conference Digest of the 4th Biennial IEEE Conference on Electromagnetic Field Computation, DB10, Toronto, Canada, 1990.
11. Freidberg, J. P. Ideal Magnetohydrodynamics, Chapter 6, 107-183, Plenum, New York and London, 1987.
12. Igarashi, H. and Honma, T. : BE and FE Analysis of Helically Symmetric MHD Equilibrium Configuration, to be published in Int. J. App. Electromagnetics in Materials, 1990.
13. Igarashi, H. and Honma, T. An Iterative Boundary Element Analysis of Helically Symmetric MHD Equilibria, Abstracts of IABEM-90, 54-55, Rome, Italy, 1990.

A Subdomain Method for FEM-BEM Coupling in Electromagnetism

F. Delincé *, P. Dular *, A. Genon, H. Hedia, W. Legros, A. Nicolet

University of Liège, Dept of Electrical Engineering, Institut Montefiore - Sart Tilman B 28 - 4000 Liège, Belgium

* These authors are Research Assistants with the Belgian National Fund for Scientific Research

ABSTRACT

The finite element and the boundary element methods each have their specific advantages in electromagnetism. To benefit from both methods we adopted a coupling method. The knowledge of the vector potential together with the tangential field on the boundary of the subdomains is obtained as a byproduct of the computation. It allows the introduction of special features such as the control of the total current in the conductors, permanent magnets, thin magnetic plates, the computation of Joule losses with the Poynting vector and the computation of forces with the Maxwell stress tensor.

INTRODUCTION

The coupling between the finite element and the boundary element methods is most helpful for the computation of two-dimensional magnetic fields.

The finite element method allows the modelization of complex material behaviour such as magnetic saturation.

The boundary element method is appropriate for dealing with open boundary problems (infinite domains) frequently encountered in electromagnetism.

The notion of subdomain appears naturally in the direct boundary element method. This notion is extended to the finite element method in order to couple it with the boundary elements. The interconnection of subdomains is made by imposing transmission conditions on the boundary value of the vector potential and its normal derivative i.e the tangential induction.

This procedure yields the value of the vector potential and of the tangential field on the boundaries of all the subdomains.
It allows the introduction of special features in the computation.

At the processing level :

- Control of the current inside conductors with the line integral of the magnetic fields (Ampere's law) and coupling with electric circuits.
- Permanent magnets as special transmission conditions.
- Thin magnetic plates (membrane conditions)

At the postprocessing level :

- Computation of Joule losses with the Poynting vector.
- Computation of forces and torques with the Maxwell stress tensor.

BOUNDARY ELEMENT-FINITE ELEMENT COUPLING

The equation of the two-dimensional magnetostatics is the Poisson equation (1) :

$$\Delta A = \mu J \qquad (1)$$

where :
A is the vector potential which has only one component
J is the current density
μ is the magnetic permeability (=1/ν the inverse of the magnetic reluctivity)

In linear magnetic non conducting media, the direct boundary element method (Brebbia[1]) is based on the following relation :

$$cA = \int_\Gamma A\ \partial G/\partial n\ d\Gamma - \int_\Gamma G\ \partial A/\partial n\ d\Gamma \qquad (2)$$

where :
- G is the free space Green function of the two-dimensional Laplace operator;
- c = 0.5 on a smooth boundary;
- $\partial./\partial n$ is for the normal derivative.

Integrals are taken on the boundary Γ of the subdomains and the method involves only A and $\partial A/\partial n$ (tangential induction) on the boundaries.

The finite element formulation (Silvester[2]) is based on the Galerkin method and we have for the domain Ω of boundary Γ :

$$\int_\Omega (\nu \text{ grad } A \text{ grad } w - Jw) d\Omega - \int_\Gamma w\ \nu \partial A/\partial n\ d\Gamma = 0 \qquad (3)$$

where ν is the magnetic reluctivity, J the current density and w the weighting function.

The boundary term is usually used to apply a Neuman boundary condition. Here it is used at the interface between the finite element and the boundary element domains to couple the methods. This can be extended to the boundary between two finite element subdomains. The tangential field is taken as an unknown on the boundary and the boundary term of (3) is introduced. It leads to the same solution for the vector potential A as the classical formulation but it also provides the value of the tangential field $\nu\partial A/\partial n = H_t$ on the boundary.

As a demonstration, consider the boundary between two finite element domains (figure 1).

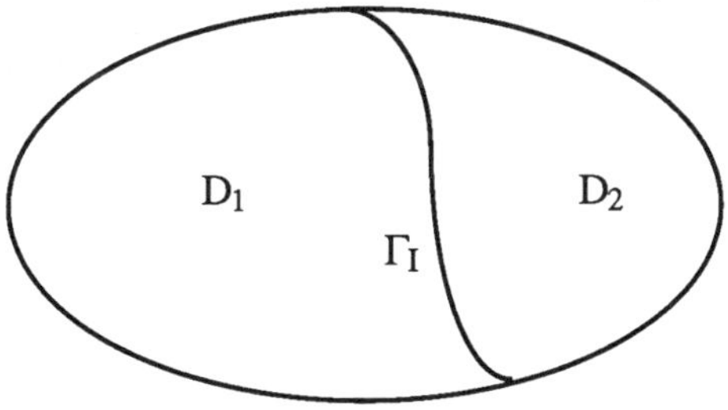

Figure 1 : interface between two subdomains

In the classical finite element formulation the interface Γ_I does not exist and D_1 and D_2 form a single domain. In this case, the finite element equation for a node i on the boundary is represented as follows :

$$\sum_{j \in D_1 \cup D_2} \alpha_{ij}^{D_1 \cup D_2} A_j = 0 \tag{4}$$

It involves only the nodal values A_j of the potential vector. In the subdomain method we have two equations and the tangential field is involved :

$$\sum_{j \in D_1} \alpha_{ij}^{D_1} A_j + \beta H_{ti} = 0 \tag{5a}$$

$$\sum_{j \in D_2} \alpha_{ij}^{D_2} A_j - \beta H_{ti} = 0 \tag{5b}$$

We know that $\alpha_{ij}^{D_1 \cup D_2} = \alpha_{ij}^{D1} + \alpha_{ij}^{D2}$. The superscripts D_1, D_2... indicate the domain of integration involved in the computation of the coefficients. The opposite signs of βH_{ti} in equations (5a) and (5b) insure that their sum leads to equation (4). It implies that the subdomain method is equivalent to the classical method and gives the same value to the potential (If numerical errors are neglected). The particular value of the coefficient β allows the physical interpretation of H_{ti} as the nodal value of the tangential field.

First and second order isoparametric elements are used for both the finite element and the boundary element methods.

VALUES ON BOUNDARIES

The value of the vector potential is automatically computed on the boundaries of the subdomains.

The tangential field is also easy to obtain. It is directly computed on the boundaries of the finite element domain. In the boundary element domain it is easily computed as $H_t = \frac{1}{\mu}\frac{\partial A}{\partial n}$ because $\partial A/\partial n$ is directly obtained and μ is constant on the domain. The normal induction B_n is another value of interest on the boundary. It is computed from the vector potential by :

$$
\begin{aligned}
B_n \quad &= \text{curl } A.\vec{n} = \frac{\partial A}{\partial y} \,.\, n_x - \frac{\partial A}{\partial x} \,.\, n_y \\
&= \left(- \frac{\partial A}{\partial y}\frac{\partial y}{\partial \xi} - \frac{\partial A}{\partial x}\frac{\partial x}{\partial \xi} \right) / \, J(\xi) \\
&= - \frac{\partial A}{\partial \xi} / \, J(\xi)
\end{aligned}
\tag{6}
$$

Where ξ is the local coordinate of the form functions used for the boundary element discretization and $J(\xi)$ the corresponding jacobian.

CONSTRAINTS FOR EXCITATIONS

Using Ohm's law, the case of eddy currents can be dealt with by introducing

$$
J = - \sigma(U + \partial A/\partial t) \tag{7}
$$

as the expression for the current density in equation (3). The conductor is characterized by its electrical conductivity σ. The time derivative of A expresses the inductive effects since U=grad V is due to an externally imposed electromotive force and can be interpreted as the terminal voltage of the conductor (per unit of length). If U is given for all the conductors of the problem, the discretisation of equation (3) leads to a differential system that can be solved. Unfortunately, in most cases, it is the total current I in a conductor which has to be imposed. As the finite element formulation involves the tangential field on the boundary, the total current I can be expressed as the line integral of the magnetic field (Ampere's law) :

$$I = \int_{\Gamma} \nu\, \partial A/\partial n\, d\Gamma \tag{8}$$

From a more general point of view we can systematically relate to every conductor an equation like (8) and two global degrees of freedom : the total current I and the terminal voltage U. To do so, a circuit equation is necessary for each conductor. The simplest cases correspond to imposing U or I for an individual conductor but any case of conductor interconnection and electric circuit coupling can be considered (Bourmanne[3]).

For practical implementation of the numerical magnetic field computation software, the two following cases are considered : a set of n parallel conductors with an imposed current I and a set of n serial conductors with an imposed voltage U. We refer to the current and the terminal voltage of the conductor i respectively as I_i and U_i.
In the case of series conductors, the following circuit equations are added :

$$\sum_{i=1}^{n} U_i = U \tag{9}$$

$$I_i = I_{i+1} \qquad \text{for } i = 1, ..., n-1 \tag{10}$$

They imply that the total terminal voltage is the sum of all the individual terminal voltages and that the current is the same in all conductors. One advantage of this method is that it automatically gives the current and the voltage for each conductor. This information is then used to compute global values such as the power or the impedance.

With sinusoidal excitations, the boundary element can also be used to compute the eddy current in conductors having linear magnetic properties. It involves the Green function of the corresponding Helmholtz equation. In this case the process of control of the current remains exactly the same.

TRANSMISSION CONDITIONS

The normal transmission conditions at the boundary between two magnetic materials is the continuity of A and $\nu\partial A/\partial n$. These conditions may be transformed to allow for other situations.

Permanent magnet

The permanent magnets are characterized by their coercitive field $\vec{H}_c$. The continuity of the tangential field becomes :

$$\left(\nu \frac{\partial A}{\partial n} + H_{ct}\right)_1 = \left(\nu \frac{\partial A}{\partial n} + H_{ct}\right)_2 \tag{11}$$

where the tangential components of the coercitive fields in the different materials are involved.

Thin magnetic plates

A thin magnetic plate of thickness e and permeability μ_3 separating two media of permeability respectively μ_1 and μ_2 is considered. In the indirect boundary element method the interface between two media is replaced by a surface current. If this technique is applied to the thin plate it is replaced by equivalent surface currents K1 on one side and K2 on the other which are very close to one another.

The approximation is to consider those two layers as a superposition of a single layer of current Ks and a double layer of current Kd at the middle of the plate. Returning to the direct formulation the expressions of Kd and Ks can be used to express the discontinuity of A and $\partial A/\partial n$ (tangential induction) due respectively to the equivalent double layer and single layer of current (Bamps[4]):

$$\nu_1 \partial A/\partial n_1 = \nu_2 \partial A/\partial n_2 \tag{12a}$$

$$((\mu_1 + \mu_2 - 2\,\mu_3) \,/\, (\mu 1 + \mu 2))\,(\partial A/\partial n_1 + \partial A/\partial n_2)\; e/2 = A_1 - A_2 \tag{12b}$$

Subscripts 1 and 2 are for media 1 and 2 respectively.

Those transmission conditions allow a very simple way to take into account thin magnetic plates.

POSTPROCESSING

Joule losses

In the sinusoidal cases the Joule losses can be computed using the Poynting theorem (Antunes[5]) :

$$P = \int_\Gamma \vec{S}.\vec{n}\, d\,\Gamma = \int_\Gamma 1/2(\vec{E} \times \vec{H}^*).\vec{n}\, d\,\Gamma = \int_\Gamma \frac{J}{2\sigma} H_t^*\, d\,\Gamma \tag{13}$$

The current density J is related to the vector potential and to the terminal voltage by equation (7). H_t^* is the complex conjugate of the tangential field.

Forces and torques

The Maxwell stress tensor method is used to compute the forces and torques (Aldefeld[6]). This method allows us to define at the boundary of objects an equivalent normal force density $\vec{F}_n$ and an equivalent tangential force density $\vec{F}_t$ as :

$$\vec{F}_n = \left(\frac{B_n^2}{2\mu_0} - \frac{\mu_0}{2} H_t^2\right) \vec{n} \tag{14}$$

$$\vec{F}_t = B_n \, H_t \, \vec{t} \tag{15}$$

The total force acting on an object of boundary Γ is given by

$$\vec{F} = \int_\Gamma \left(\vec{F}_n + \vec{F}_t\right) d\Gamma \tag{16}$$

The total torque is given by

$$\vec{T} = \int_\Gamma \vec{r} \times \left(\vec{F}_n + \vec{F}_t\right) d\Gamma \tag{17}$$

Where $\vec{r}$ is the position vector relative to the considered center of rotation.

Line integrals appear as a powerful tool in the postprocessing of the subdomain method. Another advantage is that the inside of domains is implicitly taken into account independently of their nature (seat of eddy currents, magnetic saturable...) and of the numerical method (BEM or FEM).

EXAMPLES

Magnetic circuit

Figure 2 shows the field lines in a simple magnetic circuit (width of the circuit $d = 10^{-2}$m, total air gap 10^{-3}m, excitation = 1 A turn, relative permeability μ = 10000).

The theoretical force (per unit of length) acting on one half of the circuit is

$$F = 2d \, \frac{B^2}{2\mu_0} \tag{18}$$

where B is the induction in the air gap and is equal to 1.213 10^{-3}T. The theoretical force is then F = 1.170 10^{-2}N and the computed force using FEM-BEM coupling is F = 1.097 10^{-2}N. (note that the theoretical computation of the force is made taking into account assumptions that are not necessary when using the numerical computation).

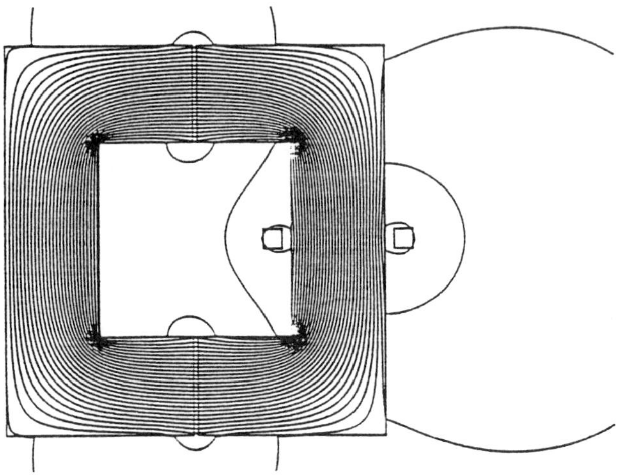

Figure 2 : field lines in a magnetic circuit

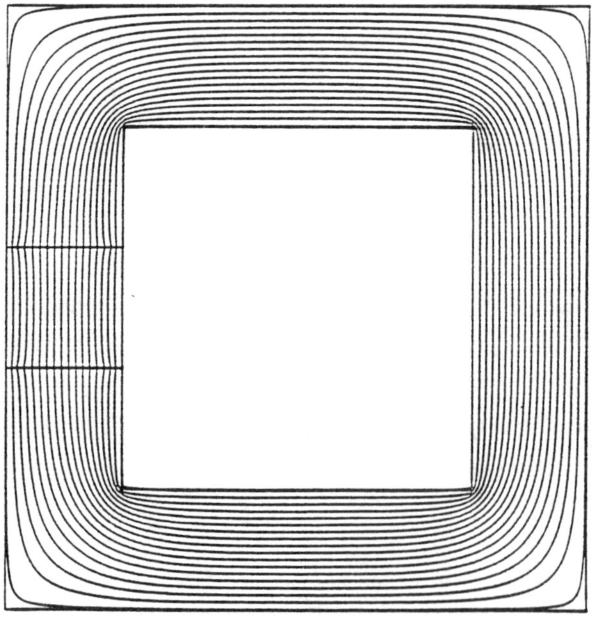

Figure 3 : field lines in a closed magnetic circuit with a permanent magnet

Closed magnetic circuit and permanent magnet

Figure 3 shows the field lines in a closed magnetic circuit with a permanent magnet ($B_r = \mu H_c = 1.T$). The width of this circuit is 0.01 m and its length about 0.16 m. The relative permeability of the whole circuit is $\mu = 1000$.The theoretical value for the induction is B = 0.0625 T and the computed one is B = 0.0694T (Note that the theoretical computation overestimates the length of the circuit and then produces an underestimated B).

Thin magnetic plate

As a test computation for the thin magnetic plates the simple case of a coil 0.1m above a thin magnetic plate of length 0.5m , thickness e = 1mm (aspect ratio 1/500) and relative permeability 10000 (Figure 3) is taken.Inductors are at a distance of 0.2m and the current is 1A. The field lines in this structure are shown on figure 5. This computation has been made with second order finite elements and special transmission conditions. The discontinuity of the potential vector and the screening effect of the plate are obvious.

Eddy currents

The test problem for eddy currents deals with two bars in a slot (Figure 6). In analytical computations the magnetic permeability of the surrounding material is supposed to be infinite while it is assumed to be equal to $1000\mu_0$ for the numerical computation. The height of an individual bar is 30 mm (60 mm for both) and the width is 5 mm. The conductivity is $5\ 10^7$ S/m and the current is I = 0.5 A in each bar.

Figure 7 compares the analytical and numerical amplitudes of current density at 50 Hz.

Figure 8 shows the Joule losses in each bar (analytical and numerical) versus frequency. The meshing of figure 6 was used for all the computations. This explains the larger error at a higher frequency where the smaller skin depth causes steeper current density variation.

CONCLUSION

The boundary element-finite element coupling method is a most effective tool for two-dimensional magnetic field computation. The notion of subdomain, that appears naturally in the direct boundary element method, has been extended to the coupling method and to the pure finite element method. It appears to provide useful information at the boundaries of the domains. This information has been used to control the excitations of the conductors, to implement special features such as permanent magnets and thin magnetic plates, and to compute forces and energy dissipations at the postprocessing level.

An exhaustive application software called LUCIE, integrating all these features, has been devised. It is based on the coupling method but allows pure BEM or pure FEM treatments. This can be useful to validate computations by comparing various models.The systematic treatment of the subdomains and of the transmission conditions have allowed a very structured and modular organization of the software that guarantees its quality.Future developments are the extension of the LUCIE software to three-dimensional computations.

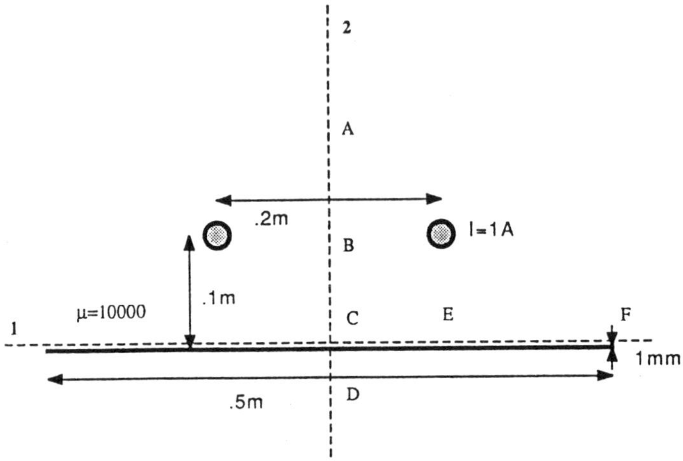

Figure 4 : thin magnetic plate

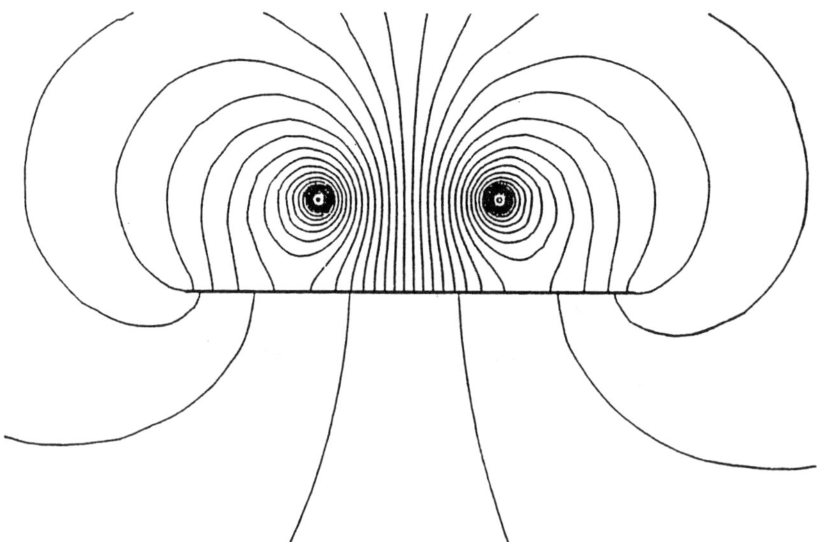

Figure 5 : field lines around a thin magnetic plate

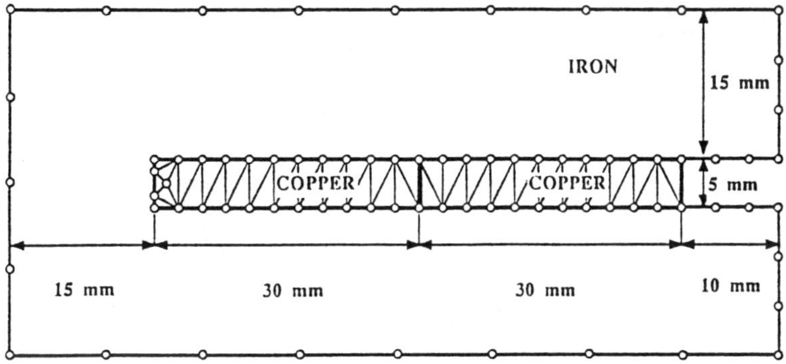

Figure 6 : two conducting bars in a slot

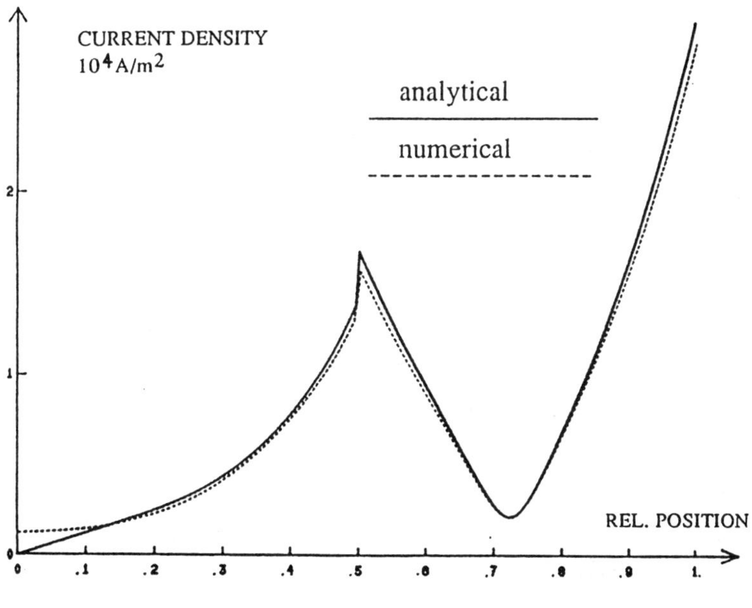

Figure 7 : amplitude of current density in the bars

D

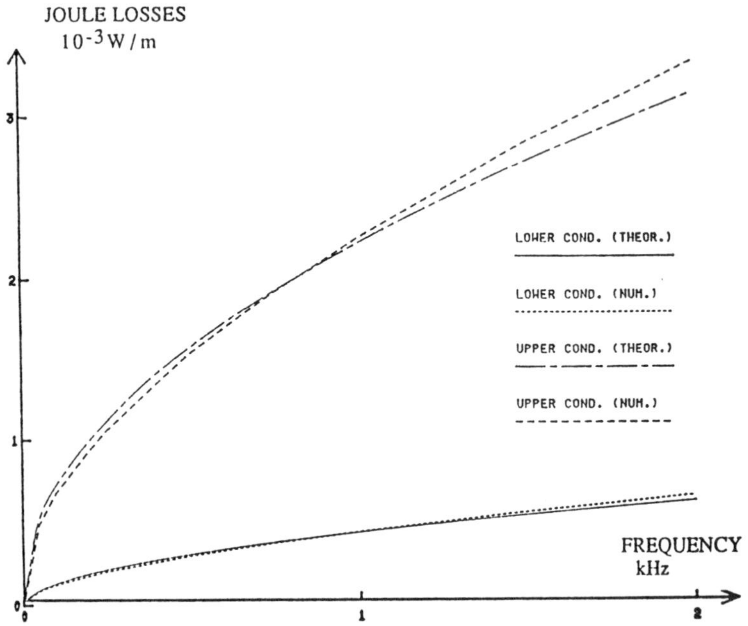

Figure 8 : Joule losses in the bars

REFERENCES

1. Brebbia, C.A., The Boundary Element Method for Engineers, Pentech Press, London, 1984.

2. Silvester, P.P., Ferrari, R.L., Finite Elements for Electrical Engineers, Cambridge University Press, Cambridge, 1990.

3. Bourmanne, P., Delincé, F., Genon, A., Legros, W., Nicolet, A, Skin Effect and Proximity Effect in Multiconductor Systems with Applied Currents and Voltages, 35th MMM Conference, San Diego, USA, October 1990.

4. Bamps, N., Bourmanne, P., Delincé, F., Genon, A., Legros, W., Nicolet, A, Comparison of Various Methods for the Modeling of Thin Magnetic Plates, 35th MMM Conference, San Diego, USA, October 1990.

5. Antunes, C.F.R.L., Bryant, C., Freeman, E.M.,Lowther, D.A., Postprocessing of 2D Eddy Current Problems; IEEE Transactions on Magnetics, vol.24, n°1, January 1988, pp 385-388.

6. Aldefeld, B., Forces in Electromagnetic Devices, Compumag 1978, Grenoble, pap. 8.1.

Modelization of Movement in Electromagnetic Devices with the Boundary Element Method

N. Bamps, F. Delincé *, P. Dular *, A. Genon, W. Legros, A. Nicolet

University of Liège, Dept of Electrical Engineering, Institut Montefiore - Sart Tilman B 28-4000 Liège, Belgium

* These authors are Research Assistants with the Belgian National Fund for Scientific Research

ABSTRACT

The paper presents the coupling between the finite element and the boundary element methods as an efficient technique to solve the problem of the modelization of movement in two-dimensional electromagnetic field computations. On the one hand, the finite elements are used inside the magnetic and conducting parts where they permit to take into account the complex behaviour of materials such as saturable ferromagnetic materials. On the other hand, the boundary elements are used around the moving parts where they prevent the deformation of the meshes.

The formulation of the eddy current problems leads to an automatic computation of the motional induced electromotive force (Legros[1]).

INTRODUCTION

Movement is of crucial importance in electromechanical devices such as motors and relays. A correct treatment of the movement involves a coupling between the mechanical model and the electromagnetic model. On the one hand, electromagnetic phenomena produce forces that mechanically act on the devices producing, among others, movement.

On the other hand, mechanical movement acts on the electromagnetic model : it transforms the geometry of the problem and it produces phenomena such as the motional induced electromotive force.

In this paper, we only consider rigid moving parts. In the two-dimensional model, the only degrees of freedom associated to such parts are 2 degrees of translation and 1 degree of rotation.

The mechanical model consists of the simple movement equation where electromagnetic force and torque are introduced. For the electromagnetic field modelization we use an existing program called LUCIE, which is based on the coupling of the finite element method and of the boundary element method.

This powerful method allows a rigorous treatment of open boundary problems with BEM together with a correct modelization of the saturable ferromagnetic materials. The modularity of the LUCIE software has allowed an easy and workable introduction of movement in the modelization.

The philosophy behind the use of the coupled method to model the movement is to use the finite element method inside the rigid moving parts that can be the seat of phenomena such as saturation and eddy currents.

The boundary elements are used for linear magnetic and non conducting media wherein rigid parts can move. The advantages of this formulation are that there is no mesh deformation in the finite element domains and that we need no explicit terms for the motional induced electromotive force, which is automatically taken into account by the variation of the boundary element method coefficients.

NUMERICAL METHODS : BOUNDARY ELEMENT - FINITE ELEMENT COUPLING

The equation for the two-dimensional magnetostatics is :

$$\Delta A = \mu J \qquad (1)$$

- A is the vector potential which has only one component;
- J is the current density;
- μ is the magnetic permeability (= $1/\nu$ inverse of the magnetic reluctivity).

In linear magnetic non conducting media, the direct boundary element method (Brebbia[2]) is based on the relation (2) :

$$cA = \int_\Gamma A\, \partial G/\partial n\, d\Gamma - \int_\Gamma G\, \partial A/\partial n\, d\Gamma \qquad (2)$$

where :
- G is the free space Green function of the two-dimensional Laplace operator;
- c = 0.5 on a smooth boundary;
- $\partial./\partial n$ is for the normal derivative.

Integrals are taken on the boundary Γ of the subdomains and the method involves only A and $\partial A/\partial n$ (tangential induction) on the boundaries.

The finite element formulation (Silvester[3]) is based on the Galerkin method (3) and for domain Ω of boundary Γ we have:

$$\int_\Omega (\nu \text{ grad } A \text{ grad } w - Jw)d\Omega - \int_\Gamma w\, \nu\partial A/\partial n\, d\Gamma = 0 \qquad (3)$$

where ν is the magnetic reluctivity, J the current density and w the weighting function.

The boundary term is usually used to apply a Neuman boundary condition.

Here it is used at the interface between finite element and boundary element domains to couple the methods. This can be extended to the boundary between two finite element subdomains. The tangential field is taken as an unknown on the boundary and the boundary term of equation (3) is introduced. It gives the same solution for vector potential A as the classical formulation but also furnishes the value of the tangential field $\nu\partial/\partial n = H_t$ on the boundary.

EDDY CURRENTS

Using Ohm's law, the case of eddy currents can be treated by introducing (Bourmanne[4]) :

$$J = -\sigma(U + \partial A/\partial t) \tag{4}$$

as the expression for current density in equation (3). The conductor is characterized by its electrical conductivity σ. The time derivative of A expresses the inductive effects since U = grad V is due to an externally imposed electromotive force and can be interpreted as the terminal voltage of the conductor (per unit of length). If U is given for all the conductors of the problem, the discretization of equation (3) leads to a differential system that can be solved. Unfortunately, in most cases, it is the total current I in a conductor that has to be imposed. As the finite element formulation involves the tangential field on the boundary, the total current I can be expressed as the line integral of the magnetic field (Ampere's law) :

$$I = \int_\Gamma \nu\, \partial A/\partial n \, d\Gamma \tag{5}$$

The numerical method used to discretize the transient eddy current problem is a semi-discrete Galerkin method. The first step is a spatial discretization of the domains leaving a continuous time variation of the degrees of freedom. This spatial discretization is a classical finite element meshing. It leads to a system of ordinary differential equations. The second step is a time discretization of the degrees of freedom. A finite difference scheme is generally used. In our particular case, a backward Euler method is used. It leads to an algebraic system of equations.

MOTIONAL INDUCED ELECTROMOTIVE FORCE

One of the major difficulties related to the numerical modelization of the movement is to take into account the motional induced electromotive force.

The problem can be analyzed as follows.

We have related current density $\vec{J}$ in a conductor to the electric field $\vec{E}$ through Ohm's law :

$$\vec{J} = \sigma\vec{E} \tag{6}$$

However this law is only valid for a conductor at rest. To determine the relation between current density and the electric field in a restless conductor, we have to consider a referential associated to the conductor and to use the non-relativistic field transformation formulae. We obtain (Landau[5]) :

$$\vec{J} = \sigma\left(\vec{E} + \vec{v}X\vec{B}\right) \tag{7}$$

This modified Ohm's law for restless conductors relates the current density in the conductor to the electric field $\vec{E}$ and to the induction $\vec{B}$ defined in a referential at rest with respect to which the conductors are moving.

If the electromotive force acting on a closed linear circuit is considered, it can be shown that the first term of the right hand member of equation (7) corresponds to a variation of the magnetic flux due to the time variation of the magnetic field itself when the position of the circuit does not change.

The second term of the right hand member of equation (7) corresponds to the magnetic flux variation due to the displacement of the conductor in a magnetic field constant in time.

Those results can be summed up by Faraday's induction law. The induced electromotive force in a circuit is equal to the total magnetic flux variation through the circuit whatever the origin of the variation : time variation of the field or movement of the circuit.

In the numerical treatment of the eddy currents the flux variation on a time step is computed as the difference of nodal values of the vector potential at two successive moments.

If the node belongs to a moving object, the flux variation is computed with two vector potential nodal values taken at different points in space but corresponding to the same "particle" of the moving body. The time derivative of the potential defined in this way is the "substantial" or "material" derivative.

If objects are displaced during the computation without any other change to the numerical methods, flux variations will be correctly computed. This automatically takes into account movements of conductors or of magnetic parts because, according to Faraday's induction law, it involves the total flux variation at one point of the conductor.

The coupling between the boundary elements and the finite elements naturally solves the problems related to the modelization of the movement.

FORCES AND TORQUES COMPUTATION

Two kinds of movement can be considered : forced movement under external forces and natural movement due to the electromagnetic forces. In this later case, the determination of the electromagnetic forces is an important point of the movement modelization. The knowledge of the vector potential A and of the tangential field H_t on the boundaries given by the coupling method suggests using the Maxwell stress tensor method (Aldefeld[6]).

This method permits to define, at the boundary of objects, an equivalent normal force density $\vec{F}_n$ and an equivalent tangential force density $\vec{F}_t$ as :

$$\vec{F}_n = \left(\frac{B_n^2}{2\mu_0} - \frac{\mu_0}{2} H_t^2 \right) \vec{n} \tag{8}$$

$$\vec{F}_t = B_n \, H_t \, \vec{n} \tag{9}$$

The tangential field H_t and the normal induction B_n are well defined even on surfaces of discontinuity. H_t is directly given by the coupling method since B_n is computed from the vector potential by :

$$\begin{aligned} B_n &= \text{curl } A.\vec{n} = \frac{\partial A}{\partial y} \, . \, n_x - \frac{\partial A}{\partial x} \, . \, n_y \\ &= \left(- \frac{\partial A}{\partial y} \frac{\partial y}{\partial \xi} - \frac{\partial A}{\partial x} \frac{\partial x}{\partial \xi} \right) / \, J(\xi) \\ &= - \frac{\partial A}{\partial \xi} / \, J(\xi) \end{aligned} \tag{10}$$

where ξ is the local coordinate of the form functions used for the boundary element discretization and $J(\xi)$ the corresponding jacobian.

The real total force acting on an object of boundary Γ is given by :

$$\vec{F} = \int_\Gamma \left(\vec{F}_n + \vec{F}_t \right) d\Gamma \tag{11}$$

The real total torque is given by :

$$\vec{T} = \int_\Gamma \vec{r} \times \left(\vec{F}_n + \vec{F}_t \right) d\Gamma \tag{12}$$

where $\vec{r}$ is the position vector relative to the considered centre of rotation considered.

MOVEMENT EQUATION

In the case of a natural movement, the displacement of an object is computed by solving the equation of the movement involving the magnetic forces.

Consider for instance a translation along the axis $\vec{e}_x$.

The following quantities are defined :

- position : x;
- speed : v = dx/dt;
- mass : M;
- elastic constant of a spring : k;
- friction coefficient : f;
- other forces including the magnetic forces : F(t).

The equation of the movement is :

$$M\frac{dv}{dt} = F(t) - fv - kx \tag{13}$$

The time discretization of this equation by the backward Euler method gives (Stoer[7]) :

$$v_{t+\Delta t} = \frac{Mv_t + F(t)\,\Delta t - kx_t\,\Delta t}{k\,\Delta t^2 + f\,\Delta t + M} \tag{14}$$

$$x_{t+\Delta t} = \frac{Mx_t + Mv_t\,\Delta t + F(t)\,\Delta t^2 + fx_t\,\Delta t}{k\,\Delta t^2 + f\,\Delta t + M} \tag{15}$$

Equations (14) and (15) relate the position $x_{t+\Delta t}$ and the speed $v_{t+\Delta t}$ at time $t+\Delta t$ to the position x_t and to the speed v_t at time t.

The movement is determined if an initial position x_0 and an initial speed v_0 are specified. Limits of movement can also be specified to avoid non physical interpenetration of moving parts. The other degrees of freedom are dealt with in a similar way (mutatis mutandis).

INTRODUCTION OF THE MOVEMENT IN THE SOFTWARE

The basis of the modelization of the movement in electromagnetic devices is an existing 2D magnetic field computation software called LUCIE. The central feature of the software is the use of the coupled FEM-BEM. The modular structure of the software is designed to make further developments and modifications as easy as possible.

The algorithm used to model the movement is the following :

1. Compute the electromagnetic state of the system.
2. Compute the magnetic forces and torques for this state.
3. Compute the displacements of the moving bodies with equations (14) and (15).
4. Modify the geometry in the database of the software using the computed displacements.
5. Start again at step 1.

This procedure involves some implicit hypotheses. The time varying electromagnetic force is considered as constant on a time step in equations (14) and (15), the displacements (i.e. the geometry variations) must not be too large on a time step.

Practical considerations show that an adaptative time step is necessary in order to keep the computational time within reasonable limits while remaining as close as possible to the hypotheses.

RESULT : MODELIZATION OF A RELAY

The modelization of an electromechanical relay requires a transient modelization because there are no periodic phenomena like in a rotating machine (Juffer[8]). Moreover, the movement plays a fundamental role in the electromagnetic modelization of the closure of the relay.

Figure 1 shows the geometry of the 2D model of a relay.

The magnetic core is composed of an upper moving part and of a lower static part. The coil has 1430 turns and its resistance is 18 Ω. The mass of the moving part is 112g and the elastic constant of the spring supporting the moving part is k = 137 N/m. Friction is neglected (f=0). The depth of the structure is 1.6 cm and, for the modelization, all the characteristics of the relay are translated into equivalent characteristics for a standard depth of 1 meter. A special treatment to force a uniform current density is applied to the coil to take into account the small section of the wire that prevents eddy currents (It allows only a global inductive effect).

The relay is fed by a step of voltage of 12 V at the terminals of the coil.

The coil and the saturable magnetic parts are modelized with finite elements. The surrounding air is modelized with boundary elements (Figure 1).

Figure 2 shows the field lines in the open relay.

Figure 3 shows the time variation of the position of the moving part.

Figure 4 compares the computed (continuous line) and the measured (dashed line) time variation of the current in the coil.

The motional induced electromotive force is responsible for the decrease of the current just before the closure is completed.

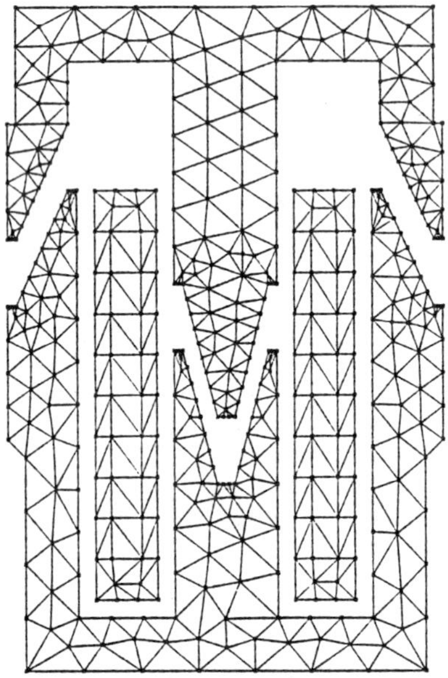

Figure 1 : meshing of an electromechanical relay

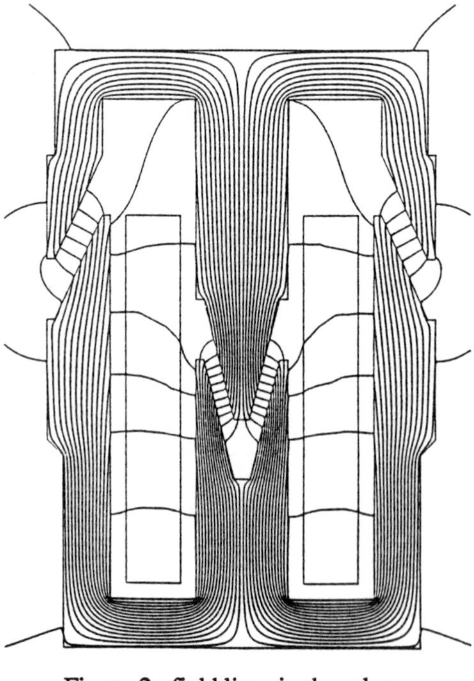

Figure 2 : field lines in the relay

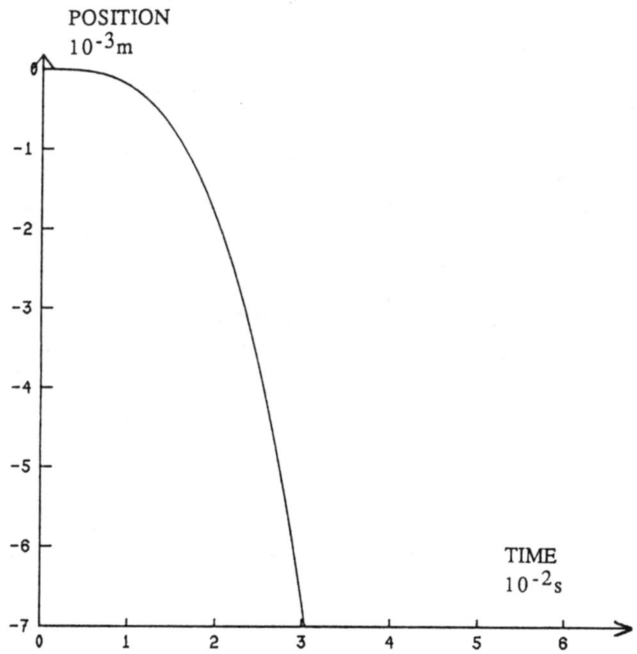

Figure 3 : displacement of the moving part

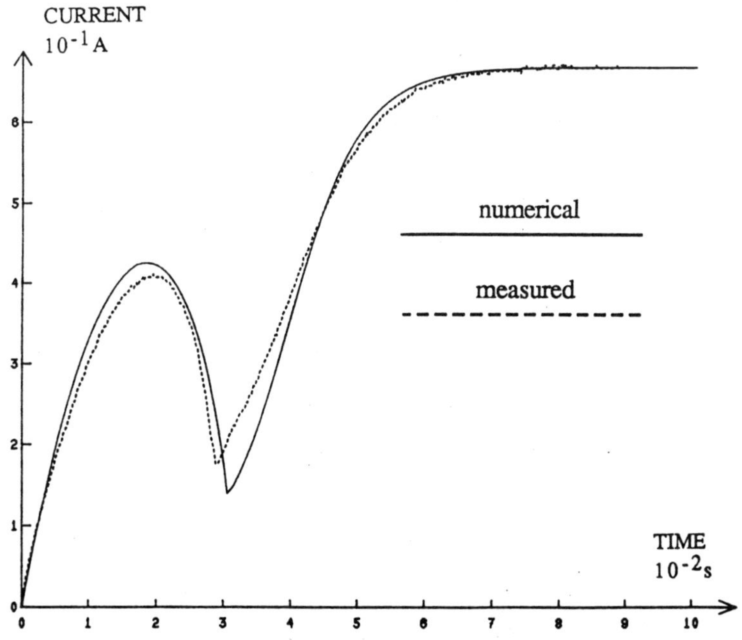

Figure 4 : current in the coil of the relay

CONCLUSIONS

The coupling between the finite elements and the boundary elements appears to be a natural and workable method to model the movement in electromagnetic devices.

The rigid moving parts involving materials with complicated behaviour are easily modelled with the finite elements. Most of the time, the media wherein rigid parts move is linear magnetic and non-conducting (air) and very often open. Its treatment with boundary elements is therefore natural and avoids all the problems of mesh deformations.The paper shows that the motional induced electromotive force is easily taken into account in the eddy current formulation. The basis of the modelization is the 2D magnetic field computation software called LUCIE using the coupling between the boundary element and the finite element methods. Its modular structure has allowed an efficient and safe modification of the software.

The algorithm of the movement modelization shows that the modifications inside the program have been kept to a minimum. Most of the computations proper to the movement are gathered in an independent module.

The numerical computations show the accuracy of the method. Further developments will be the extension to three-dimensional computations.

REFERENCES

1. Legros, W., Nicolet, A., Paganini, M., Numerical Modelization of Transient Current in Relay, IEEE Transactions on magnetics, Vol. 25, n°5 September 1989, pp. 3593-3595.

2. Brebbia, C.A., The Boundary Element Method for Engineers, Pentech Press, London, 1984.

3. Silvester, P.P., Ferrari, R.L., Finite Elements for Electrical Engineers, Cambridge University Press, Cambridge, 1990.

4. Bourmanne, P., Delincé, F., Genon, A., Legros, W., Nicolet, A, Skin Effect and Proximity Effect in Multiconductor Systems With Applied Currents and Voltages, 35th MMM Conference, San Diego, USA, October 1990.

5. Landau, L., Lifchitz, E., Electrodynamique des milieux continus, Edition Mir, Moscow, 1969.

6. Aldefeld, B., Forces in Electromagnetic Devices, Compumag 1978, Grenoble, pap. 8.1.

7. Stoer, J., Bulirsch, R., Introduction to Numerical Analysis, Springer Verlag, Berlin and New York, 1980.

8. Juffer, M., Transducteurs électromécaniques, Traité d'électricité, Vol IX, Presses Polytechniques Romandes, Lausanne, 1985.

Comparison of Two Different BEM Formulations for Nonlinear Eddy Currents

A. Kost

Institut für Elektrische Maschinen, Technische Universität Berlin, Einsteinufer 11, DW 1000 Berlin 10, Germany

ABSTRACT

A method to calculate 3D nonlinear eddy currents by the Boundary Element Method has been developed in 1989 by the author. It is based on a formulation for the vectors of the electric ($\vec{E}$) and magnetic ($\vec{H}$) field strength whereas the nonlinearity is focused in a part of the magnetic induction ($\vec{B}_{nonl}$). For several practical nonlinear applications it could be interesting to have a formulation by the vectorpotential ($\vec{A}$) and scalar potential (φ) because especially in the 2D–case the fields often can be described by only one vectorpotential component. Such a formulation is deducted in this paper, where a modified Lorentz gauge is introduced. It is finally compared with the above mentioned one. Both formulations are very well adapted for parallel computation.

INTRODUCTION

The ($\vec{E}$, $\vec{H}$, $\vec{B}_{nonl}$)–BEM formulation by Kost[1] takes the full time and field dependence of the permeability into account. It is further based on a periodical time dependence of the field vectors, which has not to be sinusoidal. This assumption is justified by the fact that the exciting fields

are often periodic with time.
The method could be improved by Kost and Shen[2] by applying the Fast Fourier Transform (FFT) for different points and the BEM for different frequencies in every step of an iterative procedure which was introduced to solve the nonlinear problem.

In the beginning of the iterative process the nonlinear part ($\vec{B}_{nonl}$) is zero. The first result is then a time harmonic field supposing the exciting field would also be time harmonic. In the next step the forcing function contains higher harmonics due to the nonlinearity of the magnetization curve and the result also contains them.

The formulation of this paper follows the same ideas. Special attention has been paid to avoid, if possible, spatial derivatives due to accuracy.
Also the derived ($\vec{A}$, φ, $\vec{H}_{nonl}$)–formulation is well suited for parallel computing.

PROBLEM FORMULATION WITH POTENTIALS AND MODIFIED LORENTZ GAUGE

There are many examples where a special interest exists to find out the eddy current distribution as well as their losses and accompanying field distribution in conductive and iron containing parts of electrical machines and related designs. Also in the field of electromagnetic compatibility (EMC) conductive iron materials are used, especially performed by plates or double plates, where a 3D nonlinear eddy current problem is given. So the analysis can be referred to a general iron body excited by a time periodic current density distribution $\vec{J}^{(e)}$ outside itself (see figure 1).

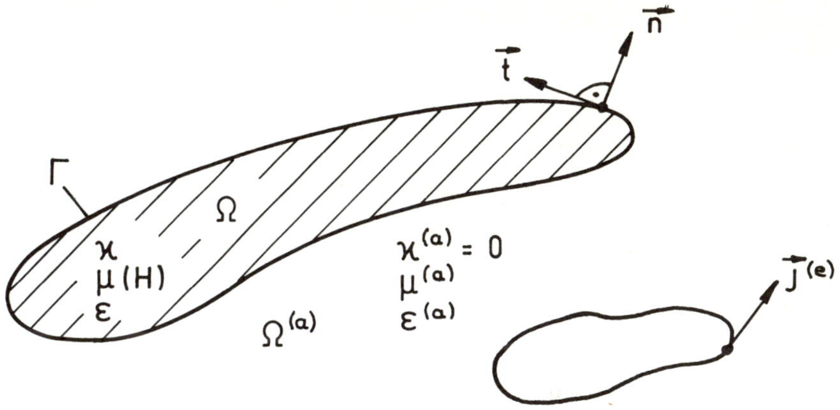

Figure 1. Conductive iron body excited by current density $\vec{J}^{(e)}$

As usually with eddy current problems the displacement current can be neglected, and the starting points are Maxwell's equations

$$\text{curl } \vec{E} = -\frac{\partial \vec{B}}{\partial t} \tag{1}$$

$$\text{curl } \vec{H} = \begin{cases} \vec{J} \text{ in } \Omega \\ \vec{J}^{(e)} \text{ in } \Omega^{(a)} \end{cases} \tag{2}$$

$$\text{div } \vec{B} = 0 \tag{3}$$

$$\text{div } \vec{D} = 0 \tag{4}$$

The material equations are

$$\vec{B} = \begin{cases} \mu(H)\, \vec{H} \text{ in } \Omega \\ \mu^{(a)}\, \vec{H} \text{ in } \Omega^{(a)} \end{cases} \tag{5}$$

$$\vec{D} = \epsilon\,\vec{E} \tag{6}$$

$$\vec{J} = \kappa\,\vec{E} \tag{7}$$

The meanings of the vectors and material properties in eq. (1) to (7) are as follows:

$\vec{E}$: electric field strength
$\vec{H}$: magnetic field strength
$\vec{D}$: electric displacement density
$\vec{B}$: magnetic flux density
$\vec{J}$: current density
μ: permeability
ϵ: permittivity
κ: conductivity

Introducing now the magnetic vectorpotential $\vec{A}$ by

$$\vec{B} = \text{curl}\,\vec{A} \tag{8}$$

in eq. (1) leads to the introduction of the scalar potential φ:

$$\text{curl}\,\vec{E} = -\,\text{curl}\,\frac{\partial\vec{A}}{\partial t} \tag{9}$$

$$\text{curl grad}\,\varphi \equiv 0 \tag{10}$$

$$\vec{E} = -\frac{\partial\vec{A}}{\partial t} - \text{grad}\,\varphi \tag{11}$$

(11) in (2) $\rightarrow$

$$\text{curl } \vec{H} = -\kappa \frac{\partial \vec{A}}{\partial t} - \kappa \text{ grad } \varphi \tag{12}$$

Introducing the reluctivity ν

$$\nu = \frac{1}{\mu} \tag{13}$$

in (12) leads to

$$\text{curl } (\nu \text{ B}) = -\kappa \frac{\partial \vec{A}}{\partial t} - \kappa \text{ grad } \varphi \tag{14}$$

$$\text{curl } (\nu \text{ curl } \vec{A}) = -\kappa \frac{\partial \vec{A}}{\partial t} - \kappa \text{ grad } \varphi \tag{15}$$

A second differential equation for $\vec{A}$ and φ arises from eq. (4):

$$\text{div } \vec{E} = 0 = -\frac{\partial}{\partial t} \text{div } \vec{A} - \text{div grad } \varphi = -\frac{\partial}{\partial t} \text{div } \vec{A} - \Delta\varphi \tag{16}$$

Now usually the Lorentz gauge

$$\text{div } \vec{A} = -\frac{\kappa}{\nu} \varphi \tag{17}$$

is used in order to get the telegraph differential equation for both the vector as well as the scalar potential in the linear case. In the nonlinear case however this type of equation cannot be reached and moreover several difficulties would arise in the further deduction of a BEM formulation. To avoid them another gauge is introduced:

$$\text{div } \vec{A} = -\frac{\kappa}{\nu^0} \varphi \tag{18}$$

which shall be called modified Lorentz gauge. The time– and position–independent reluctivity ν^0 is defined by splitting off a linear part in the function H(B), see fig. 2, by

$$\vec{H} = \vec{H}_{lin} + \vec{H}_{nonl} = \nu \text{ curl } \vec{A} = \nu^0 \vec{B} + \vec{H}_{nonl}\ ; \quad \nu^0 \neq f(x,y,z,t) \tag{19}$$

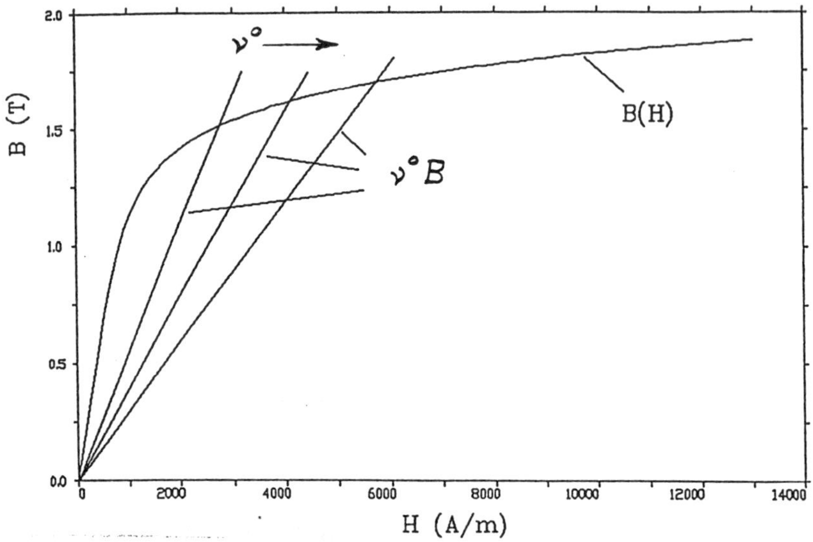

Figure 2. Splitting off a linear part of the magnetization curve $\nu^0 = 1/\mu^0$

Hereby the following two equations arise from eq. (15) and (16):

$$\nu^0 \text{ curl curl } \vec{A} + \kappa \frac{\partial \vec{A}}{\partial t} + \kappa \text{ grad } \varphi + \text{curl } \vec{H}_{nonl} = 0 \tag{20}$$

$$\Delta \varphi - \frac{\kappa}{\nu^0} \frac{\partial \varphi}{\partial t} = 0 \tag{21}$$

The three–dimensional vectors $\vec{A}$, $\vec{H}_{nonl}$ and the scalar potential φ are a function of position x,y,z and time t. They have to satisfy boundary conditions in the surface Γ which are formulated in eq. (52) to (54).

INTEGRAL FORMULATION WITH VECTOR AND SCALAR POTENTIAL

The following formulation is based on the second Green's theorem derived for vectors by Stratton[3]:

$$\int_{\Omega} (\vec{\Phi} \text{ curl curl } \vec{P} - \vec{P} \text{ curl curl } \vec{\Phi}) \, d\,\Omega =$$

$$\int_{\Gamma} (\vec{P} \times \text{curl } \vec{\Phi} - \vec{\Phi} \times \text{curl } \vec{P}) \, \vec{n} \, d\,\Gamma \tag{22}$$

where $\vec{P}$ and $\vec{\Phi}$ are vector functions of position being continuous together with their first and second derivatives inside, outside and on the surface Γ which separates the two regions Ω and $\Omega^{(a)}$. $\vec{n}$ is the normal unit vector at the source point on Γ.

A weighted residual technique following Brebbia[4] is now applied on eq. (20):

$$\int_{\Omega} \vec{\Phi} \, [\nu^{0} \text{ curl curl } \vec{A} + \kappa \frac{\partial \vec{A}}{\partial t} + \kappa \text{ grad } \varphi + \text{curl } \vec{H}_{\text{nonl}}] \, d\,\Omega = 0 \tag{23}$$

with $\vec{\Phi}$: weighting vector function.
In order to replace $\vec{\Phi}$ curl curl $\vec{A}$ by $\vec{A}$ curl curl $\vec{\Phi}$ the second Stratton's theorem is operated on the first addend of eq. (23):

$$\int_{\Omega} \nu^{0} \vec{A} \text{ curl curl } \vec{\Phi} \, d\,\Omega + \int_{\Gamma} \nu^{0} (\vec{A} \times \text{curl } \vec{\Phi} - \vec{\Phi} \times \text{curl } \vec{A}) \, \vec{n} \, d\,\Gamma$$

$$+ \int_{\Omega} \vec{\Phi} \, \kappa \frac{\partial \vec{A}}{\partial t} \, d\,\Omega + \int_{\Omega} \vec{\Phi} \, \kappa \text{ grad } \varphi \, d\,\Omega + \int_{\Omega} \vec{\Phi} \text{ curl } \vec{H}_{\text{nonl}} \, d\,\Omega = 0 \tag{24}$$

The vector function $\vec{\Phi}$ of any size is first slightly determined by

$$\vec{\Phi} = \Phi\,\vec{a} \tag{25}$$

with an arbitrary constant unit vector $\vec{a}$ with div $\vec{a} = 0$, curl $\vec{a} = 0$. Then the first volume integral in eq. (24) can be changed:

$$\nu^0 \int_\Omega \vec{A}\ \text{curl curl}\ (\Phi\,\vec{a})\ d\,\Omega = \nu^0 \int_\Omega \vec{A}\ \text{grad}\ (\vec{a}\ \text{grad}\ \Phi)\ d\,\Omega$$

$$-\nu^0 \int_\Omega \vec{A}\ \Delta\Phi\ \vec{a}\ d\,\Omega \tag{26}$$

With $\vec{A}$ grad $\alpha = \text{div}\ (\vec{A}\ \alpha) - \alpha\ \text{div}\ \vec{A}$, (27)

Gauss' theorem and the modified Lorentz gauge eq. (26) can be changed to

$$\nu^0 \int_\Omega \vec{A}\ \text{curl curl}\ (\Phi\,\vec{a}) = \vec{a}\ \nu^0 \int_\Gamma (\vec{A}\ \vec{n})\ \text{grad}\ \Phi\ d\,\Gamma$$

$$-\vec{a}\ \nu^0 \int_\Omega \vec{A}\ \Delta\Phi\ d\,\Omega + \vec{a}\ \kappa \int_\Omega \varphi\ \text{grad}\ \Phi\ d\,\Omega \tag{28}$$

Further contributions in eq. (24) are changed by

$$\vec{n}\ (\vec{A} \times \text{curl}\ \vec{\Phi}) = \vec{a}\ [(\vec{n} \times \vec{A}) \times \text{grad}\ \Phi] \tag{29}$$

$$\vec{n}\ (\vec{\Phi} \times \text{curl}\ \vec{A}) = -\,\vec{a}\ \Phi\ [\vec{n} \times \text{curl}\ \vec{A}] \tag{30}$$

The following equation is very important, because it avoids the spatial derivative of $\vec{H}_{nonl}$ in eq. (24) and therefore enhances the accuracy of the computation:

$$\int_\Omega \vec{\Phi}\ \text{curl}\ \vec{H}_{nonl}\ d\,\Omega = \int_\Omega \text{div}\ (\vec{H}_{nonl} \times \vec{\Phi}) d\,\Omega + \int_\Omega \vec{H}_{nonl}\ \text{curl}\ \vec{\Phi}\ d\,\Omega$$

$$= \vec{a} \int_\Gamma \Phi\ (\vec{n} \times \vec{H}_{nonl}\ )\ d\,\Gamma + \vec{a} \int_\Omega (\vec{H}_{nonl} \times \text{grad}\ \Phi)\ d\,\Omega \tag{31}$$

So eq. (24) is transferred to

$$-\int_{\Omega} \vec{A}\, \Omega\Phi \,d\,\Omega + \frac{\kappa}{\nu^0}\int_{\Omega} \Phi \frac{\partial \vec{A}}{\partial t}\,d\,\Omega + \int_{\Gamma} (\vec{A}\,\vec{n})\,\mathrm{grad}\,\Phi\,d\,\Gamma$$

$$+\int_{\Gamma} (\vec{n}\times\vec{A})\times\mathrm{grad}\,\Phi\,d\,\Gamma + \int_{\Gamma}\Phi\,(\vec{n}\times\mathrm{curl}\,\vec{A})\,d\,\Gamma$$

$$+\frac{1}{\nu^0}\int_{\Gamma}\Phi\,(\vec{n}\times\vec{H}_{nonl})\,d\,\Gamma + \frac{1}{\nu^0}\int_{\Gamma}(\vec{H}_{nonl}\times\mathrm{grad}\,\Phi)\,d\,\Omega$$

$$+\frac{\kappa}{\nu^0}\int_{\Omega}[\varphi\,\mathrm{grad}\,\Phi + \Phi\,\mathrm{grad}\,\varphi]\,d\,\Omega = 0 \qquad (32)$$

The last volume integral in eq. (32) can also be transformed to the boundary:

$$\varphi\,\mathrm{grad}\,\Phi + \Phi\,\mathrm{grad}\,\varphi = \mathrm{grad}\,(\varphi\,\Phi) \qquad (33)$$

$$\int_{\Omega}\mathrm{grad}\,(\varphi\,\Phi)\,d\,\Omega = \int_{\Gamma}\varphi\,\Phi\,\vec{n}\,d\,\Gamma \qquad (34)$$

Applying the weighted residual technique also on eq. (21) leads to

$$\int_{\Omega}\Delta\varphi\,\Phi\,d\,\Omega - \frac{\kappa}{\nu^0}\int_{\Omega}\frac{\partial\varphi}{\partial t}\Phi\,d\,\Omega = 0 \qquad (35)$$

and the second Green's theorem applied on the first integral in eq. (35):

$$\int_{\Omega}\varphi\,\Delta\,\Phi\,d\,\Omega - \frac{\kappa}{\nu^0}\int_{\Omega}\frac{\partial\varphi}{\partial t}\Phi\,d\,\Omega - \int_{\Gamma}\varphi\,\mathrm{grad}\,\Phi\,\vec{n}\,d\,\Gamma + \int_{\Gamma}\Phi\,\mathrm{grad}\,\varphi\,\vec{n}\,d\,\Gamma = 0$$
$$(36)$$

ITERATIVE SOLUTION PROCESS

With $\mathcal{L}$ being a linear operator and $\mathcal{N}$ a nonlinear one, eq. (32) can be written

$$\mathcal{L}\,(\vec{A},\,\varphi,\,\Phi) + \mathcal{N}\,(\vec{H}_{nonl},\,\Phi) = 0 \tag{37}$$

where $\mathcal{N}$ is representing the volume and surface integral depending on $\vec{H}_{nonl}$ in eq. (32) and $\mathcal{L}$ the rest of the equation. In the following iterative solution process the nonlinear parts are now handled as known forcing functions. So step (K) of the process is given by

$$\mathcal{L}\,(\vec{A},\,\varphi,\,\Phi)^{(K)} + \mathcal{N}\,(\vec{H}_{nonl},\,\Phi)^{(K-1)} = 0 \tag{38}$$

Because of the time periodic exciting currents all field quantities must also be time periodic:

$$\vec{A}^{(K)} = \sum_{n=0}^{N} \text{Re}\left\{ \underline{\vec{A}}^{(K)}_{2n+1}\,(x,y,z)\; e^{j(2n+1)\omega t} \right\} \tag{39}$$

$$\varphi^{(K)} = \sum_{n=0}^{N} \text{Re}\left\{ \underline{\varphi}^{(K)}_{2n+1}\,(x,y,z)\; e^{j(2n+1)\omega t} \right\} \tag{40}$$

$$\vec{H}^{(K-1)}_{nonl} = \sum_{n=0}^{N} \text{Re}\left\{ \underline{\vec{H}}^{(K-1)}_{nonl\;2n+1}\,(x,y,z)\; e^{j(2n+1)\omega t} \right\} \tag{41}$$

BEM–FORMULATION WITH TIME–PERIODIC VECTOR AND SCALAR POTENTIAL

Introducing the time dependence of eq. (39) into the integral equation (32) the first two volume integrals there can be written

$$-\int_{\Omega} \underline{\vec{A}}^{(K)}_{2n+1}\,\Delta\Phi\,d\,\Omega + \int_{\Omega} j\,(2n+1)\,\omega\frac{\kappa}{\nu^0}\,\underline{\vec{A}}^{(K)}_{2n+1}\,\Phi\,d\,\Omega =$$

$$-\int_{\Omega} \underline{\vec{A}}^{(K)}_{2n+1}\,(\Delta\Phi - j\,(2n+1)\,\omega\frac{\kappa}{\nu^0}\,\Phi)\,d\,\Omega \tag{42}$$

The weighting function Φ chosen to be the fundamental solution of

$$\Delta\Phi + \beta_{2n+1}^{2}\,\Phi = -\,\delta_{i} \qquad (43)$$

with $\beta_{2n+1}^{2} = -\,j\,(2n+1)\,\omega\,\dfrac{\kappa}{\nu^{0}}$

δ_{i} : Dirac–delta function

$$\Phi = \Phi_{2n+1} = \frac{1}{4\pi r}\exp\left(-\,j\,\beta_{2n+1}\,r\right) \qquad (44)$$

r : distance between source point and computation point

finally leads from eq. (32) to the boundary integral equation

$$\frac{\Omega_i}{4\pi}\underline{\vec{A}}^{(K)}_{2n+1_i} = \int_{\Gamma}\Big[-\big(\underline{\vec{A}}^{(K)}_{2n+1}\,\vec{n}\big)\,\mathrm{grad}\,\Phi_{2n+1} - \big(\vec{n}\times\underline{\vec{A}}^{(K)}_{2n+1}\big)\times\mathrm{grad}\,\Phi_{2n+1}$$

$$-\big(\vec{n}\times\mathrm{curl}\,\underline{\vec{A}}^{(K)}_{2n+1}\big)\,\Phi_{2n+1}\Big]\,d\Gamma$$

$$-\frac{\kappa}{\nu^{0}}\int_{\Gamma}\underline{\varphi}^{(K)}_{2n+1}\,\Phi_{2n+1}\,\vec{n}\,d\Gamma - \frac{1}{\nu^{0}}\int_{\Gamma}\big(\vec{n}\times\underline{\vec{H}}^{(K-1)}_{non l\,2n+1}\big)\,\Phi_{2n+1}\,d\Gamma$$

$$-\frac{1}{\nu^{0}}\int_{\Omega}\big(\underline{\vec{H}}^{(K-1)}_{non l\,2n+1}\times\mathrm{grad}\,\Phi_{2n+1}\big)\,d\Omega \qquad (45)$$

and from eq. (36) to the boundary integral equation

$$\frac{\Omega_i}{4\pi}\underline{\varphi}^{(K)}_{2n+1_i} = -\int_{\Gamma}\underline{\varphi}^{(K)}_{2n+1}\,\mathrm{grad}\,\Phi_{2n+1}\,\vec{n}\,d\Gamma + \int_{\Gamma}\Phi_{2n+1}\,\mathrm{grad}\,\underline{\varphi}^{(K)}_{2n+1}\,\vec{n}\,d\Gamma \qquad (46)$$

where Ω_i denotes the solid angle subtended by Γ at computation point i.

One of the boundary conditions in Γ is the fact that the current cannot flow outside the conductive region, supposing that the displacement current has been neglected:

$$\kappa\,\vec{n}\left(-\frac{\partial\vec{A}}{\partial t} - \mathrm{grad}\,\varphi\right) = 0 \qquad (47)$$

Hereby the spatial derivative grad $\underline{\varphi}^{(K)}_{2n+1}$ in eq. (46) can be avoided and replaced by the normal component of $\vec{A}$:

$$\vec{n}\ \text{grad}\ \underline{\varphi}^{(K)}_{2n+1} = -\, j(2n+1)\ \omega\ (\vec{n}\ \underline{\vec{A}}^{(K)}_{2n+1}) \tag{48}$$

$$\frac{\Omega_i}{4\pi}\ \underline{\varphi}^{(K)}_{2n+1_i} = -\int_\Gamma \underline{\varphi}^{(K)}_{2n+1}\ \text{grad}\ \Phi \cdot \vec{n}\ d\Gamma$$

$$-\int_\Gamma \Phi_{2n+1} \cdot j(2n+1)\ \omega\ (\vec{n}\ \underline{\vec{A}}^{(K)}_{2n+1})\ d\Gamma \tag{49}$$

There is a volume integral to be evaluated in eq. (45) but of an integrand which is known from the step (K–1) before. So the unknowns of step (K) can be found on the boundary Γ alone, for i lying on Γ.
To do this equivalent equations for the exterior of the nonlinear body (symbol (a)) can be derived from eq. (45), (46):

$$\frac{\Omega_i}{4\pi}\ \underline{A}^{(K)(a)}_{2n+1_i} = \int_\Gamma [(\underline{A}^{(K)(a)}_{2n+1}\ \vec{n})\ \text{grad}\ \Phi^{(a)}_{2n+1} + (\vec{n} \times \underline{A}^{(K)(a)}_{2n+1}) \times \text{grad}\ \Phi^{(a)}_{2n+1}$$

$$+ (\vec{n} \times \text{curl}\ \underline{A}^{(K)(a)}_{2n+1})\ \Phi^{(a)}_{2n+1}]\ d\Gamma$$

$$+ \int_\Gamma \mu\ \underline{\vec{J}}^{(e)}_{2n+1}\ \Phi^{(a)}_{2n+1}\ d\Gamma \tag{50}$$

$$\frac{\Omega_i}{4\pi}\ \underline{\varphi}^{(K)(a)}_{2n+1_i} = \int_\Gamma \underline{\varphi}^{(K)(a)}_{2n+1}\ \text{grad}\ \Phi^{(a)}_{2n+1}\ \vec{n}\ d\Gamma$$

$$-\int_\Gamma \Phi^{(a)}_{2n+1}\ \text{grad}\ \underline{\varphi}^{(K)(a)}_{2n+1_i}\ \vec{n}\ d\Gamma$$

$$+ \underline{\varphi}^{(e)}_{2n+1} \tag{51}$$

with the fundamental solution $\Phi^{(a)} = 1/4\pi r$.

Eq. (50) and (51) also contain the harmonics of the exciting current density $\vec{J}^{(e)}$ and potential $\varphi^{(e)}$.

Together with the remaining boundary conditions for the continuity of the potentials and the tangential component of the magnetic field strength in Γ:

$$\underline{\vec{A}}_{2n+1} = \underline{\vec{A}}^{(a)}_{2n+1} \tag{52}$$

$$\underline{\varphi}_{2n+1} = \underline{\varphi}^{(a)}_{2n+1} \tag{53}$$

$$\vec{n} \times \nu \operatorname{curl} \underline{\vec{A}}_{2n+1} = \vec{n} \times \nu^{(a)} \operatorname{curl} \underline{\vec{A}}^{(a)}_{2n+1} \tag{54}$$

eq. (45), (49), (50)...(54) state after introducing boundary elements a system of equations for the unknowns of the tangential and normal components of the vector potential, the tangential components of the flux density and the scalar potential on the conductor surface.

COMPARISON WITH ($\vec{E}$, $\vec{H}$, $\vec{B}_{nonl}$)–FORMULATION AND CONCLUSION

The equation systems are comparable in size for the general 3D case.
Fig. 3 shows a flux diagram of the parallel computation of one step of the iterative procedure. Compared with the correspondent flux diagram for the ($\vec{E}$, $\vec{H}$, $\vec{B}_{nonl}$)–formulation, showed by Kost and Shen[2], a disadvantage lies in the fact that a numerical derivative curl $\vec{A}^{(K-1)}$ has to be done in the ($\vec{A}$, φ, $\vec{H}_{nonl}$)–formulation of this paper, a source of additional numerical error.
On the other hand the number of unknowns can be reduced for several problems in the ($\vec{A}$, φ, $\vec{H}_{nonl}$)–formulation as an advantage.

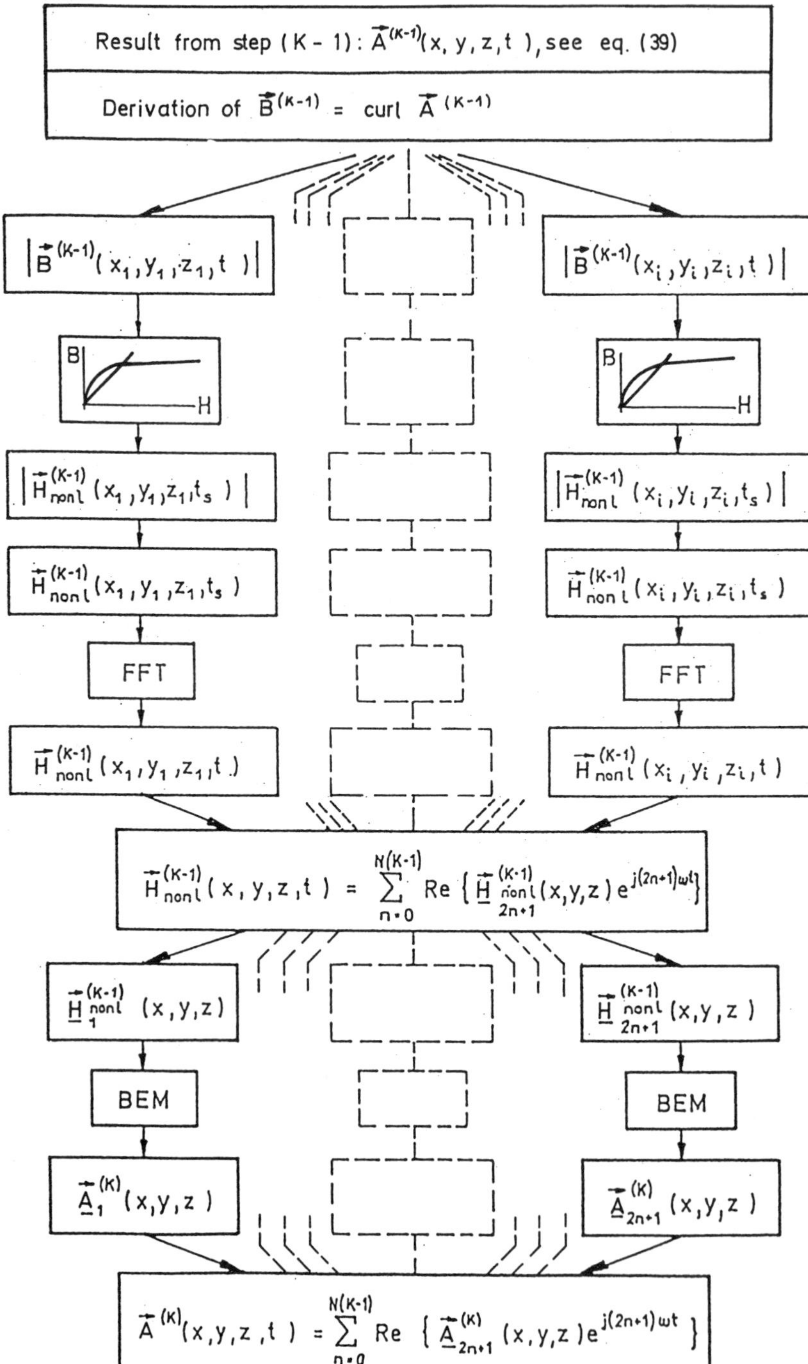

Fig. 3: Parallel computing of step (K)

So in conclusion at this moment the ($\vec{E}$, $\vec{H}$, $\vec{B}_{nonl}$)–formulation seems to be more appropriate in the general 3D case, but the ($\vec{A}$, φ, $\vec{H}_{nonl}$)–formulation may have advantages in special applications, especially in the 2D case. This tendency can only be consolidated by further investigations and examples.

REFERENCES

1. Kost A. (1989)
Calculation of Three–Dimensional Nonlinear Eddy Currents in an Electrical Machine by the Boundary Element Method
Proceedings of the IV. International Conference on Boundary Element Technology, Windsor, Canada (1989), pp. 227–238

2. Kost A. and Shen J. (1989)
Parallel Computation of 3–D Nonlinear Eddy Currents by the Boundary Element Method (BEM) and the Fast Fourier Transform (FFT)
COMPEL–The Int. Journal for Comp. and Math. in Electrical and Electronic Eng., Vol. 9, Suppl. A (1990), pp. 181–184

3. Stratton J.A. (1941)
Electromagnetic Theory
Mc Graw – Hill Book Company
New York and London (1941)

4. Brebbia C.A., Telles J.C.F., Wrobel L.C.
Boundary Element Techniques
Springer Verlag
Berlin, Heidelberg, New York, Tokyo 1984

Procat System: Recent Advances and Future Developments in Numerical Simulation of Cathodic Protection Problems

W.J. Mansur, J.C.F. Telles, J.A.F. Santiago
COPPE/UFRJ, Civil Engineering Department, Cx P 68506, CEP: 21945, Fax (5521) 290-6626, Rio de Janeiro-RJ., Brazil

ABSTRACT

This paper is concerned with improvements in efficiency of the PROCAT SYSTEM, and with future developments still to be incorporated in order to make numerical simulation of cathodic protection (CP) problems more realistic.

A summary of the direct Boundary Element Method and a brief description of PROCAT features is first shown. Subsequently the J/bi-CG iterative algorithm to solve linear systems of equations is presented, and the CPU time economy obtained is illustrated through the analysis of the impressed current CP system of a semisubmersible platform.

Finally two algorithms to consider dynamic polarization curves dependent on time history of cathodic regions are described.

INTRODUCTION

The cathodic protection computer system (PROCAT) discussed here has been developed at COPPE/Federal University of Rio de Janeiro for PETROBRÁS, the Brazilian State Oil Company. Previous papers on the PROCAT System (Telles et al.[1], Wrobel et al.[2], Mansur et al.[3], Telles et al.[4]) described the code in detail and showed several applications to practical engineering problems.

Version one of PROCAT SYSTEM was developed from 1983 to 1987. In the years that followed its conclusion, users observed that modelling complex real CP systems can in some cases demand large CPU computer time. This fact motivated the study on iterative solvers described by Araujo and Mansur[5] and Araujo et al.[6], who selected the bi conjugate Gradient Method with Jacobi preconditioning, J/bi-CG procedure, (among those studied so far) as the best for BEM analyses. The economy in BEM modelling of CP systems is substantial as will be shown in this paper.

The other point which will be discussed here is concerned with a better representation of the cathodic protection phenomena. For offshore structures, the behaviour of cathodic surfaces can be better modelled by time dependent polarization curves, different at each surface point, which describe the non-linear relation between potential and current distribution. Dynamic polarization curves are influenced by scale deposition which is strongly influenced by the current-potential history of the surface. This history may be quite complex, being impossible to cover all of the existing situations by experiments. However, information about the polarization curve can be retrieved from potentiostatic polarization data (or galvanostatic polarization data), obtained from experiments in which a formation potential (or formation current density) is maintained constant in time (see Nisancioglu[7]). These curves can be used together with adequate time marching schemes to simulate the history dependent phenomenon; i.e., the time varying potential and current density of actual cathodic protection systems. In this Work the Boundary Element Method is used in conjunction with two different time marching schemes to analyse cathodic protection systems of offshore structures. These are, fictitious potential and fictitious time procedures.

PROBLEM EQUATIONS

In order to find the corrosion rate at any point on a metal surface, submerged in an electrolyte and subjected to cathodic protection, corrosion engineers need the electrochemical potential ϕ in the electrolyte adjacent to the point, i.e., the potential distribution over the interface electrolyte/structure plays a key role in the design and analysis of cathodic protection systems. Herein, particular emphasis is given to offshore structures and pipelines in which the seawater represents an approximately homogeneous electrolyte. Such problems have been seen to be adequately modeled within this conducting domain Ω (electrolyte) by the Laplace equation (Munn[8]) :

$$\nabla^2\phi(x) = 0 \qquad x \in \Omega \tag{1}$$

subject to appropriate boundary conditions of the form

$$\left.\begin{array}{lll} \phi(x) = \bar{\phi}(x) & x \in \Gamma_1 & (i) \\ i(x) = \bar{i}(x) & x \in \Gamma_2 & (ii) \\ i(x) = a\,\phi(x)+b & x \in \Gamma_3 & (iii) \\ i(x) = f(\phi) & x \in \Gamma_4 & (iv) \end{array}\right\} \tag{2}$$

where $\Gamma = \Gamma_1 + \ldots + \Gamma_4$ is the boundary of Ω, i(x) is the current density, a and b are known constants, and $f(\phi)$ is a nonlinear function of the potential (polarization curve) that may also depend on the time history of the system.

In addition, from Ohm's law

$$i(x) = k\,\frac{\partial\phi(x)}{\partial n(x)} = kq(x) \qquad x \in \Gamma \tag{3}$$

in which k is the conductivity of the electrolyte and n is the outward normal to the boundary Γ.

A boundary integral equation, equivalent to Equations(1) and (2), can therefore be written as (Brebbia, Telles and Wrobel)[9] :

$$C(\xi)\phi(\xi) + \int_\Gamma q^*(\xi,x)\phi(x)d\Gamma(x) = \int_\Gamma \phi^*(\xi,x)q(x)d\Gamma(x) \tag{4}$$

where $C(\xi)$ depends on the boundary geometry at the point ξ,$\phi^*(\xi,x)$ is the free space fundamental solution to the Laplace equation and $q^*(\xi,x) = \partial\phi^*(\xi,x)/\partial n(x)$. Notice that impressed current anodes can also be simulated as concentrated or linearly distributed internal sources. To this end, one simply has to add to the right of the Equation(4) a term presented by Telles et al.[1] that is computed in closed form.

The application of Equation(4) to external problems in infinite regions (e.g., seawater) requires a careful examination of the behaviour of the functions involved over the boundary at infinity (Telles, Mansur and Wrobel)[10]. In this case, a term ϕ' representing a constant potential at infinity should be added to the right side of Equation(4), with the contour Γ representing only the internal boundaries of the region. The inclusion of the unknown potential ϕ' makes it possible to impose a self-equilibrated current density between the anodes and cathodes of the problem. This is automatically obtained by satisfying the condition

$$\int_\Gamma q(x)\, d\Gamma(x) = 0 \qquad x \in \Gamma \tag{5}$$

which ensures that there is no loss of current to infinity.

In order to solve numerically Equation(4) (and Equation(5), if required), the boundary Γ is discretized into a number of boundary elements within which the variables ϕ and **q** (unknown or prescribed) are interpolated as functions of their nodal values. Using a collocation technique, it is then required that Equation(4) be satisfied for each nodal point ξ of the problem.

This results in a system of algebraic equations that can be solved directly if the boundary conditions are of the type (*i*), (*ii*), or (*iii*) in Equation(2) or iteratively for the nonlinear case (*iv*).

FEATURES OF PROCAT

The PROCAT system consists of two modules, which can be used to study two-dimensional, axisymmetric, and three-dimensional cathodic protection problems. In the two-dimensional/axisymmetric module, the boundary of the region under study is represented by straight superelements that are internally subdivided into a predetermined number of constant elements of equal size.

There are six types of superelements in the three-dimensional module as follows (see Figure 1) :

(*A*) *Plane Triangular Superelement* - This superelement is very useful in order to define plane surfaces. It also presents a facility for automatic subdivision into a number of constant triangular elements, homotetic to the superelement.

(*B*) *Rectangular Cylindrical Superelement* - This superelement was developed to represent cylindrical cavities, such as platform legs, pipelines and cylindrical tanks. The rectangular cylindrical superelement can be automatically subdivided into a predetermined number of constant rectangular elements of cylindrical surface.

(*C*) *Triangular Cylindrical Superelement* - This superelement, as the previous one, is also used to model cylindrical cavities. In this case, however, the possibility of modelling tubular joints is taken into account.

(*D*) *Tubular Superelement* - This superelement, also employed in the modelling of cylindrical cavities, subdivides the cylindrical surface into a series of ring-shaped elements, with no functional variation (constant) along the circumference and quadratic interpolation function along its length. This superelement has demonstrated great computational efficiency, as well as facility for data input/output, when applied to large problems.

(*E*) *Circular Plane Superelement* - This superelement is of particular importance for the closing (beginning or end) of cylindrical cavities. Each superelement corresponds to a constant plane element, with circular geometry and nodal point at the centroid.

(*F*) *Spherical Cavity Superelement* - The spherical cavity superelement is used to represent approximately galvanic anodes as spherical cavities.

This superelement may contain various constant spherical cavity elements, the nodal point of each element being the point of the spherical cavity that is farthest from the axis of the cylinder to which the element belongs.

A more detailed description of the elements implemented is given by Mansur et al.[3] The PROCAT system can also be used to study the CP of buried pipelines in nonhomogeneous solids by considering the medium as piecewise homogeneous, using the subregions technique. In this case, the interface between homogeneous subregions should also be discretized and adequate compatibility conditions are automatically introduced by the program.

If the problem presents one or more planes of symmetry, only the boundary of the region located on one side of the respective symmetry plane need be discretized. This is due to an automatic process of reflection and condensation of variables that permits the analysis of a condensed version of the final system of equations, with no need for discretization of the symmetry planes.

The polarization curve of the cathode is defined by a series of pairs of points (ϕ,i) that are, internally, linearly interpolated. In this way, any type of nonlinear function can be analyzed by the program. The final system of equations is solved by the Gauss elimination technique. When the system is large, PROCAT uses an out-of-core block solver based on the one published by Das[11], which enables the problem to be solved in relatively small computers.

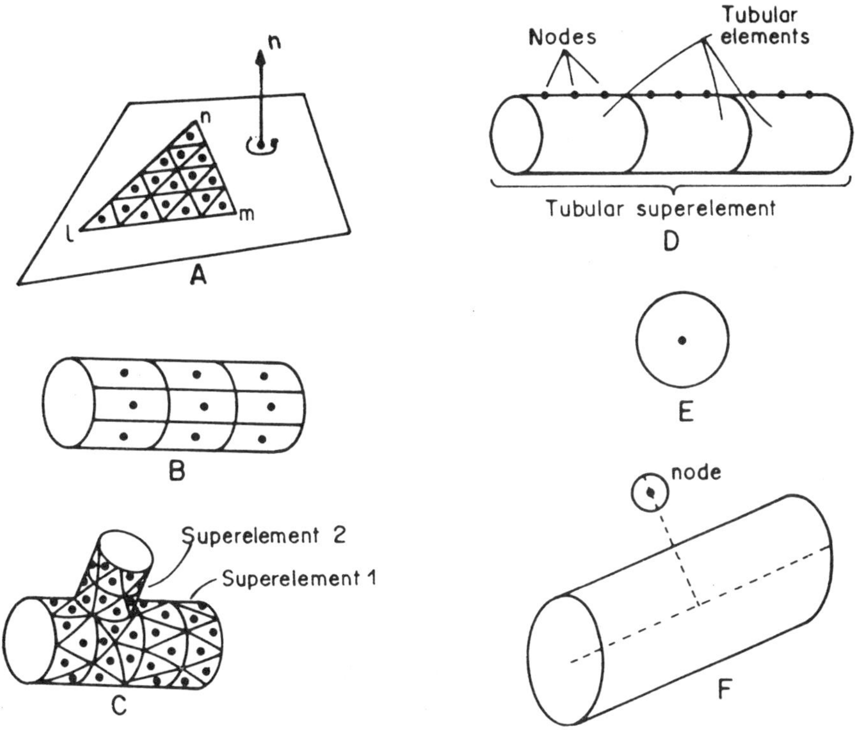

Figure 1. Superelements available in the PROCAT system

Nonlinear problems are solved by using an efficient Newton-Raphson technique (Telles et al.[2]), which has proved to converge very fast for general problems. At each iteration, the system of equations is solved in-core or out-of-core, depending on the number of unknowns.

IMPROVEMENT IN EFFICIENCY

The most substantial recent improvement available in the PROCAT system consists of the J/bi-CG iterative procedure[5,6] to solve large systems of equations.

This iterative method can be used to find solutions of the system of algebraic equations obtained from the integral Equation(4), which after the discretization process, can be written as[9] :

$$\mathbf{H}\,\phi = \mathbf{G}\,\mathbf{i} \tag{6}$$

or (after boundary conditions are introduced)

$$\mathbf{A}\,\mathbf{x} = \mathbf{b} \tag{7}$$

The J/bi-CG algorithm to solve Equation(7) is defined by the following Equations(8) to (14) :

$${}^{n+1}\mathbf{x} = {}^{n}\mathbf{x} + \lambda_n\,{}^{n}\mathbf{p} \tag{8}$$

$${}^{n}\mathbf{p} = \begin{cases} \mathbf{r}_0 & , \quad n = 0 \\ {}^{n}\mathbf{r} + \alpha_n\,{}^{n-1}\mathbf{p} & , \quad n \geq 1 \end{cases} \tag{9}$$

$${}^{n}\mathbf{r} = {}^{n-1}\mathbf{r} - \lambda_{n-1}\,\mathbf{A}\,{}^{n-1}\mathbf{p} \tag{10}$$

$${}^{n}\bar{\mathbf{p}} = \begin{cases} {}^{0}\bar{\mathbf{r}} = {}^{0}\mathbf{r} & , \quad n = 0 \\ {}^{n}\bar{\mathbf{r}} + \alpha_n\,{}^{n-1}\bar{\mathbf{p}} & , \quad n = 1,2,.. \end{cases} \tag{11}$$

$${}^{n}\bar{\mathbf{r}} = {}^{n-1}\bar{\mathbf{r}} - \lambda_{n-1}\,\mathbf{A}^{T}\,{}^{n-1}\bar{\mathbf{p}} \tag{12}$$

$$\lambda_{n-1} = \frac{\left({}^{n-1}\bar{\mathbf{r}}\ ,\ {}^{n-1}\mathbf{r}\right)}{\left({}^{n-1}\bar{\mathbf{p}}\ ,\ \mathbf{A}\,{}^{n-1}\mathbf{p}\right)} \tag{13}$$

$$\alpha_n = \frac{\left({}^{n}\bar{\mathbf{r}}\ ,\ {}^{n}\mathbf{r}\right)}{\left({}^{n-1}\bar{\mathbf{r}}\ ,\ {}^{n-1}\mathbf{r}\right)} \tag{14}$$

The iterative procedures described above can be stopped whenever the parameter ε defined by Expression(15) is smaller than a predetermined tolerance TOL

$$\varepsilon = \left\{ \frac{\left({}^{n}\mathbf{r}^{*} \;,\; {}^{n}\mathbf{r}^{*} \right)}{\left(\mathbf{b}^{*} \;,\; \mathbf{b}^{*} \right)} \right\}^{1/2} \tag{15}$$

with

$$\begin{aligned} {}^{n}\mathbf{r}^{*} &= \mathbf{Q}^{-1}\, {}^{n}\mathbf{r} \\ \mathbf{b}^{*} &= \mathbf{Q}^{-1}\, \mathbf{b} \end{aligned} \tag{16}$$

where $\mathbf{Q}$ is the Jacobi preconditioning matrix, whose inverse premultiplies Equation(7), prior to its solution.

IMPROVEMENT IN MODELLING

Among the modern references, the work by Nisancioglu[7] is that which presents the most complete investigation concerning the time evolution of polarization curves.

The model recognizes the influence of the following parameters in the polarization process :

$$i = f_{\phi_f} \Big(\phi,\ t,\ P,\ T,\ v_{\infty}, \text{metal surface conditions, seawater chemistry} \Big) \tag{17}$$

where P is pressure, T is temperature and v_{∞} is the seawater velocity close to the cathodic surface.

The dynamic polarization curve represented by Equation(17) can be determined from the potentiostatic data which takes into account implicitly P, T, v_{∞} and the seawater chemistry. Thus Equation(17) can be written as :

$$i = f_{\phi_f} \; (\phi, t) \tag{18}$$

where ϕ_f is the formation potential.

Solution Methods

Time history of a cathodic protection system can be obtained with two different numerical procedures called fictitious time and fictitious potential. The procedures have the common initial stage described below.

Initial stage - Compute initial current and potential distribution by solving Laplace's Equation (Equation(1)) subjected to the following boundary conditions :

$$\phi = \bar{\phi} \qquad \text{or} \qquad i = \bar{i} \qquad \text{on anodic regions} \tag{19}$$

$$i = f_{\phi_f}(\phi,\ t = 0) \qquad \text{on cathodic regions} \tag{20}$$

It must be observed that the polarization curve indicated by Equation(20) is the same for the whole cathodic surface.

The current density and potential obtained in this stage, for a boundary node j, are denoted by i_j^o and ϕ_j^o respectively.

Fictitious time method (FTM) - The potential over each boundary element is supposed not to vary during a time interval Δt, selected for each stage.

I - Each boundary node j is kept under a constant formation potential ϕ_{fj}^1 equal to ϕ_j^o computed at the previous stage, during the time interval Δt.

II - At $t_j^1 = \Delta t$, each boundary node j has its own polarization curve corresponding to ϕ_{fj}^1. Equation(1) is then solved again, being the new boundary conditions on a typical cathodic boundary node j given by

$$i_j = f_{\phi_{fj}^1}\left(\phi_j \ ,\ t_j^1\right) \tag{21}$$

III - Current density and potential at each node. obtained in II are now respectively denoted by i_j^1 and ϕ_j^1.

IV - Compute a fictitious time $\hat{t}_j^1$ by solving the following non-linear equation (by a modified Newton-Raphson method) :

$$i_j^1 = f_{\phi_j^1}\left(\phi_j^1 \ ,\ \hat{t}_j^1\right) \tag{22}$$

V - Set $t_j^2 = \hat{t}_j^1 + \Delta t$, and $\phi_{fj}^2 = \phi_j^1$ for each cathodic boundary node j.

VI - Repeat steps I, II, III, IV and V up to the final time of the analysis.

Fictitious potential method (FPM) - The potential over each boundary element is supposed not to vary during a time interval Δt selected for each step, as for the FTM method presented.

I - See I of FTM

II - See II of FTM

III - See III of FTM

IV - Compute a fictitious potential $\hat{\phi}_j^1$ by solving the following non-linear equation

$$i_j^1 = f_{\hat{\phi}_j^1}\left(\hat{\phi}_j^1 \; , \; t_j^1\right) \tag{23}$$

V - Set $t_j^2 = t_j^1 + \Delta t$ and $\phi_{fj}^2 = \hat{\phi}_j^1$ for each cathodic boundary node j.

VI - Repeat steps II, III, IV and V up to the final time of the analysis.

The methods presented above can represent in a discrete form the time and space behaviour of cathodic protection systems subjected to time varying current density and potential. The solution is approximated and the smaller the time step, the better the representation is. Dynamic polarization curves obtained from potentiostatic data[7] are required in order to apply the algorithms presented.

APPLICATIONS

Semisubmersible Platform

In this application J/bi-CG has been employed in a three-dimensional potential analysis which consists of using the BEM to verify the cathodic protection system of the semisubmersible[4] platform shown in Figure 2.

The impressed current anodes used to protect the platform were considered as concentrated sources in the BEM analysis.

The geometrical model and superelement discretization adopted are shown in Figure 3. Triangular superelements were subdivided into triangular elements not shown.

Cylindrical constant and tubular quadratic boundary elements, were employed to discretize cylindrical parts of the boundary.

Two numerical analyses have been carried out, in which two distinct types of boundary conditions have been considered. In the problem entitled 'semisubmersible 1' the boundary conditions were of the type (*iii*) indicated in Equation(2). Thus, in this analysis the coefficients of the system matrix **A** are given by

$$\mathbf{A}_{ij} = \mathbf{H}_{ij} - a\ \mathbf{G}_{ij} \tag{24}$$

In the problem entitled 'semisubmersible 2' boundary conditions of type (*iii*) indicated in Equation(2) were prescribed on some of boundary elements and the potential ϕ was prescribed on the remainder. This second analysis did not simulate any actual cathodic protection problem, rather it aimed to force the system matrix **A** to have part of its coefficients given by Expression(24) and the remainder of its coefficients given by (see Equation(6))

$$\mathbf{A}_{ij} = \mathbf{G}_{ij} \tag{25}$$

Results for the two analyses presented on Table I, show that in this case the different nature of the coefficients of matrix **A** have not increased iterative solver solution cost.

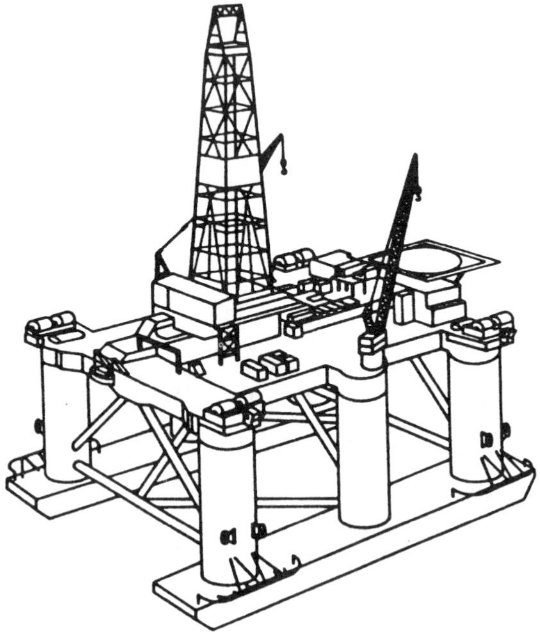

Figure 2. Semisubmersible platform model

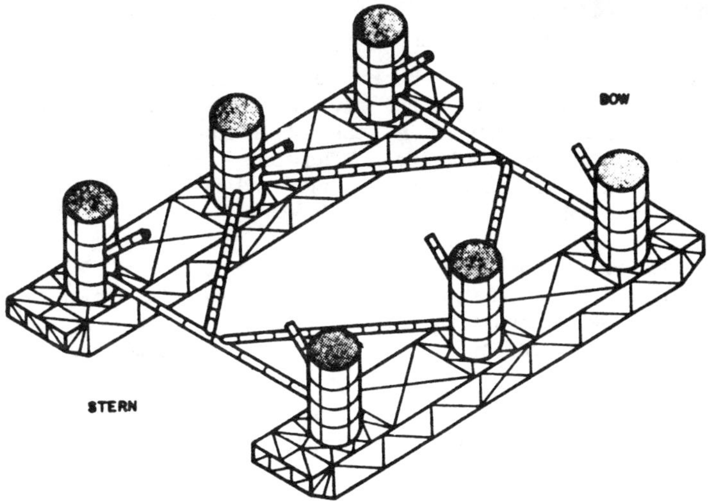

Figure 3. Discretization of submerged structure

Table I. Number of iterations and CPU time for semisubmersible offshore platform.

Number of equations equal to 901

PROBLEM TYPE	SEMI-SUB.1	SEMI-SUB.2	SEMI-SUB.1	SEMI-SUB.2
	CPU time (sec)		Number of iterations	
Gauss	3450	3450	——	——
J/bi-CG	523	448	37	32

Cylindrical Tank

This application has been chosen to be the simplest possible to illustrate the behaviour of some time dependent variables, which can be either physical (potential and current density) or inherent to the numerical procedure adopted (fictitious time or fictitious potential). More realistic engineering applications will be carried out in future works.

The example chosen simulates a cylindrical tank with infinite axis. Therefore the analysis is two-dimensional as shown in Figure 4. Internal and external radius are respectively equal to 100cm and 500cm and the conductivity of the confined electrolyte was taken equal to $0.04\Omega^{-1}cm^{2}$.

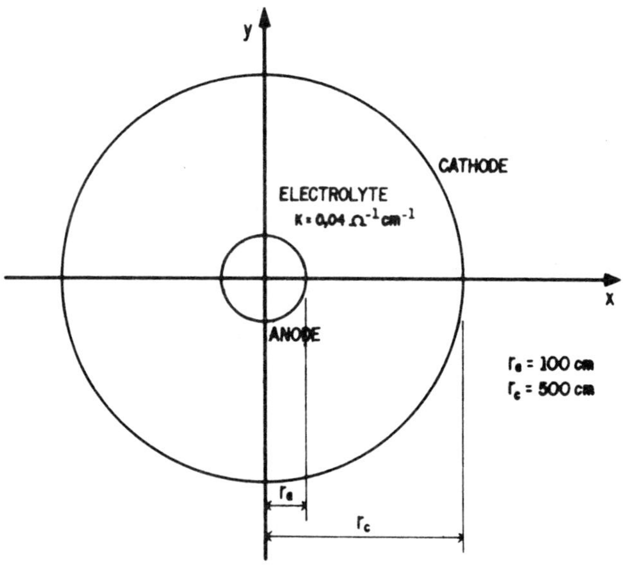

Figure 4. Circular tank

The mesh employed in this analysis is shown in Figure 5. The internal wall is anodic with an impressed current density equal to $-0.30mA/cm^{2}$. The external wall is cathodic with known dynamic polarization curves.

Behaviour of the variables on the cathode with time advance is presented in Figures 6 and 7.

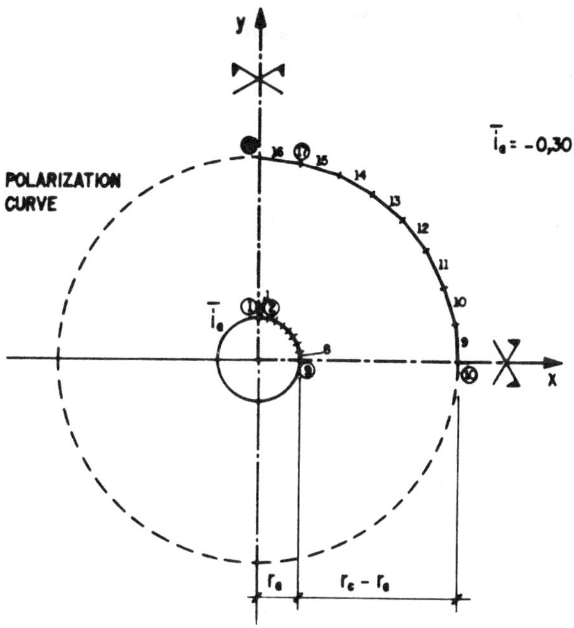

Figure 5. Boundary element discretization

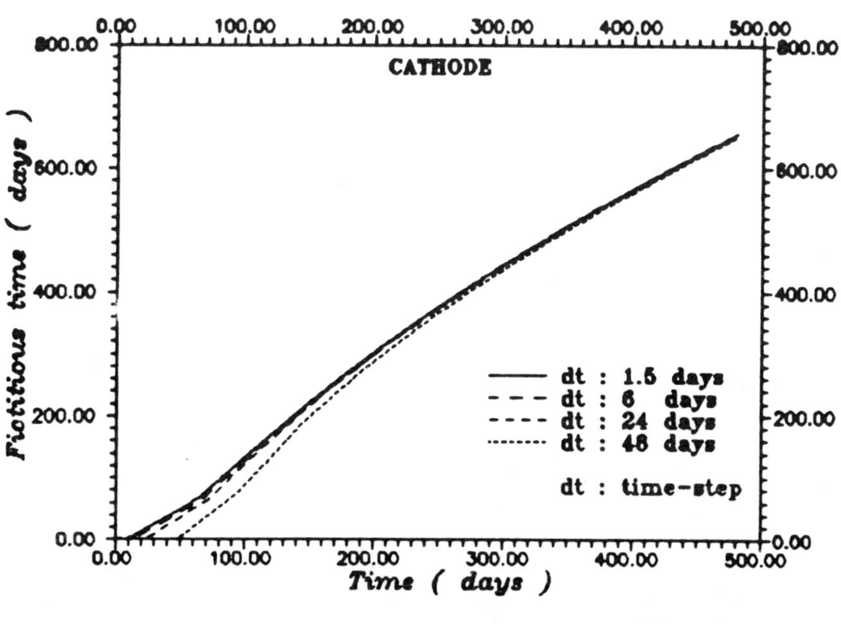

Figure 6. Behaviour of the fictitious time with time advance

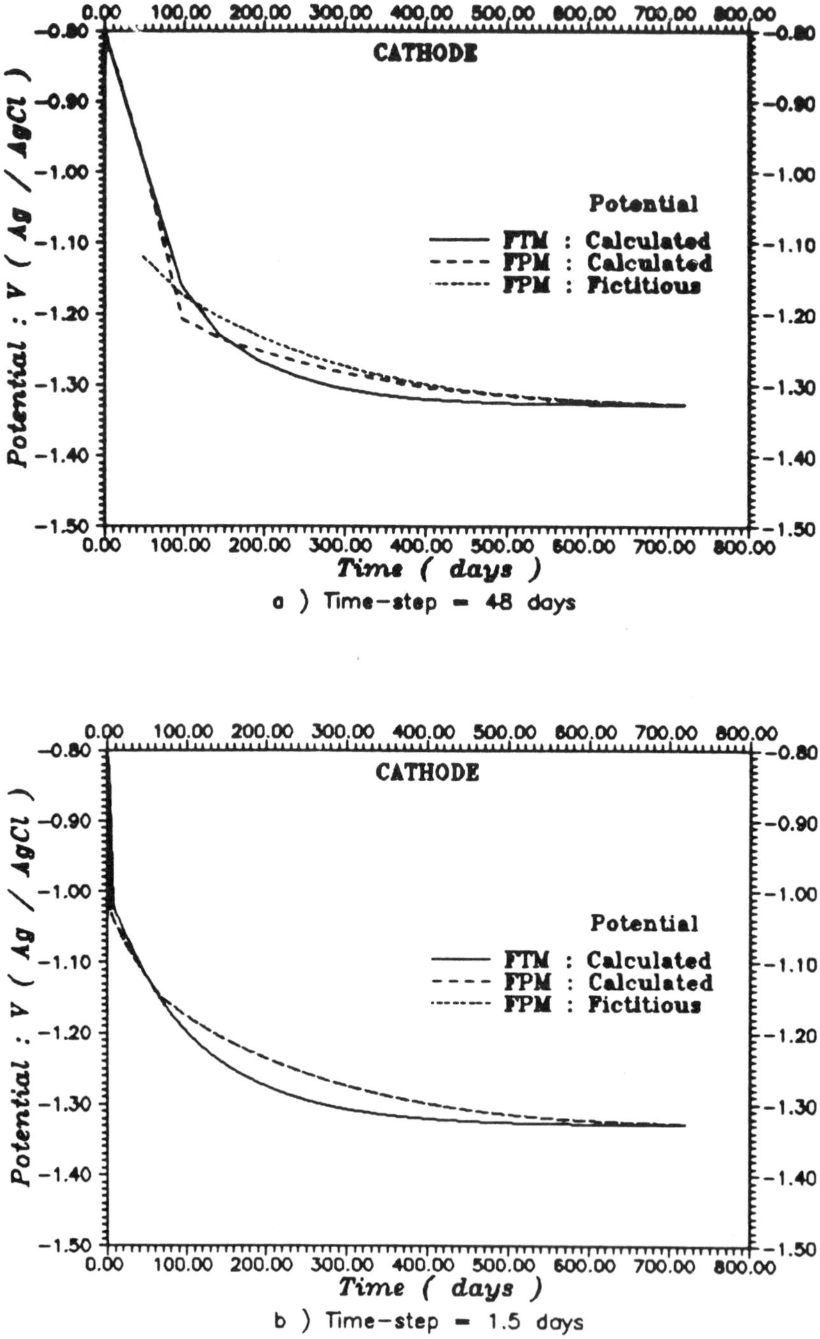

a) Time-step = 48 days

b) Time-step = 1.5 days

Figure 7. Behaviour of the potential with time advance

CONCLUSIONS

The discussion presented in this paper which concerns numerical simulation of cathodic protection problems, focused on two aspects :

(*a*) improvement in computer efficiency
(*b*) improvement in physical modelling of cathodic protection phenomena.

The proposed J/bi-CG algorithm for solving BEM algebraic systems of equations has shown to be quite efficient. In the analysis of the cathodic protection system of a semisubmersible platform, the CPU computer time of J/bi-CG for solving a system of equations which had 901 unknowns was about six times less than that required by the direct Gauss Elimination.

In order to improve the physical modelling, fictitious time and fictitious potential hypothesis have been studied, aiming to simulate actual dynamic polarization curves in cathodic protection systems of offshore structures. Physical characteristics of the seawater have been represented by dynamic polarization curves for constant polarization potential fitted to potentiostatic in situ experiments. The application carried out in the paper illustrates the difference in time response of these two hypothesis (fictitious time and fictitious potential) for a simple problem. Experiments are now being carried out in order to establish which of the two approaches employed can represent better the actual behaviour of cathodic protection systems.

REFERENCES

1 . Telles J.C.F., Wrobel L.C., Mansur W.J. and Azevedo J.P.S. Boundary Elements for Cathodic Protection Problems, Boundary Elements VII, vol. 1, Proc. of the 7th International Conference on BEM, Villa Olmo, Springer-Verlag, Berlin, 1985.

2 . Wrobel L.C., Telles J.C.F., Mansur W.J. and Azevedo J.P.S. A Boundary Element System for Cathodic Protection Design, Offshore Engineering, vol. 5, Proc. of the 5th Int. Symp. on Offshore Engng., Rio de Janeiro, Pentech Press, London, 1985.

3 . Mansur W.J., Wrobel L.C., Telles J.C.F., Azevedo J.P.S. and Santiago, J.A.F. The PROCAT System for Cathodic Protection Design, Boundary Element Technics: Applications in Fluid Flow and Computational Aspects, Proceedings of Betech 87 Conf., Rio de Janeiro, CML Publications, Southampton 1987.

4 . Telles J.C.F., Mansur, W.J., Wrobel L.C., and Marinho M.G. Numerical Simulation of a Cathodically Protected Semisubmersible Platform Using the PROCAT System, Corrosion, vol. 46, no. 6, pp. 513-518, 1990.

5 . Araujo F.C. and Mansur W.J. Iterative Solvers for BEM Systems of Equations, Advances in Boundary Elements: Computations and Fundamentals, Proc. of the 11th Int. Conference on Boundary Element Methods, Cambridge, Springer-Verlag, Berlin, 1989.

6 . Araujo F.C., Mansur W.J. and Malaghini J.E.B. Biconjugate Gradient Acceleration for Large BEM Systems of Equations, Boundary Elements XII: Applications in Stress Analysis, Potential and Diffusion, Proc. of the 12th International Conf. on Boundary Element Methods, Sapporo, Springer- Verlag, Berlin, 1990.

7 . Nisancioglu K. Predicting the Time Dependence of Polarization Cathodically Protected Steel in Seawater, Corrosion, vol. 43, no. 2, pp. 100-111.

8 . Munn R.S. A Mathematical Model for a Galvanic Anode Cathodic Protection System, Materials Performance, vol. 21, no. 8, pp. 29-36.

9 . Brebbia C.A., Telles J.C.F. and Wrobel L.C. Boundary Element Techniques, Springer-Verlag, Berlin, 1984.

10 . Telles J.C.F., Mansur W.J., Wrobel L.C. On Boundary Elements for External Potential Problems, Mechanics Research Communications, vol. 11, no. 6, pp. 373-377.

11 . Das P.C. A Disc Block Elimination Technique used for the Solution of Non-Symmetrical Fully Populated Matrix System Encountered in the Boundary Element Methods, Pentech Press, London, 1978.

SECTION 3: POTENTIAL TYPE PROBLEMS

Transient Saltwater Intrusion in Confined Aquifers using a B-Spline Boundary Element Formulation

J.J.S.P. Cabral *, L.C. Wrobel

Computational Mechanics Institute, Wessex Institute of Technology, Ashurst Lodge, Ashurst, Southampton, SO4 2AA, England

* On leave from Water Recources and Environmental Engineering Group, Federal University of Pernambuco, Recife, PE, Brazil

ABSTRACT

Continuity of the derivatives of the main variable is an important feature to obtain an accurate representation of moving boundaries with iterative or time-marching schemes. A recently developed BEM formulation using cubic B-splines provides up to C^2 continuity between adjacent elements. This formulation is applied in this work to saltwater intrusion problems in confined aquifers.

INTRODUCTION

Moving boundary problems generally demand an accurate computation of both direction and modulus of the velocity at points located on the moving line or surface. When using a discretization procedure with Lagrangian interpolation (even of higher order) oscillations may develop due to the appearance, during the motion, of points of geometric tangent discontinuity between adjacent elements at which the velocity is not unique.

A recently developed BEM formulation based on uniform and non-uniform cubic B-splines [1],[2], capable of preserving continuity of the tangent and curvature (C^2 continuity) between elements, is applied in the present work to study the motion of the interface between fresh and salt water in the context of saltwater intrusion problems.

Boundary elements have been applied to saltwater intrusion studies by Awater [3] using a formulation in the complex plane, with application to a lake reclamation problem. Liu *et al.* [4] developed a formulation using real variables and linear elements for the case of confined aquifers while Kemblowski [5], [6] applied a similar formulation to saltwater upconing problems. Taigbenu *et al.* [7] presented a different approach involving a modified potential and areal modelling. Cabral and Wrobel [8] expanded the formulation of Liu *et al.* [4] by including drains and trenchs for simulating water exploitation in confined coastal aquifers while Cabral and Cirilo [9] have used a similar approach to model a semiconfined aquifer.

Cubic B-splines are used in this work both for geometric representation and as interpolation functions. The numerical results obtained with the present formulation are compared with experimental and analytical solutions showing very accurate results with coarse discretizations.

PHYSICAL CONSIDERATIONS

In many practical cases the thickness of the transition zone between freshwater and saltwater is relatively small compared to the aquifer dimensions. It is thus a common approximation to neglect the transition zone and assume a sharp interface separating the two regions. Bear [10] has given some examples of real aquifers in which experimental measurements have shown the salt concentration to vary sharply at a determined location, clearly establishing a region of low salt concentration (freshwater) and a region of high salt concentration (saltwater).

A further approximation can be introduced into the mathematical modelling of the problem by assuming a constant potential value in the saltwater region and analysing only the freshwater region. This approach, although simpler, may lead to unsatisfactory results [11].

In view of the above, the present paper adopts a sharp interface model including the two regions by employing a standard BEM sub-regions technique [12].

MATHEMATICAL FORMULATION

Combination of Darcy's law and the mass balance equation for a two-dimensional flow through a homogeneous isotropic porous medium gives, for a vertical model,

$$K\left(\frac{\partial^2 u}{\partial x^2}+\frac{\partial^2 u}{\partial y^2}\right)=S\frac{\partial u}{\partial t} \qquad (1)$$

where K is the hydraulic conductivity, S the specific storage and the potential u is the piezometric head

$$u = y + \frac{p}{\gamma}$$

in which p is pressure, γ is specific weight and y is elevation.

For practical purposes, transient flow can be treated as a sequence of successive steady states in a step-by-step procedure. So, the problem can be modelled using Laplace's equation

$$\frac{\partial^2 u}{\partial x^2}+\frac{\partial^2 u}{\partial y^2}=0 \qquad (2)$$

In the present paper, this equation is applied to each region independently, and compatibility and equilibrium conditions employed to relate the potential and normal flux on the interface between regions [12].

The following integral equation, equivalent to (2), can be obtained after some transformations [12]

$$c^i u^i + \int_\Gamma u q^* d\Gamma = \int_\Gamma u^* q d\Gamma \tag{3}$$

where $c^i = \theta/2\pi$, θ is the internal angle at the point i on the boundary Γ, u^* is the fundamental solution of Laplace's equation, $q = \partial u/\partial n$ and $q^* = \partial u^*/\partial n$.

Discretizing the boundary Γ into N elements, we obtain

$$c^i u^i + \sum_{j=1}^{N} \int_{\Gamma_j} u q^* d\Gamma = \sum_{j=1}^{N} \int_{\Gamma_j} u^* q d\Gamma \tag{4}$$

The values of the potential and its normal derivative using cubic B-splines as interpolation functions are given at any point by

$$u(\omega) = E_0(\omega)a_{i-1} + E_1(\omega)a_i + E_2(\omega)a_{i+1} + E_3(\omega)a_{i+2} \tag{5}$$

$$q(\omega) = E_0(\omega)b_{i-1} + E_1(\omega)b_i + E_2(\omega)b_{i+1} + E_3(\omega)b_{i+2} \tag{6}$$

This expression applies to the segment between nodes i and $i+1$; a_i are the coefficients (control points) of the potential representation and b_i the coefficients of the normal derivative; $E_k(\omega)$ are the blending functions for cubic B-splines given in [1]; and ω is a local parameter which varies between 0 and 1 within each segment. To represent corners, multiple knots are used and non-uniform blending functions must be applied as reported in [2].

The present formulation employs isoparametric boundary elements for which the coordinates x and y of each point are also expressed by B-spline functions

$$x = E_0 X_{i-1} + E_1 X_i + E_2 X_{i+1} + E_3 X_{i+2} \tag{7}$$

$$y = E_0 Y_{i-1} + E_1 Y_i + E_2 Y_{i+1} + E_3 Y_{i+2} \tag{8}$$

where X and Y are the coordinates of the control points.

Application of the BEM equation (4) to all nodal points, incorporating approximations (5) to (8) within each boundary element, generates a system of algebraic equations which, in the present case, will have the form

$$\mathbf{Ha} = \mathbf{Gb} \tag{9}$$

Evaluation of the coefficients of matrices **H** and **G** is discussed in detail in [1] and [2].

Boundary conditions

The boundary conditions of the problem specify the potential or flux on the boundary, but generally the coefficients a_i and b_i are not known. So one must modify the system and substitute the coefficients by their relation with the potential and flux.

Writing expression (5) for all collocation points and collecting the result in matrix form gives

$$\begin{Bmatrix} u_1 \\ u_2 \\ \cdot \\ \cdot \\ u_n \end{Bmatrix} = \begin{bmatrix} E_{11} & E_{12} & \dots & E_{1n} \\ E_{21} & \dots & \dots & E_{2n} \\ \cdot & \cdot & \cdot & \cdot \\ \cdot & \cdot & \cdot & \cdot \\ E_{n1} & E_{n2} & \dots & E_{nn} \end{bmatrix} \begin{Bmatrix} a_1 \\ a_2 \\ \cdot \\ \cdot \\ a_n \end{Bmatrix} \tag{10}$$

Each row of matrix $\mathbf{E}$ of nodal values of blending functions has three non-zero terms (four blending functions but one of them is null at each node), and all the other terms are zero.

Inverting equation (10) and substituting for vector $\mathbf{a}$ in (9), adopting a similar procedure for vector $\mathbf{b}$, produces

$$\mathbf{HE}^{-1}\mathbf{u} = \mathbf{GE}^{-1}\mathbf{q} \tag{11}$$

The next step in the formulation is to combine an equation of the above type for each region through the application of interface conditions. For each node along the interface between the freshwater and saltwater zones the following conditions are used:

- Equilibrium of pressure on both sides of the interface, leading to

$$u_s = \frac{u_f}{\chi} + \frac{\chi - 1}{\chi} y \tag{12}$$

  where $\chi = \rho_s/\rho_f$, ρ is the density and the subscripts f and s are related to the fresh and salt waters.

- Compatibility of fluxes, which gives [13]

$$q_f = -\frac{\chi}{\alpha} q_s \tag{13}$$

  where $\alpha = \mu_s/\mu_f$ and μ is the dynamic viscosity.

TRANSIENT PROCEDURE

After solving the BEM system of equations, the normal derivatives at the interface are used to compute the interface motion [13]. Assuming $x = \lambda(y,t)$, the following equation can be written:

$$\frac{\partial \lambda}{\partial t} = -\frac{1}{\sin\beta}\frac{K}{\epsilon} q_f \tag{14}$$

in which β is the angle between the interface and the horizontal axis, and ϵ is the effective porosity.

Applying a finite difference approximation to represent the interface motion equation (14) becomes

$$\lambda^{t+\Delta t} = \lambda^t - \frac{\Delta t\, K}{\epsilon \sin \beta^t} \left[\theta q_f^{t+\Delta t} + (1-\theta) q_f^t\right] \tag{15}$$

The above equation involves a linearization since the sine of β is calculated at time t although the equation is written for time $t + \Delta t$. The use of small time steps is generally sufficient to produce accurate results without iteration.

COMPUTATIONAL IMPLEMENTATION

Quasi-singularity

It is a common feature of groundwater analysis that the horizontal dimension of aquifers is generally much larger than the vertical one. Numerical problems may occur if, for example, a source point on the top surface of the aquifer is very close to a boundary element on the bottom surface or interface.

Furthermore, on dealing with moving interface problems, quasi-singularities may develop after some time steps have ellapsed due to a source point becoming very close to a neighbouring element or due to a very large length ratio between adjacent elements.

To avoid this problem, Telles' algorithm [14] for quasi-singular integrals was applied. In this algorithm, according to the smallest relative distance from the element to the source point, a cubic transformation is used and a higher concentration of integration points obtained in the region of the element nearest to the source point, improving the accuracy of standard Gaussian integration schemes.

Interpolated elements

When implementing Dirichlet boundary conditions (specified potential) for salt-water intrusion in confined aquifers, there are two unknown fluxes to be calculated at the interface tip.

The singularity of matrix **G** can be avoided using the so-called *Interpolated element* [15] in which the source point is assumed somewhere inside the element rather than at the extreme points as usual. This procedure differs from the non-conforming elements [12] because in the interpolated element algorithm, the unknowns remain at the extreme points and only the collocation point is moved. So, one avoids the extrapolation of functions that could introduce inaccuracies.

Several tests [15] have shown that placing the source point at a quarter of the element length generally gives optimum results for the interpolated element implementation. In this work, the source point was assumed at a position corresponding to a B-spline parameter ω equal to 0.4 if the singularity is at the beginning of the element and at $\omega = 0.6$ if it is at the end of the element; this gives a position near a quarter element size.

For computing the coefficients of matrices **H** and **G** with the singularity of the collocation point inside the element, Telles' algorithm [14] was used with more integration points than in the case of singularity at the extreme. Accurate results were obtained with 24 integration points for the case of singularity inside the element and 10 integration points for the cases of non-singularity and singularity at the extreme point of the element.

Discharge computation

The discharge rate through each boundary element can be obtained by integrating the flux along the element length

$$Q_{elem} = \int_{\omega=0}^{\omega=1} -K\frac{\partial u}{\partial n}d\Gamma \tag{16}$$

Applying expressions (6) to (8) into equation (16), one obtains

$$Q_{elem} = -K\int_{\omega=0}^{\omega=1} [E_0(\omega)b_{i-1} + E_1(\omega)b_i + E_2(\omega)b_{i+1} + E_3(\omega)b_{i+2}] \mid G \mid d\omega \tag{17}$$

which can be evaluated using standard Gauss quadrature.

Time stepping

Cubic B-splines may provide superior spatial representation but present the same limitations of Lagrangian elements regarding temporal representation.

In the present implementation, small time steps are adopted at the beginning of the process; as time progresses, this value is increased based on the geometrical and physical characteristics of the problem and on mass continuity concepts, *i.e.* the equilibrium between the given rate of inflow and the calculated rate of outflow.

APPLICATIONS

Segregation by gravity action

This simple theoretical problem initially considers a vertical barrier separating fresh and salt waters in a horizontal confined aquifer of constant thickness. At time $t = 0$, the barrier is removed and the interface begins to tilt in a wedge-like shape due to the saltwater motion, forcing the freshwater upwards.

Several researchers have studied this problem applying theoretical, experimental and numerical techniques. Keulegan [16] presented an analytical solution based on the Dupuit approximation which is valid for large times, in the form

$$\ell(t) = \left(\frac{\Delta\rho\, K_f\, H}{\epsilon\, \rho_f}\right)^{\frac{1}{2}} t^{\frac{1}{2}}$$

in which ℓ is the location of the interface toe, H is the aquifer thickness and $\Delta\rho = \rho_s - \rho_f$. Gardner *et al.* [17] modified Keulegan's expression to better represent the motion at small times, obtaining the equation

$$\ell(t) = \left(\frac{3t^{*2}}{4 + 3t^*}\right)^{\frac{1}{2}} H$$

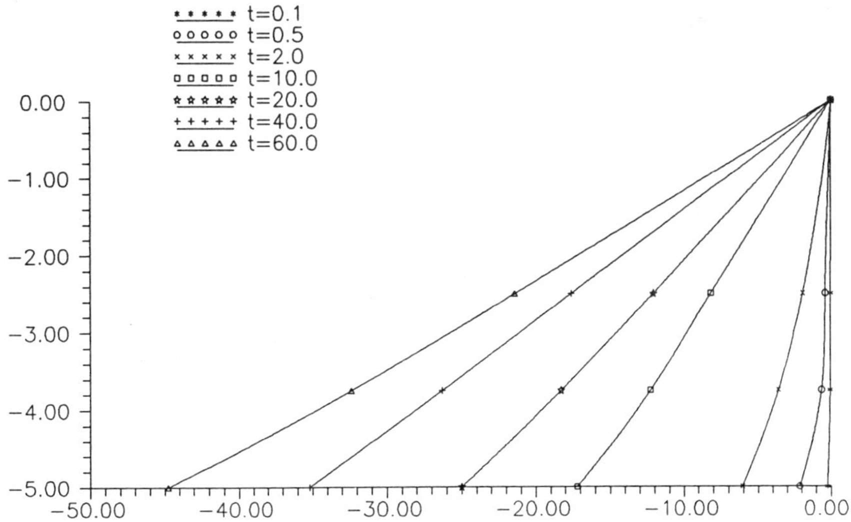

Figure 1: Segregation by gravity action: interface position at different times

where t^* is a dimensionless time defined as

$$t^* = \frac{\Delta\rho\, K_f}{\epsilon\, \rho_f\, H}\, t$$

Figure 1 shows the interface position at different times, as obtained by the present BEM formulation. The physical and geometrical characteristics of the aquifer are: $\rho_f = 1.0g/cm^3, \rho_s = 1.025g/cm^3, \mu_f = \mu_s = 1.2\times10^{-3}g/(cm\, s), K_f = 39.024m/day, K_s = 40m/day$ and $H = 10m$. The boundary element discretization employed 28 nodes along the external boundary and 7 nodes to model the interface. The initial time step was equal to $0.1day$; during the motion, this value was changed to $0.2day, 0.5day$ and finally $2.0day$.

Figure 2 compares the BEM results for the toe position with those obtained from the formulas of Keulegan [16] and Gardner *et al.* [17]. Very good agreement between the solutions can be observed in the figure.

An interesting feature of this problem is that a Neumann boundary condition ($q = 0$) is applied all along the external boundaries, so the solution is not uniquely defined. To complete the problem specification and in order not to destroy the symmetry of the solution, a prescribed potential ($u = 0$) is applied at the middle point of the interface.

Recession of saltwater tongue by increased recharge

The present BEM formulation was next applied to simulate Bear and Dagan's

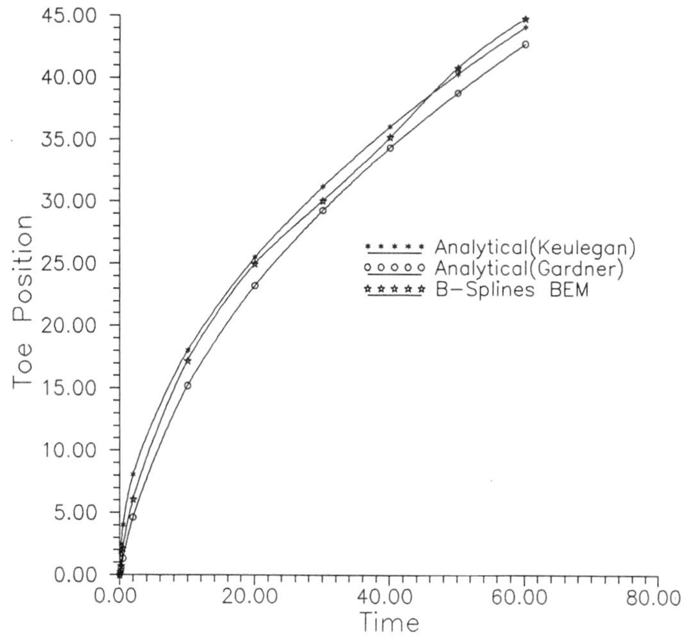

Figure 2: Segregation by gravity action: toe position at different times

experiment with a Hele-Shaw cell [18]. Figure 3 shows a schematic representation of experiment 3 of [18] in which the initial recharge rate of $Q = 3.9cm^2/s$ is increased to $Q = 18.8cm^2/s$ for $t > 0$. Other characteristics of the problem are $\rho_f = 1.0g/cm^3, \rho_s = 1.029g/cm^3, K = 69cm/s, \epsilon = 1.0, \mu = 1.2 \times 10^{-4}g/(cm\, s)$ and $H = 27cm$.

For the numerical analysis, an arbitrary position of the interface was initially assumed and a study undertaken to determine the initial steady-state configuration. Then, another analysis was performed, with the second recharge rate, to compute the transient behaviour of the saltwater tongue.

The interface tip presents a singular behaviour in this case. Bear [10] and Liu *et al.* [4] have shown that the interface should intersect the outflow face at a right angle at all times. From the BEM point of view, the potential is known at the intersection point but there are two unknown fluxes; thus, an interpolated element was used to overcome the problem. The horizontal derivative at this point is always null and, for calculating the location of the interface tip, an extrapolation was performed assuming a parabolic shape for the interface (following van Dam [19]).

Figure 4 depicts the interface location at different times, compared to the experimental results [18]. It can be seen that the numerical results differ quite considerably from the experimental ones. However, as quoted by Liggett and Liu [13], comparisons between other theoretical and numerical results [20], [21] and

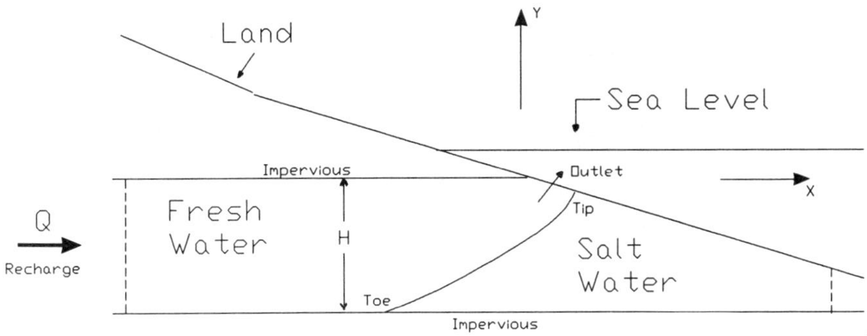

Figure 3: Recession of saltwater tongue by increasing recharge

the experimental ones of Bear and Dagan [18] were not close either, and there is reason to believe that the experimental error may be significant. The present results are very similar to that of Liggett and Liu [13] although a much coarser discretization is employed herein.

In figure 5, a comparison is made between the two steady-state BEM solutions (for the two values of Q) and analytical results using Glover's equation as presented by van Dam [19]. The agreement is very reasonable in this case.

CONCLUSIONS

For transient flow in porous media with a moving boundary or interface, the normal derivative at points along the moving surface can be accurately computed using elements that provide derivative continuity. The results obtained suggest that, for the same level of accuracy, one can generally use far less degrees of freedom with B-spline elements than other kinds of elements.

ACKNOWLEDGMENT

The first author would like to acknowledge the financial support of CAPES/Ministry of Education, Brazil.

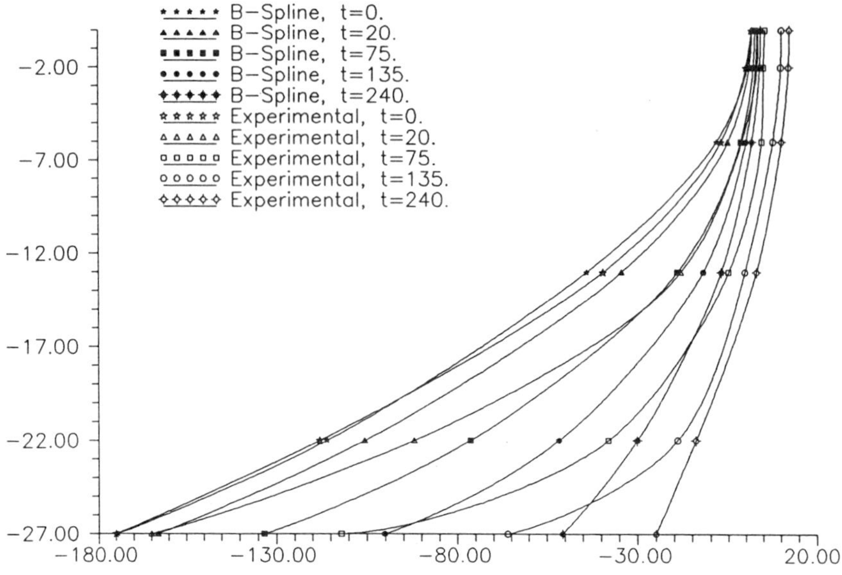

Figure 4: Transient behaviour of interface

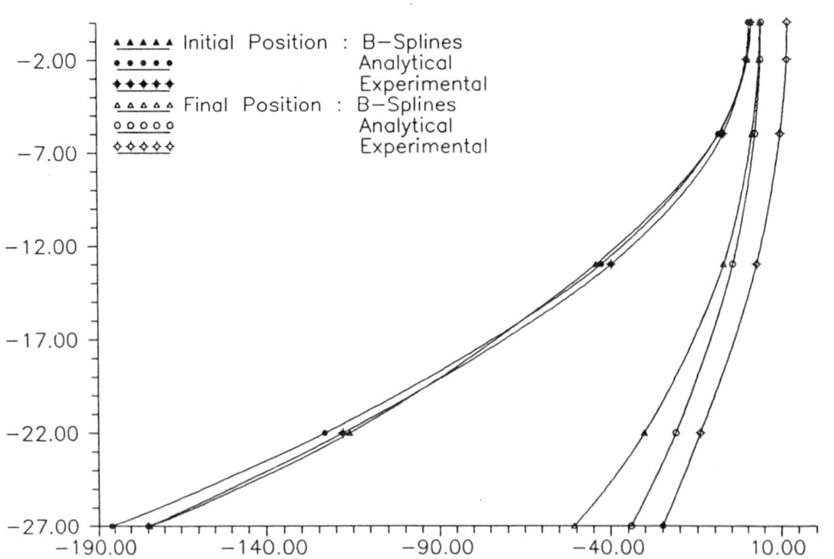

Figure 5: Steady-state positions of interface

References

[1] Cabral,J.J.S.P., Wrobel,L.C. and Brebbia,C.A., A BEM formulation using B-splines: I- uniform blending functions, *Engineering Analysis*, Vol. 7, pp. 136-144, 1990.

[2] Cabral,J.J.S.P., Wrobel,L.C. and Brebbia,C.A., A BEM formulation using B-splines: II- multiple knots and non-uniform blending functions, *Engineering Analysis*, Vol. 8, pp. 51-55, 1991.

[3] Awater,R., The transient behaviour of a fresh/salt water interface, *New Developments in Boundary Element Methods*, Computational Mechanics Publications, Southampton and Springer-Verlag, Berlin, 1980.

[4] Liu,P.L-F., Cheng,A.H-D., Liggett,J.A. and Lee,J.H., Boundary integral equation solutions to moving interface between two fluids in porous media, *Water Resources Research*, Vol. 17, pp. 1445-1452, 1981.

[5] Kemblowski,M., Saltwater upconing under a river - a boundary element solution, *BEM VI*, Computational Mechanics Publications, Southampton and Springer-Verlag, Berlin, 1984.

[6] Kemblowski,M., Saltwater-freshwater transient upconing - an implicit boundary element solution, *Journal of Hydrology*, Vol. 78, pp. 35-47, 1985.

[7] Taigbenu,A.E., Liggett,J.A. and Cheng,A.H-D., Boundary integral solution to seawater intrusion into coastal aquifers, *Water Resources Research*, Vol. 20, pp. 1150-1158, 1984.

[8] Cabral,J.J.S.P. and Wrobel,L.C., A numerical analysis of saltwater intrusion in coastal aquifers, *Brazilian Engineering Journal, Water Resources Division*, Vol. 3, pp. 29-52, 1985 (in Portuguese).

[9] Cabral,J.J.S.P. and Cirilo,J.A., Saltwater-freshwater interface motion in leaky aquifers, *BETECH 87*, Computational Mechanics Publications, Southampton, 1987.

[10] Bear,J., *Hydraulics of Groundwater*, McGraw-Hill, New York, 1979.

[11] de Lange,W.J., Application of the boundary integral element method to analyse the behaviour of a freshwater-saltwater interface calculating three-dimensional groundwater flow, *9th Salt Water Intrusion Meeting*, Delft, 1986.

[12] Brebbia,C.A., Telles,J.C.F. and Wrobel,L.C., *Boundary Element Techniques*, Springer-Verlag, Berlin, 1984.

[13] Liggett,J.A. and Liu,P.L-F., *The Boundary Integral Equation Method for Porous Media Flow*, Allen and Unwin, London, 1983.

[14] Telles,J.C.F., A self-adaptive coordinate transformation for efficient numerical evaluation of general boundary element integrals, *International Journal for Numerical Methods in Engineering*, Vol. 24, pp. 959-973, 1987.

[15] Marques,E., Coupling of the finite element method and the boundary element method: an application to potential problems, *M.Sc. Thesis*, COPPE/UFRJ, Rio de Janeiro, 1986 (in Portuguese).

[16] Keulegan,G.H., Ninth progress report on model laws for density currents: An example of density current flow in permeable media, *Report No. 3411*, National Bureau of Standards, Washington, D.C., 1954.

[17] Gardner,G.H.F., Downie,J. and Kendall,H.A., Gravity segregation of miscible fluids in linear model, *Society of Petroleum Engineers Journal, Transactions AIME*, Vol. 225, pp. 95-104, 1962.

[18] Bear,J. and Dagan,G., Moving interface in coastal aquifers, *Journal of the Hydraulic Division, ASCE*, Vol. 90, pp. 193-216, 1964.

[19] van Dam,J.C., The shape and position of the salt water wedge in coastal aquifers, *Hamburg Symposium in Relation of Groundwater Quantity and Quality*, IAHS, Publication No. 146, 1983.

[20] Shamir,V. and Dagan,G., Motion of the seawater interface in coastal aquifers: a numerical solution, *Water Resources Research*, Vol. 7, pp. 644-657, 1971.

[21] Sá da Costa,A.A. and Wilson,J.L., A numerical model of seawater intrusion in aquifers, *Report No. 247*, Ralph M. Parsons Laboratory, MIT, 1979.

Sound Radiation of a Cantilever Plate - Comparisons of BEM Results with Experimental Data

O. von Estorff *, A. Homm**, F. Bartels **

* *Industrieanlagen-Betriebsgesellschaft mbH, Einsteinstr. 20, D-8012 Ottobrunn, Germany*

** *Fraunhofer-Forschungsgruppe für Hydroakustik, Waldparkstr. 41, D-8012 Ottobrunn, Germany*

ABSTRACT

The sound radiation of a vibrating cantilever plate is investigated by a combined Finite Element / Boundary Element approach. First, in a FE–analysis the mode shapes of the plate are determined. Second, for three selected frequencies the acoustic pressure distribution around the plate is calculated using an indirect boundary element formulation. The results are compared with experimental data in order to show the accuracy and applicability of the numerical approach.

INTRODUCTION

The investigation of acoustic problems is a very complex and pretentious task. In particular in fields such as the automotive and aerospace industries, environmental engineering, hydroacoustics (sonar), and safety analyses of underwater structures, the knowledge of the acoustic wave propagation is of considerable importance.

While in practice the majority of acoustic investigations are still performed experimentally, in recent years also numerical techniques, namely the Finite Element Method (FEM) [1] and the Boundary Element Method (BEM) [2], [3], have been developed and applied to these kind of problems.

At the beginning, mainly the FEM was employed and it was found that its main advantage was the flexibility to handle complex geometries and various boundary conditions. Moreover, the robustness of this well known methodology and the fact that conventional structural FE codes could be used to treat acoustic problems [4], [5] were reasons why people started with this approach. However, there is one stringent limitation encountered with the FEM: its inability to model unbounded domains. To overcome this problem the introduction of infinite "elements" were required, which, however, are not yet established for the general case (fullspace). Nevertheless, the FEM is an effective tool when closed regions, such as passenger compartments or rooms of a building, are under consideration [6], [7].

For the treatment of unbounded domains, the boundary element method is increasingly coming to be recognized as a method that is superior to the FEM. In particular, for the investigation of problems, such as noise radiation in air or scattering of underwater pressure waves, the BEM is ideally suited as it automatically takes into account wave radiation to infinity (SOMMERFELD's condition). Antes [8], Amini and Wilton [9], and Ciskowski and Brebbia [10] discussed comprehensively the application of the boundary element technique to acoustic problems. While, e.g., Coyette, Dejehet, Guisset and Fyfe [11], and Seybert and Cheng [12] concentrated on the solution of harmonic problems using frequency domain formulations, Antes and von Estorff [13], Mansur [14], Antes and Meise [15], and Groenenboom [16] investigated transient problems by means of a BE–method working directly in the time domain.

Only very little work has been done in the field of validation of the acoustic models. The sound propagation in a passenger compartment of a car was investigated, e.g., by Burfeindt and Zimmer [17] making use of the FE analogy described in [5], and by Guy and Bhattacharya [18] employing an analytical formulation. Ochmann and Wellner [19], on the other hand, calculated the sound radiation of an L–shaped structure by boundary elements and compared their results with measured values. A fluid– structure interaction problem, namely a plate under water, was recently tested by Coyette and Fyfe [20], but only the change of the resonant frequencies was considered.

In the present paper the sound generated by a cantilever plate is investigated (exterior problem). The main objective of this study is to compare the numerically obtained results with experimental data in order to enhance the confidence in boundary element solutions for wave radiation problems. The comparison is performed in two steps. First, several mode shapes of the cantilever plate (surrounded by air) are calculated using finite elements, and the results are compared with the tested configuration. Second, the acoustic field generated by the vibrating plate (exterior problem) is considered. By means of an indirect boundary element approach, realized in the computer program SYSNOISE [21], the acoustic pressure at various points in the vicinity of the plate is analyzed and compared with the experimental results.

NUMERICAL MODELING

For the numerical determination of the mode shapes of the considered plate a finite element analysis seemed to be most appropriate. During the course of this study, linear shell elements will be employed to model the plate, and the eigenvalue problem will be solved by using the Subspace Iteration Method. Details of the FEM approach are extensively discussed in [1], therefore only the indirect BE approach, used to calculate the radiating noise, will be treated in this paper.

Assuming time dependence with the circular frequency ω, a mathematical model of the propagation of small–amplitude acoustic waves in a homogeneous acoustic medium is the linear Helmholtz equation

$$\nabla^2 p + k^2 p = -\gamma \quad \text{in } \Omega \tag{1}$$

where p is the acoustic pressure and $k = \omega/c$ is the wavenumber; c denotes the speed of the sound waves and γ the amplitude of a harmonic excitation in the interior of the domain Ω. Moreover, the velocity vector $\mathbf{v}$ of a fluid particle is connected to the space–derivatives of the scalar pressure by

$$\mathbf{v} = \frac{\nabla p}{-i \rho \omega} \tag{2}$$

where ρ is the density of the acoustic fluid.

Assuming the general case of a mixed boundary value problem, the appropriate boundary conditions for equation (1) can be defined as (see Fig. 1):

$$p = \bar{p} \quad \text{on } \Gamma_1 \tag{3}$$

$$p_{,n} = -i \rho \omega \bar{v}_n \quad \text{on } \Gamma_2 \tag{4}$$

where $\mathbf{n}$ is the outward normal to the boundary Γ and the comma indicates the spatial derivative; given values are marked by "–".

Instead of eq. (1) also an integral representation can be used to describe the propagation of the acoustic waves in Ω. In the indirect BEM, or so–called singularity method, this integral equation relates fictious singularity distributions along the boundary Γ which have to be determined in such a way that the acoustic field caused by the given boundary conditions is enforced.

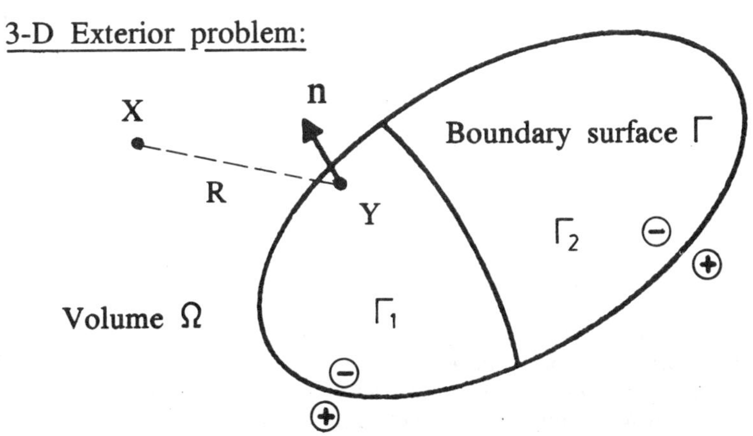

Fig. 1: Definition of the boundaries (indirect BEM)

For the acoustic problem under consideration an integral equation of the form

$$p(X) = \int_\Gamma \left[\mu(Y) \frac{\partial G(X,Y)}{\partial n_Y} - \sigma(Y)\, G(X,Y)\right] d\Gamma(Y) \qquad (5)$$

can be used. It relates two singularities which are the jump of pressure between the two boundaries of the regions + and − (see Fig. 1),

$$\mu = p^+ - p^- \qquad (6)$$

and the jump of the pressure normal derivatives between these regions

$$\sigma = p^+_{,n} - p^-_{,n} \;. \qquad (7)$$

The kernel G(X,Y) is known as the fundamental solution of the Helmholtz equation for a point source. It is defined by

$$G(X,Y) = \frac{\exp(-i\,kR)}{4\pi\, R} \qquad (8)$$

where R is the distance between the field point X and the surface point Y. According to eqs. (3) and (4) the boundary conditions for the singularities are

$$\mu = 0 \quad \text{on } \Gamma_1 \qquad (9)$$

$$\sigma = 0 \quad \text{on } \Gamma_2 \qquad (10)$$

Using these conditions two integral equations can be derived in order to determine, **in a first step**, the unknown singularities μ and σ along the boundary. Letting the field point X reach Γ_1, one obtains

$$\bar{p}(X) = -\int_{\Gamma_1} \sigma(Y)\, G(X,Y)\, d\Gamma_1 + \int_{\Gamma_2} \mu(Y) \frac{\partial G(X,Y)}{\partial n_Y} d\Gamma_2 \qquad (11)$$

assuming that X is located on Γ_2 yields

$$-i\rho\omega\, \bar{v}_n(X) = -\int_{\Gamma_1} \sigma(Y) \frac{\partial G(X,Y)}{\partial n_X} d\Gamma_1 + \int_{\Gamma_2} \mu(Y) \frac{\partial^2 G(X,Y)}{\partial n_X\, \partial n_Y} d\Gamma_2 \quad (12)$$

These integral equations can be solved either by a collocation scheme or, alternatively, by a variational formulation. The second of these approaches can be quite advantageous, since it leads to symmetric coefficient matrices which can easily be coupled with, for instance, finite elements [11], [21]. Once the singularities μ and σ are known, equation (5) can be used, **in a second step**, to determine the acoustic pressure p at any point X within the acoustic medium.

EXPERIMENTAL SETUP

To measure the eigenfrequencies and to visualize the mode shapes of the cantilever plate the system shown in Fig. 2 a) has been set up. It consists of a horizontal aluminum plate of a/b/t = 305/254/12.7 mm which is held at one edge by a heavy circular steel disk to simulate a clamped end. The material data of the plate are YOUNG's modulus E = 70000 N/mm^2, Shear–Modulus G = 27000 N/mm^2, and density ρ = 2660 kg/m^3.

The eigenfrequencies characterizing the vibration behavior of the plate can be determined by measuring the shock response of the plate with accelerometers. By means of a fourier transformation of the measured, transient accelerations (using a dual channel FFT analyzer), a frequency response spectrum of the plate can be found, where sharp peaks show the resonance frequencies of the system.

The structural modes of the plate can be visualized by the CHLADNI method [22]. For that purpose the horizontally adjusted plate has to be evenly covered with sand and one of its edges must be excited by an electromechanical shaker running with the plate resonant frequency of interest. The vibrations cause the sand to concentrate at the lines of zero vertical displacements. These are the nodal lines of the corresponding vibration mode [see Fig. 2 a)].

For the measurements of the acoustic field generated by the plate, an additional aspect has to be considered: it is difficult to excite the plate, e.g. with a shaker, and not to disturb the acoustic field. Therefore, a reciprocity approach is employed, where, instead of measuring the sound emitted by the vibrating plate directly (via a microphone), the plate is excited by the constant sound of a loudspeaker which is set up in place of the microphone. The method makes use of the fact that the vibration amplitude of the plate, measured by an accelerometer, is proportional to the directivity of the sound.

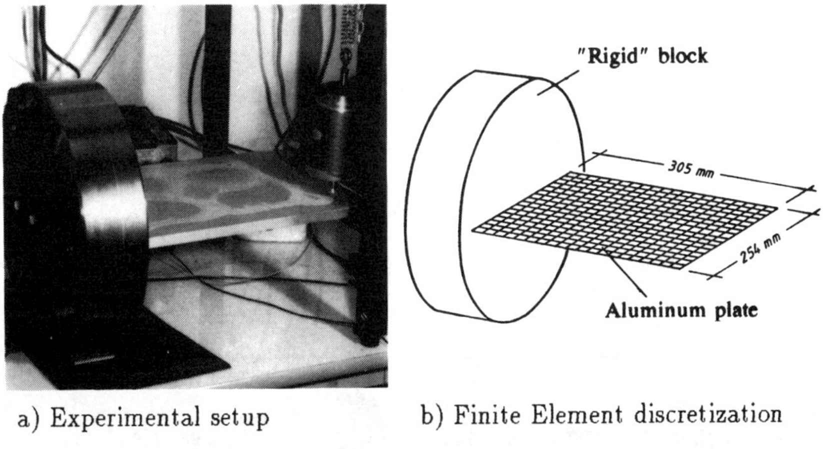

a) Experimental setup b) Finite Element discretization

Figure 2: Configuration to determine the eigenfrequencies and modes.

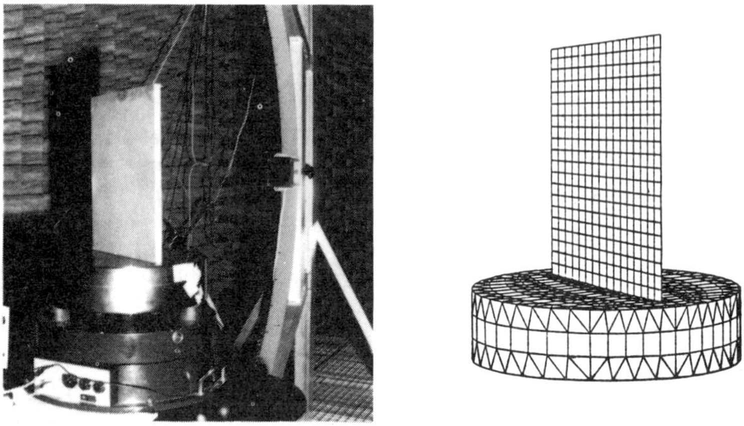

a) Experimental setup b) Boundary Element discretization

Figure 3: Configuration to determine the acoustic radiation.

The experimental setup for the radiation measurements is shown in Fig. 3 a). The accelerometer is mounted at one upper corner of the plate and the mounting block holding the plate is resting on a revolving table. The loudspeaker is mounted on a circular guide–rail at a distance of 0.95 m from the center of the plate. With this configuration the angle of elevation of the loudspeaker can be varied in a wide range so that a nearly complete sphere around the vibrating plate can be scanned.

NUMERICAL AND EXPERIMENTAL STUDIES

Vibration behavior of the plate

The dynamic behavior of the aluminum plate under consideration (material properties given in the previous section) can be characterized best by the frequency spectrum shown in Fig. 4. From this curve three eigenfrequencies were selected to compare calculated and measured values.

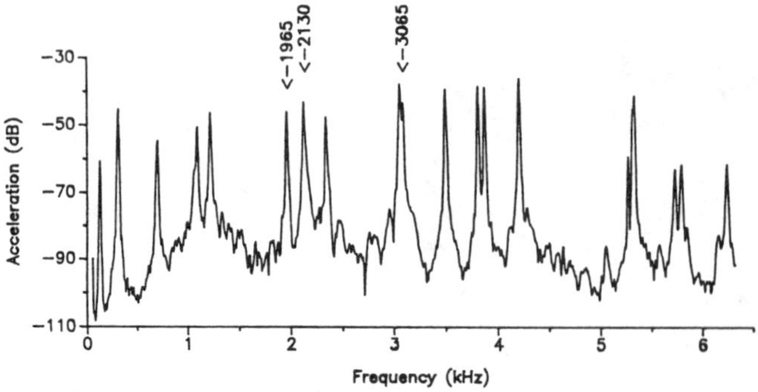

Figure 4: Frequency spectrum of the plate (measured).

	7th Mode	8th Mode	11th Mode
Calculation (FEM)	1982 Hz	2089 Hz	3173 Hz
Experiment	1965 Hz	2130 Hz	3062 Hz

Table 1: Comparison of the eigenfrequencies.

The plate was discretized by 320 finite shell elements, as shown in Fig. 2 b), and the Subspace–Iteration–Method was used to determine the eigenfrequencies and mode shapes of the system. Table 1 contains the calculated and measured resonant frequencies and it can be seen that the results match very well, i.e. the maximum difference was only 3.5 %. Additionally, for the 8th eigenfrequency, f = 2130 Hz, the calculated and the experimentally visualized eigenmode are shown in Fig. 5. Again, if one compares the lines of zero vertical displacements in both cases, an excellent agreement can be observed. Although not shown here, the accuracy of the two other structural modes was also extremely good.

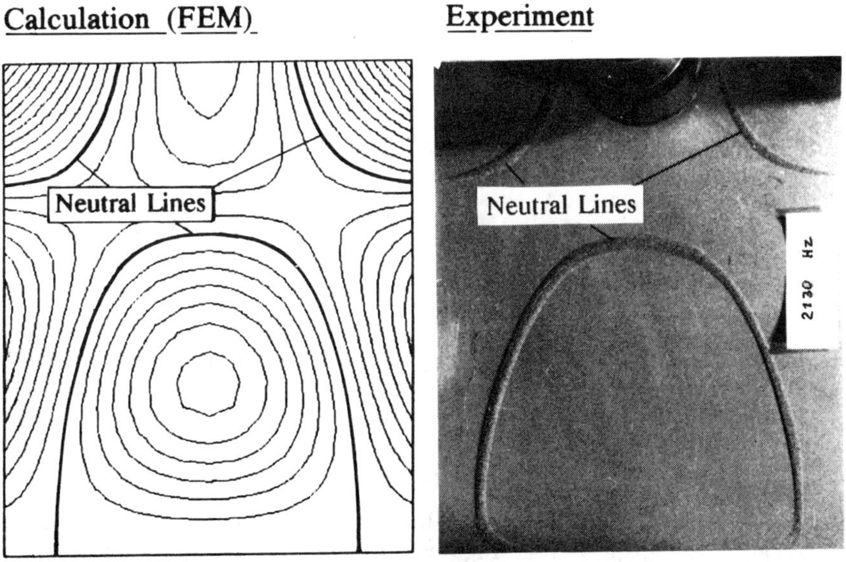

Figure 5: Comparison of the 8th structural mode.

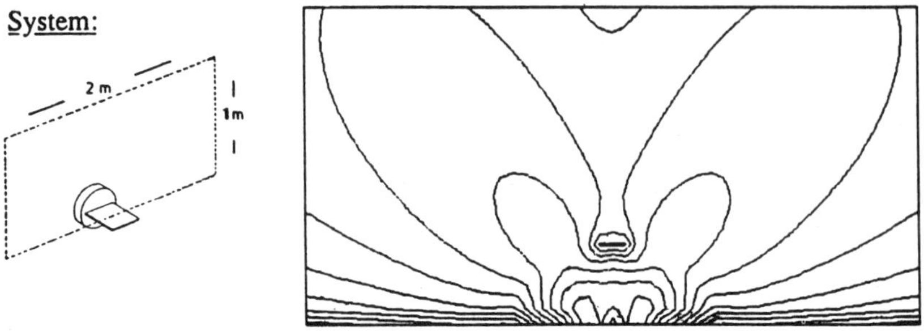

Figure 6: Sound pressure in a plane perpendicular to the plate.

Sound radiation
After the mode shapes stemming from the finite element analysis were found to be in good agreement with the experiments, the plate and its mounting block were discretized with 946 boundary elements [Fig. 3 b)]. At the elements modeling the surface of the plate the structural velocities, vertical to the mid–plane of the plate, were prescribed, separately for each of the three investigated modes. The mounting block, on the other hand, was assumed to be rigid, i.e. zero surface velocities were prescribed at the respective boundary elements. For the air surrounding the plate a density of $\rho^f = 1.189$ kg/m³ and a wave speed of $c = 343$ m/s were assumed.

First, applying the indirect BE–formulation provided in the program SYSNOISE, the unknown singularity distribution at the surface of the entire plate/block system was calculated for the three frequencies. Next, in a region of 1 m x 2 m, perpendicular to the mid–plane of the plate (see sketch in Fig. 6), 911 field points were defined. As an example, the acoustic pressure at these points, generated by the plate at a frequency of 2130 Hz, is plotted in Figure 6. The result clearly shows the typical sound characteristics (maximum/minimum) emitted by the vibrating surface of a structure.

In order to compare calculated and measured results, 180 field points were defined on a circle of radius $r = 0.95$ m around the center of the plate (see sketch in Fig. 7). The acoustic pressure at these points was calculated and plotted in a polarized manner. Thus, so–called directivity patterns were obtained which are commonly used to characterize the sound radiation behavior of a structure.

The analyzed and measured directivity plots for the three selected frequencies are displayed in Figures 7 to 9. Since neither the FE mode analysis nor the performed experiments yielded absolute results (e.g. displacements), the data had to be scaled in order to get equal maxima at 0 dB. It should be noted, however, that the performed linear analysis allows the scaling with any factor, i.e. the consideration of the results presented here is sufficient to judge about the accuracy of the BE–model.

The comparison of the three investigated directivity patterns, particularly the locations of their minima and maxima as well as the general shape of the curves, shows excellent agreement between numerical and the measured values.

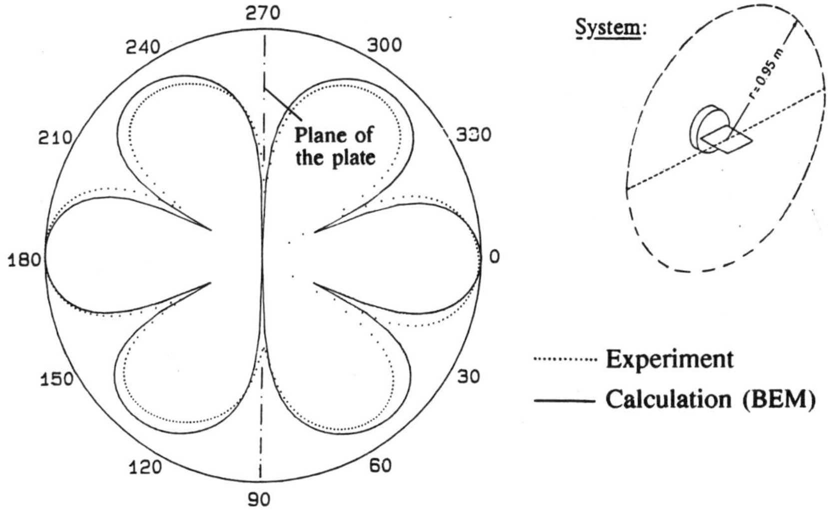

Figure 7: Comparison of the directivity pattern : f = 1965 Hz.

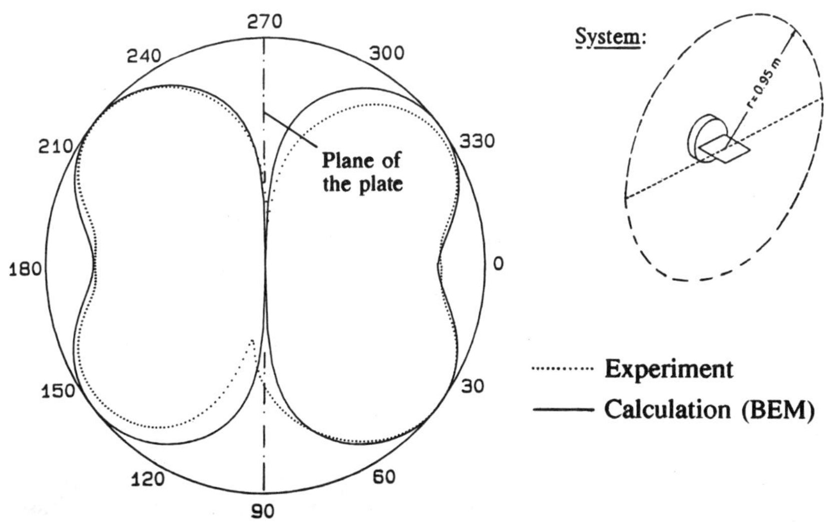

Figure 8: Comparison of the directivity pattern : f = 2130 Hz.

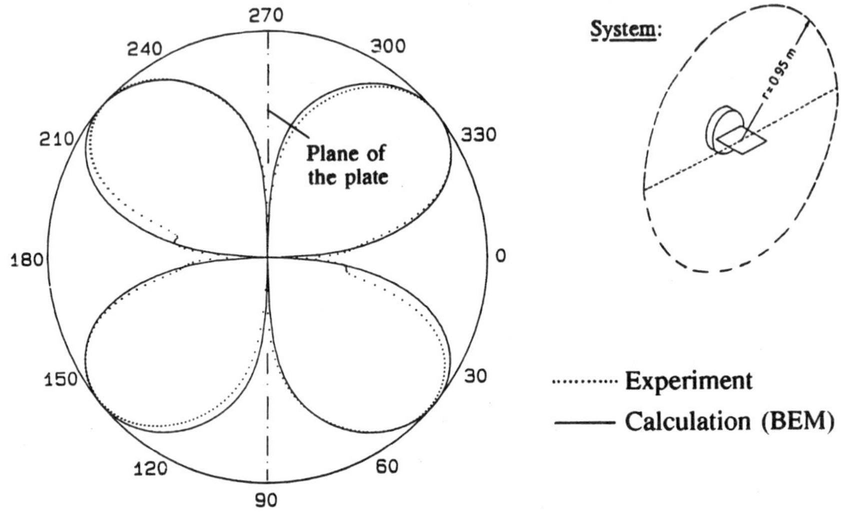

Figure 9: Comparison of the directivity pattern : f = 3062 Hz.

CONCLUSIONS

An indirect boundary element method, implemented in the computer program SYSNOISE, has been applied to study the sound radiation of a cantilever plate in air. The main objective of the investigation was to examine how detailed numerically obtained results match with experimental data.

In view of the gained experiences and the obtained results, which are not completely documented in this paper, the following conclusions can be drawn:

- The finite element calculation of the eigenfrequencies and mode shapes yielded very accurate results. (In a preceding study the first 20 resonant frequencies were compared.)
- Air/structure interaction effects also were considered but found to be negligible.
- The boundary element method is very well suited to treat sound radiation problems. All comparisons showed excellent agreement between numerical and experimental data.

The study clearly enhances the confidence in boundary element solutions for wave radiation problems. In the near future a similar configuration, namely a plate submerged in water, will be investigated in order to assess how fluid–structure interaction affects the results.

REFERENCES

1. Bathe, K.–J. Finite Element Procedures in Engineering Analysis, Prentice–Hall, Englewood Cliffs, N.J., 1982.
2. Brebbia, C.A. The Boundary Element Method for Engineers, Pentech Press, London, 1978.
3. Bausinger, R., Kuhn, G. The Boundary Element Method, (in German), Expert–Verlag, Ehningen, FRG, 1987.
4. Raasch, I. Procedures to Solve Acoustic Problems with MSC/NASTRAN, Proc. of the MSC/NASTRAN Users' Conf., Munich, 1981.
5. MacNeal, R.H., Citerley, R., Chargin, M. A New Method of Analyzing Fluid–Structure Interaction Using MSC/NASTRAN, MacNeal–Schwendler Corporation, Los Angeles, 1980.
6. Coyette, J.P., Guisset, P. Finite Element Methods for Interior Acoustics, Proc. of the 2nd Int. Conf. on Methodology and Innovations in Automotive Experimentations, Florence, 1988.
7. von Estorff, O. Numerical Methods for Automotive Acoustics, Proc. of the 3rd European Cars/Trucks Simulation Symposium, Schliersee, FRG, 1991.
8. Antes, H. Application of the Boundary Element Method in Elastodynamics and Fluiddynamics, (in German), Teubner–Verlag, Stuttgart, 1988.
9. Amini, S. , Wilton, D.T. An Investigation of Boundary Element Methods for the Exterior Acoustic Problem. Comp. Meth. in Appl. Mech. and Engng, Vol. 54, pp. 59–65, 1986.
10. Ciskowski, R., Brebbia, C.A. (Ed.). Advances in BEM in Acoustics, Computational Mechanics Publications, Southampton, 1991.
11. Coyette, J.P., Dejehet, E., Guisset, P., Fyfe, K. Numerical Analysis of Acoustic and Elasto–Acoustic Problems Using Combined Finite Element and Boundary Element Methods, Proc. of the 7th Int. Modal Analysis Conf., Las Vegas, 1989.
12. Seybert, A.F., Cheng, C.Y.R. Application of the Boundary Element Method to Acoustic Cavity Response and Muffler Analysis, J. of Vibration, Acoustics, Stress and Reliability in Design, Vol. 109, pp. 15–21, 1987.
13. Antes, H., von Estorff, O. Radiation of Transient Acoustic Waves – Investigations Using a Time Domain BEM, (in German), Ingenieur–Archiv, Vol. 59, pp. 17–31, 1989.
14. Mansur, W.J. A Time–Stepping Technique to Solve Wave Propagation Problems Using the Boundary Element Method. Ph.D. Thesis, Southampton University, 1983.

15. Antes, H., Meise, T. Scalar Wave Propagation – Calculation Capabilities of a 3–D Time Domain Boundary Element Method, in Advances in Boundary Elements, (C.A. Brebbia and J.J. Connor, Eds), pp. 343–358 in Vol. 3, Stress Analysis, Proc. 11th BEM Conf., Cambridge, USA, Springer Verlag, Berlin and New York, 1989.

16. Groenenboom, P.H.L. Wave Propagation Phenomena, Chapter 2 in Progress in Boundary Element Methods (Brebbia, C.A., Ed.), Vol. 2, Springer Verlag, Berlin and New York, 1983.

17. Burfeindt, H., Zimmer, H. Calculating Sound Pressure in Car Interiors, Proc. of the MSC/NASTRAN Users' Conf., London, 1989.

18. Guy, R.W., Bhattacharya, M.C., The Transmission of Sound Through a Cavity–Backed Finite Plate, J. of Sound and Vibration, Vol. 27, pp. 207–223, 1973.

19. Ochmann, M., Wellner, F. Analysis of the Noise Radiation of Three–Dimensional Vibrating Structures by a Boundary Element Multi Grid Method, (in German), Acustica, 1991, (in print).

20. Coyette, J.P., Fyfe, K.R. Solution of Elasto–Acoustic Problems Using a Variational Finite Element / Boundary Element Technique, Dynamic Engineering, Heverlee, Belgium, 1989.

21. SYSNOISE User's Manual, Dynamic Engineering, Heverlee, Belgium, 1988.

22. Grinsted, B. Nodal Pattern Analysis, Proceedings of the Institution of Mechanical Engineers (A), Vol. 166, pp. 309–326, 1952.

23. Junger, M.C., Feit, D. Sound, Structures, and Their Interaction, MIT–Press, 1986.

24. Meyer, E., Neumann, E.G. Physical and Technical Acoustics, (in German), Vieweg, Braunschweig, 1979.

25. Blöchl, B., Gamser, A. Experimental Determination of Acoustic Directivity Patterns of Two Propellers for Selected Resonance Frequencies, (in German), Fraunhofer–Forschungsgruppe für Hydroakustik, FHAK Report 177/89, Ottobrunn, Germany, 1989.

SECTION 4: STRESS ANALYSIS

Anisotropic Effective Stress Soil Behaviour by BEM

F. Lin, L.A. Wood

Department of Civil & Structural Engineering, South Bank Polytechnic, London, U.K.

ABSTRACT

In the present paper, soil is modelled as a non-homogeneous, transversely isotropic and linear elastic medium using boundary element method(BEM). The integral equations are formulated for soil in plane strain problems. Some engineering applications of soil-structure interaction are discussed and the results are compared with those from observation and the finite element method.

INTRODUCTION

In geotechnical engineering, the anisotropic effects of soil are an important aspect to be taken into consideration since usually soil properties are different in the horizontal and vertical directions. The analytical technique has to depend both on the soil model and on the boundary value problems. Numerical methods are the most important techniques for solving geotechnical problems. A realistic linear elastic model is having Young's modulus increase linearly with depth, for which some non-homogeneous problems are suggested in the BEM analysis. As an efficient numerical technique with elastic model, boundary element method has been widely applied for solving geotechnical problems. Especially in tunnel engineering, there are many reports by Crouch[1], Venturini[2] and Brebbia[3] using linear elastic and non-linear models.

In the previous paper[4], the isotropic soil behaviour is modelled by BEM, but it is found that in many instances the soil presents anisotropic behaviour, such as London Clay which is overconsolidated and exhibits transverse isotropy. Namely, that the stiffness in the vertical direction is different from that in the horizontal plane. The objective of this paper is to develop the boundary element method for solving the anisotropic soil behaviour and the interaction between soil and structure. At first, the fundamental solutions for the transversely isotropic soils in integral equations are formulated and the effective stress principle of soil is combined into the formulation. A Fortran program with elastic model is written for non-homogeneous transversely isotropic soil materials. The program can be applied to modelling soil structure interaction(e.g. retaining walls and tunnels in foundations) by defining different regions, and the simplest data input is set up for the problems with tunnels whose geometry and boundary conditions can be automatically generated. A graphic system is also available for checking the geometry mesh and results. In this paper, some geotechnical applications in London are analysed by BEM. Back analysis is carried out on the ground movements and the interaction during the construction of the Victoria Line(London Underground) tunnels beneath Regents Park. Another analysis is the deep excavation of a foundation in London. The numerical analysis is based on the soil parameters obtained from the laboratory testing and the propped loads from measurement. The results are checked from the data by monitoring the movement of a retaining wall and referring to the report[5], and compared with those by finite element method(FEM) as well.

INTEGRAL EQUATIONS FOR TRANSVERSELY ISOTROPIC SOILS

In soil mechanics, soil is regarded as a three phase medium of air, water and solids. When soil is saturated with water, the effective stress principle is set up by the relationship between effective stresses and pore pressure U on the shear direction s and normal direction n:

$$\begin{cases} \sigma_n = \sigma_n' + U \\ \sigma_s = \sigma_s' \end{cases} \tag{1}$$

where σ_s and σ_n are total shear and normal stresses; σ_s' and $\sigma_n'$are effective shear and normal stresses of soil.

When the anisotropic behaviour of soil is considered, the analyses can be carried out in undrained and drained conditions. Generally, the total stresses are used in undrained analyses, and effective stresses in drained analyses. Hence, the reciprocal theory can be represented as the form of total stresses in directed boundary element method,

$$\int_\Gamma (\sigma_s u_s^* + \sigma_n u_n^*) d\Gamma = \int_\Gamma (u_s \sigma_s^* + u_n \sigma_n^*) d\Gamma \tag{2}$$

where σ_s^*, σ_n^*, u_s^* and u_n^* are the corresponding fundamental solutions for displacement and stress on shear and normal directions at any point of boundary Γ.

Fundamental Solutions for Transversely Isotropic Soils

The fundamental solution for the problem of a line of concentrated force in a transversely isotropic elastic solid is given by Rizzo and Shippy[6]. If the concentrated force is placed at point i on contour Γ, the solutions at point j caused by force at i can be represented:

$$u_x = m_1 m_2 m_3 \left(\frac{R_1}{m_1} G_{11} - \frac{R_2}{m_2} G_{12}\right) P_x - m_3 (G_{21} - G_{22}) P_y$$

$$u_y = -m_1 m_2 m_3 (G_{21} - G_{22}) P_x - m_3 \left(\frac{m_1}{R_1} G_{11} - \frac{m_2}{R_2} G_{12}\right) P_y$$

$$\sigma_{xx} = m_1 m_2 m_3 C_{66} \left(\frac{1+m_1}{m_1 R_1} G_{31} - \frac{1+m_2}{m_2 R_2} G_{32}\right) P_x + \quad (3)$$

$$+ m_3 C_{66} \left(\frac{1+m_1}{R_1} G_{41} - \frac{1+m_2}{R_2} G_{42}\right) P_y$$

$$\sigma_{yy} = -m_1 m_2 m_3 C_{66} \left[\frac{R_1 (1+m_1)}{m_1} G_{31} - \frac{R_2 (1+m_2)}{m_2} G_{32}\right] P_x -$$

$$- m_3 C_{66} \left[(1+m_1) G_{41} - (1+m_2) G_{42} \right] P_y$$

$$\sigma_{xy} = m_1 m_2 m_3 C_{66} \left[\frac{1+m_1}{m_1} G_{41} - \frac{1+m_2}{m_2} G_{42}\right] P_x -$$

$$- m_3 C_{66} \left[\frac{1+m_1}{R_1} G_{31} - \frac{1+m_2}{R_2} G_{32}\right] P_y$$

where P_x and P_y are the components of the force at x and y directions in X-Y co-ordinate system; m_1, m_2 and m_3 are material constants; R_1 and R_2 are the real and positive roots from the characteristic equation for the material[3] in plane strain; G_{nk} are the functions defined by:

$$G_{1k} = \frac{1}{2} Ln \left(X^2 + \frac{Y^2}{R_k{}^2}\right)$$

$$G_{2k} = arctg \left(\frac{Y}{X R_k}\right)$$

$$G_{3k} = \frac{X}{X^2 + Y^2/R_k{}^2} \qquad k = 1,2 \quad (4)$$

$$G_{4k} = \frac{1}{R_k} \frac{Y}{X^2 + Y^2/R_k{}^2}$$

The constants m_1, m_2 and m_3 are as follows:

$$m_1 = \frac{C_{11} R_1{}^2 - C_{66}}{C_{12} + C_{66}}$$

$$m_2 = \frac{C_{11} R_2{}^2 - C_{66}}{C_{12} + C_{66}} \quad (5)$$

$$m_3 = \frac{1}{2\pi C_{66} (m_1 + m_2)}$$

where C_{nm} are the material compliances in $\{\varepsilon\} = [c]\{\sigma\}$.

When a local x-y co-ordinate system is chosen whose origin is at point j and outward normal direction is taken as the positive y direction, the transformation formulas for the forces P_x and P_y at point i are written:

$$\begin{Bmatrix} P_x \\ P_y \end{Bmatrix} = \begin{bmatrix} \cos\beta_i & -\sin\beta_i \\ \sin\beta_i & \cos\beta_i \end{bmatrix} \begin{Bmatrix} P_s \\ P_n \end{Bmatrix} \tag{6}$$

where β_i is the angle between X axis and tangent of point i on boundary Γ. The displacements u_x and u_y at point j,

$$\begin{Bmatrix} u_x \\ u_y \end{Bmatrix} = \begin{bmatrix} \cos\beta_j & -\sin\beta_j \\ \sin\beta_j & \cos\beta_j \end{bmatrix} \begin{Bmatrix} u_s \\ u_n \end{Bmatrix} \tag{7}$$

where β_j is the angle between X axis and tangent of point j on boundary Γ. Of course, co-ordinates X and Y in the function G_{nk} should be replaced by the local co-ordinates x and y using the co-ordinate transformation:

$$\begin{Bmatrix} X \\ Y \end{Bmatrix} = \begin{bmatrix} \cos\beta_j & -\sin\beta_j \\ \sin\beta_j & \cos\beta_j \end{bmatrix} \begin{Bmatrix} x - c_x \\ y - c_y \end{Bmatrix} \tag{8}$$

where c_x and c_y are the co-ordinates of point j in local x-y co-ordinate system. After substituting (8) into (4), G_{nk} are transformed into G_{nk}'. Substitution of (6) and (7) into (3), the displacements on shear and normal directions at local co-ordinate system are:

$$\begin{Bmatrix} u_s \\ u_n \end{Bmatrix} = \begin{bmatrix} u_{ss}^* & u_{ns}^* \\ u_{sn}^* & u_{nn}^* \end{bmatrix} \begin{Bmatrix} P_s \\ P_n \end{Bmatrix} \tag{9}$$

where u_{ss}^*, u_{ns}^*, u_{sn}^* and u_{nn}^* are the fundamental solutions for displacements by taking $P_s = 0$ and $P_n = 0$ in turn.

As for the fundamental solutions for stresses, the traction vector can be resolved into components in s and n direction by simple statics:

$$\begin{cases} \sigma_s = -(\sigma_{xx} - \sigma_{yy})\sin\beta_j\cos\beta_j - \sigma_{xy}(\sin^2\beta_j - \cos^2\beta_j) \\ \sigma_n = \sigma_{xx}\sin^2\beta_j - 2\sigma_{xy}\sin\beta_j\cos\beta_j + \sigma_{yy}\cos^2\beta_j \end{cases} \tag{10}$$

Substituting (6) into (3) and (8) into (4), then put (3) into (10):

$$\begin{Bmatrix} \sigma_s \\ \sigma_n \end{Bmatrix} = \begin{bmatrix} \sigma_{ss}^* & \sigma_{ns}^* \\ \sigma_{sn}^* & \sigma_{nn}^* \end{bmatrix} \begin{Bmatrix} P_s \\ P_n \end{Bmatrix} \tag{11}$$

where σ_{ss}^*, σ_{ns}^*, σ_{sn}^* and σ_{nn}^* are the fundamental solutions for stresses by taking $P_s = 0$ and $P_n = 0$ in turn[1].

Integral Equations

When the boundary Γ is divided into N elements, the integral equation (2) can be written as two equations on each element for s and n directions:

$$\begin{aligned} \sum_{\Gamma_j=1}^{n} \sigma_s u_{ss}^* + \sum_{\Gamma_j=1}^{n} \sigma_n u_{ns}^* &= \sum_{\Gamma_j=1}^{n} u_s \sigma_{ss}^* + \sum_{\Gamma_j=1}^{n} u_n \sigma_{ns}^* \\ \sum_{\Gamma_j=1}^{n} \sigma_s u_{sn}^* + \sum_{\Gamma_j=1}^{n} \sigma_n u_{nn}^* &= \sum_{\Gamma_j=1}^{n} u_s \sigma_{sn}^* + \sum_{\Gamma_j=1}^{n} u_n \sigma_{nn}^* \end{aligned} \tag{12}$$

Formula (12) includes the singular integrals when self-effection is considered on each element, and the function G_{nk}' are given after singular integration:

$$\begin{aligned}
G_{1k}' &= -(c_x - a + \frac{B_k c_y}{2A_k})F_1(-a,k) + (c_x + a + \frac{B_k c_y}{2A_k})F_1(a,k) - \\
&\quad - \frac{c_y}{A_k R_k}[F_2(-a,k) - F_2(a,k)] \\
G_{2k}' &= -\frac{c_y}{A_k R_k}[F_1(-a,k) - F_1(a,k)] + (c_x - a + \frac{B_k c_y}{2A_k})F_2(-a,k) - \\
&\quad - (c_x + a + \frac{B_k c_y}{2A_k})F_2(a,k) \\
G_{3k}' &= \frac{\cos\beta_j}{A_k}[F_1(-a,k) - F_1(a,k)] + \frac{\sin\beta_j c_y}{|c_y| A_k R_k}[F_3(-a,k) - F_3(a,k)] \\
G_{4k}' &= \frac{\sin\beta_j}{A_k R_k}[F_1(-a,k) - F_1(a,k)] - \frac{\cos\beta_j c_y}{|c_y| A_k}[F_3(-a,k) - F_3(a,k)]
\end{aligned} \tag{13}$$

where A_k and B_k are the constants about material; a is the half length of element j; $F_n(\alpha,k)$ are the function of local co-ordinates x and y:

$$\begin{aligned}
F_1(\alpha,k) &= \frac{1}{2} Ln [A_k (c_x + \alpha)^2 + B_k (c_x + \alpha) c_y + C_k c_y] \\
F_2(\alpha,k) &= arctg [\frac{c_x + \alpha + B_k c_y/(2A_k)}{c_y} A_k R_k] \\
F_3(\alpha,k) &= arctg [-\frac{c_x + \alpha + B_k c_y/(2A_k)}{c_y} A_k R_k]
\end{aligned} \tag{14}$$

The constants A_k, B_k and C_k are given by Crouch[1]:

$$\begin{aligned}
A_k &= \cos^2\beta_j + \sin^2\beta_j \frac{1}{R_k^2} \\
B_k &= (\frac{1}{R_k^2} - 1) \sin 2\beta_j \\
C_k &= \sin^2\beta_j + \cos^2\beta_j \frac{1}{R_k^2}
\end{aligned} \tag{15}$$

Non-Homogeneous Materials

When the non-homogeneous behaviour of soil is considered, the whole domain should be divided into several subregions, and boundary elements are given on each boundary and interface. The subregions can be coupled together utilising continuity of displacement and equalising tractions on the interface[4]. Therefore, some structures, e.g. retaining walls and tunnels with lining, in the foundation can be modelled as different materials.

APPLICATIONS ON GEOTECHNICAL PROBLEMS

A FORTRAN program is written in the basis of formula(12) and the interface conditions for non-homogeneous materials. It can be applied to model complex boundary conditions with simple data input. For problems involving tunnels, the boundary elements of the internal face of the tunnel and the interface elements between tunnel and soil can be generated automatically by the program. Therefore it is convenient to discuss the precision in tunnel analyses by using different divisions of tunnel elements. Moreover, a graphic system is also available for checking the boundary mesh and presenting the BEM results.

Ground Movement due to Tunnel Excavation

An analysis is carried out of ground movement and tunnel interaction behaviour. The field investigation of ground movement associated with the excavation of the twin tunnels of the London Underground at Regents Park was reported by Barratt et al[5] in 1975. Fig. 1 shows that there are two tunnels of 4.15m diameter which were excavated and then constructed with expanded concrete linings at depths of 34m and 20m. Based on the observation of the ground settlement, the boundary mesh is assumed to be the width of 75m and 4 subregions are formed. There were 8 borehole tubes at A, B, ... J which were installed in the foundation for measuring both horizontal and vertical movement at ground levels. The observation of the ground movement was obtained after 53 days of excavation, and the undrained analysis is suggested as very low permeability of London clay in the foundation.

Fig. 2 shows the distribution of ground settlements by BEM and field observation. The distributions of horizontal movement from ground level

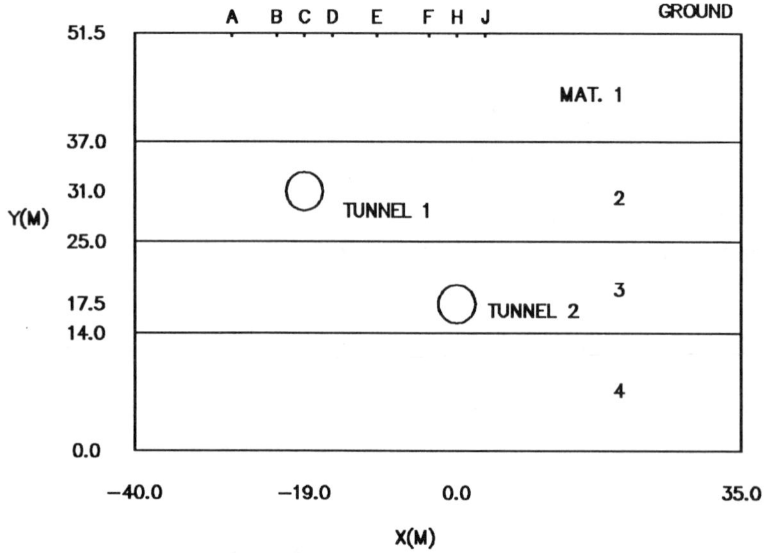

Fig. 1 Boundary Element Mesh for Regents Park

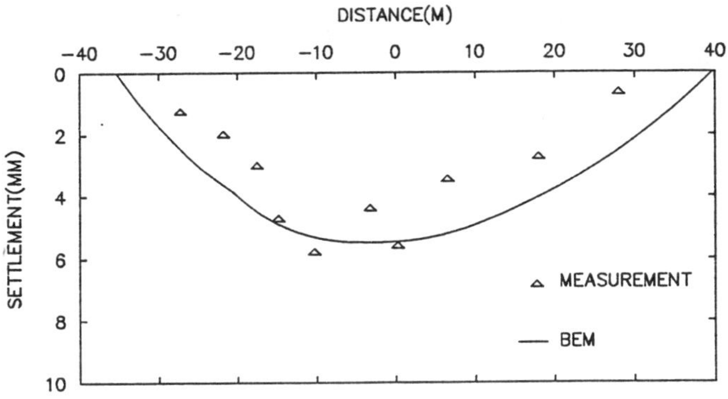

Fig.2 Distribution of ground Settlement

to foundation at boreholes B, D, F, and J by BEM and the observation are presented in Fig. 3. The distributions of vertical movement with respect to those boreholes are also shown in Fig. 4, where the negative indicates settlement and positive indicates upward movement. Referring to the movements shown in above figures, it is seen that the numerical results on vertical direction are closer to the observation than those on horizontal direction. For the horizontal movements, the modelling method depends on not only the excavation of the tunnel but also the restriction of both side boundaries where the fixed boundary elements are given. It is seen in Fig. 3 that there are outward movements near the ground level at boreholes B, D, F and J, which show the difference from the analysis results. In theory, it may be the unbalanced building loading between the boreholes D and H due to the excavation of tunnels, which can obviously dominate the shallow ground movements. The horizontal movements at deep levels are still controlled by the excavation of tunnels, where the reasonable results are presented by BEM. For the vertical movements in Fig.2 and Fig. 4, the good agreement between the analyses and the observation is obtained as the boundary conditions are more easily modelled when the boundary is assumed deep enough in the foundation.

As for the interaction between 2 tunnels, both the BEM results and the observation at borehole F in Fig. 3 show the effect of the excavation of tunnel 1, which moves forward to the tunnel 1 around its level. It is seen in Fig. 4 that there is little upward movement around tunnel 1, which shows the effect from the excavation of tunnel 2. As for the movement around tunnel 2, it has little effect from tunnel 1 as tunnel 2 is 14 m deeper than tunnel 1. The calculation is also carried out for the foundation movement when only tunnel 1 is excavated. The results are shown in Fig. 4 with dash lines, where the maximum upward movement is found about 1.7mm at boreholes B and D. Compared with the results from two tunnels, it is found that the extra settlement at borehole B due to the excavation of tunnel 2 is

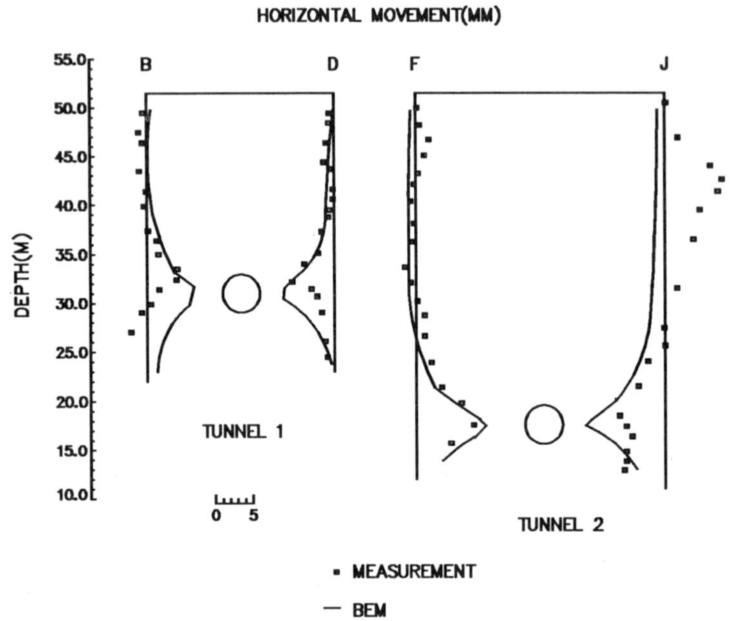

Fig.3 Distribution of Horizontal Movement

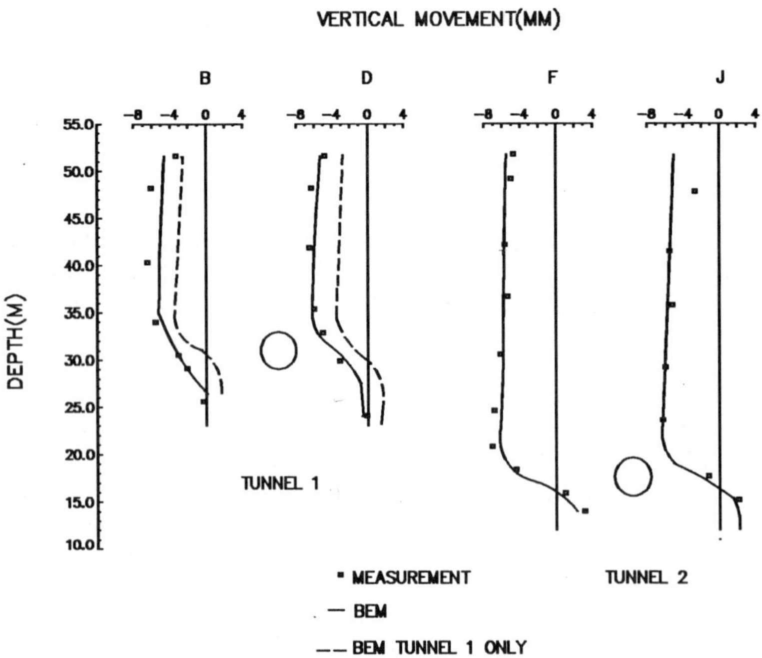

Fig. 4 Distribution of Vertical Movement

about 2.0mm on average, and that at borehole D is about 2.5mm on average. It is, therefore, that the BEM results still show the principle behaviour of soil movement on horizontal direction for the interaction of tunnels.

The Behaviour of Retaining Wall and Lined Tunnel with Deep Excavation

There is another application of using BEM to analyse a diaphragm wall with deep excavation in London clay. The excavation is carried out to 12m in depth. The diaphragm wall reaches to 18m depth from ground level and some temporary steel struts are installed at 4m depth from the ground level in order to retain the wall and adjacent buildings in stable condition during the excavation and construction. As is shown in Fig. 5 , there is an existing tunnel in the foundation, which is 14m lower than the retaining

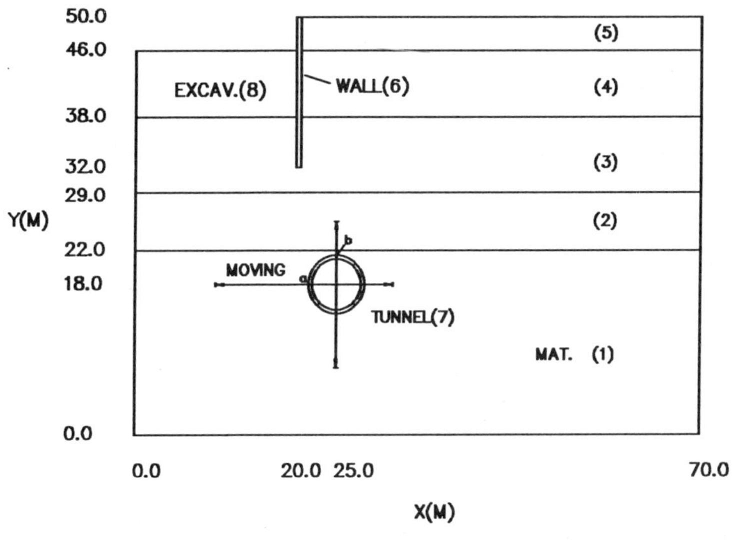

Fig. 5 BEM Mesh for Retaining Wall

wall. Undrained condition has been assumed for the interaction analysis between retaining wall and tunnel during the excavation. According to the measurement from strain gauge, the average propped force at each strut is 848.9 KN with the struts at 7.5m centres. Therefore the line load applied in the analysis to represent the strutting effect is 113.5 KN/m in plane strain condition. The wall movements were observed taking the bottom of the wall as the fixed point. In the analysis, the unbalanced load due to the excavation is applied on the wall as the external load. Fig. 6 shows the comparison of the wall movements from BEM and observation in which the displacements refer to the zero displacement at the bottom of the wall. The FEM result is also presented in Fig. 6.

In order to study the effect of the position of the tunnel relative to the excavation a parametric study has been carried out. The tunnel position has been moved in both horizontal and vertical directions as shown in Fig. 5. The relative wall movements(Δu_x), maximum shear forces(ΔQ_{max}) and bending moments on excavation level(ΔM_{exa}) of the retaining wall are listed in Table 1, and relative movements(Δu_x, Δu_y) and effective stresses ($\Delta\sigma'_s$, $\Delta\sigma'_n$) of the tunnel are listed in Table 2. The results listed in Table 1 and 2 are calculated from the unbalanced load acting on the wall. Referring to Table 1, the most sensitive movements of the wall occur when the tunnel moves from the normal position in Fig. 5 to the left, e.g. to excavation centre. The largest increment of movement is found in bottom of the wall. However the bending moments and shear forces are reduced in this case. Another behaviour is that the wall movements at all levels increase with about the same values, i.e., less deformation of the wall when the tunnel moves to a deeper position in the foundation. The relative shear forces and bending moments are also reduced. Referring to Table 2, when the tunnel is located deeper in the foundation, it moves outward from the excavation centre, and less upward movement is found. When the tunnel position moves to the right, less movements in x direction and more upward movements are found at the tunnel, and effective stresses at the tunnel are also reduced. Based on the above analysis, it is concluded that there is little interaction between the existing lined tunnel and retaining wall when a 12m deep excavation is carried out.

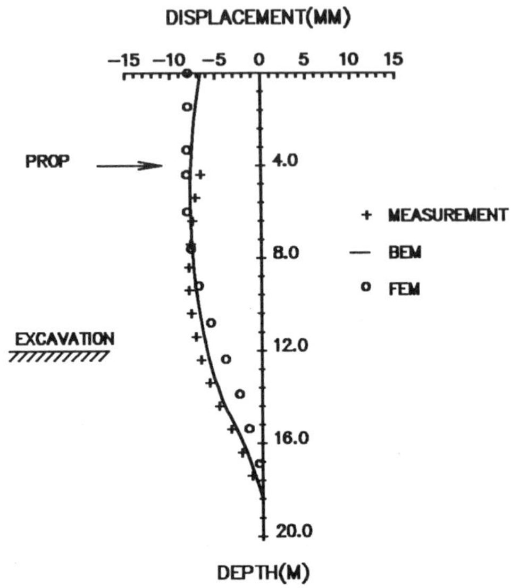

Fig. 6 Comparison of Wall Movement

In the BEM analysis, there are 233 boundary elements used which includes 24 tunnel elements. However there are 452 elements used in FEM and the data preparation is ten times more than those by BEM. It is found that enough precision is given when the number of tunnel elements chosen is more than 24. Another advantage of using BEM is the pressure, shear force and bending moment can be directly calculated with boundary elements, and it is convenient to the retaining wall analysis.

CONCLUSIONS

The integral equations are formulated for solving the non-homogeneous and transversely isotropic soils with linear elastic model in plane strain. A FORTRAN program has been written together with a graphic system to present the results.

There are two applications which are analysed by BEM. The analysis on the excavation of twin tunnels has shown the interaction between each tunnel. The reasonable distributions of the movements are obtained by BEM. In the analysis of the deep excavation, the interaction of retaining wall and lined tunnel is discussed. It is seen that BEM can precisely calculate the pressure, shear force and bending moment associated with the retaining wall. However, it is not easy to calculate them directly by FEM. Moreover, BEM results have shown good agreement with those from FEM and the field. Compared with FEM, BEM has the advantage on reducing the data preparation especially when there are some tunnels in the mesh.

REFERENCES

1. Crouch, S. L. and Starfield, A. M. Boundary Element Methods in Solid Mechanics, George Allen & Unwin, London, 1983.

2. Venturini, W. S. Boundary Element Methods in Geomechnics, Springer-Verlag, 1983.

3. Brebbia, C. A. and Walker, S. Boundary Element Techniques in Engineering, Butterworths, London, 1979.

4. Lin, F. and Wood, L. A. Boundary Element Method Applied to Two-Dimensional Non-Homogeneous Soils, Proceedings of 12th International Conference on Boundary Element Method, 1990.

5. Barratt, D. A. and Tyler, R. G. Measurements of Ground Movement and Lining Behaviour on the London Underground at Regents Park, TRRL Laboratory Report 684, Structure Department, Transport and Road Research Laboratory, Crowthorue, Berkshire, U.K., 1976.

6. Rizzo, F. J. and Shippy, D. J. A Method for Stress Determination in Plane Anisotropic Elastic Bodies, Journal of Composite Materials 4, pp36-61, 1970.

7. Wood, L. A. and Creed S. G. The Application of Boundary Elements in Offshore Engineering, Proceedings of 4th International Conference on BEM, pp391-406, Springer-Verlag, 1982.

Table 1. Comparison of Retaining Wall Behaviour by Moving Tunnel

Tunnel position	Left 15.0m	Left 7.0m	Left 5.0m	Left 2.0m	Left 1.0m	Right 1.0m	Right 2.0m	Right 4.0m
Δu_{xt} (mm)	0.037	0.020	0.016	0.002	0	0.002	0.003	0.002
Δu_{xp} (mm)	0.039	0.022	0.018	0.004	0.001	0	0.001	0
Δu_{xb} (mm)	0.067	0.044	0.039	0.019	0.009	-0.005	-0.006	-0.001
ΔQ_{max} (kn)	-1.685	-1.025	-1.005	-0.763	-0.411	0.287	0.336	0.047
ΔM_{exa} (kn-m)	-12.873	-10.319	-10.259	-7.572	-4.038	2.854	3.492	1.26

Δu_{xt}, Δu_{xb}: Relative displacements at top of wall and at bottom of wall.

Δu_{xp}: Relative displacement at prop level of wall.

Table 1. Cont.

Tunnel position	Down 1.0m	Down 2.0m	Down 4.0m	Down 7.0m	Down 10.0m	Up 7.5m
Δu_{xt} (mm)	0.017	0.031	0.051	0.066	0.068	-0.550
Δu_{xp} (mm)	0.026	0.031	0.052	0.067	0.070	-0.565
Δu_{xb} (mm)	0.023	0.042	0.069	0.092	0.099	-0.738
ΔQ_{max} (kn)	-0.319	-0.554	-0.906	-1.246	-1.484	9.339
ΔM_{exa} (kn-m)	-2.666	-4.606	-7.496	-10.348	-12.481	70.295

Table 2. Comparison of Tunnel Behaviour by Moving Tunnel

Tunnel position	Left 15.0m	Left 7.0m	Left 5.0m	Left 2.0m	Left 1.0m	Right 1.0m	Right 2.0m	Right 4.0m
Δu_{xa}(mm)	-0.178	0.006	0.069	0.012	0.001	-0.010	-0.047	-0.181
Δu_{yb} (mm)	-1.085	-0.834	-0.618	-0.222	-0.109	0.114	0.228	0.402
$\Delta\sigma_{sb}'$(kn)	-3.8	2.6	-4.3	-19.5	1.1	-0.6	0	-4.0
$\Delta\sigma_{nb}'$(kn)	-29.6	-17.4	-21.0	-7.5	8.1	-3.0	-6.0	-3.0

Δu_{xa} & Δu_{yb}: Relative displacements of tunnel in x, y directions at point a and b.

$\Delta\sigma_{sb}'$ & $\Delta\sigma_{nb}'$: Relative shear and normal effective stresses of tunnel at point b

Table 2. Cont.

Tunnel position	Down 1.0m	Down 2.0m	Down 4.0m	Down 7.0m	Down 10.0m	Up 7.5m
Δu_{xa}(mm)	0.180	0.339	0.584	0.779	0.768	-3.148
Δu_{yb} (mm)	-0.028	-0.058	-0.115	-0.187	-0.245	0.398
$\Delta\sigma_{sb}'$(kn)	-1.4	-4.6	-10.0	-15.6	-20.3	34.2
$\Delta\sigma_{nb}'$(kn)	-0.3	-2.2	-6.1	-9.1	-10.4	6.2

The Use of a Foundation Element in a Three-Dimensional Structural Analysis Code

M.A. Ramalho, W.S. Venturini

São Carlos School of Engineering, University of São Paulo, Brazil

ABSTRACT

The standard codes for structural analysis usually assume all columns of buildings have hinged or pinned down ends. Although the foundation stiffness, the three-dimensional solid of soil or rock plus the concrete elements, is very high, it cannot guarantee the initial assumption that displacements equal zero, which often means inaccuracies for efforts, displacements and second order effects. This paper shows a simple way to take into account the foundation influences using three-dimensional boundary element relations, combined with the main system of equations written for the building structure.

INTRODUCTION

Formulations proposing to combine the boundary element and finite element method have been suggested a long time. Some of those works, such as Shaw [1], Osias [2] and Zienkiewicz [3], among others, are often mentioned in the specialized literature as the first works in this field. The combination of those techniques is an important strategy to be followed to solve many problems in civil engineering, in particular those characterized by interactions between the structure and the foundation region usually assumed as an infinite domain. Specially dedicated to this kind of interaction problem, several procedures have already been suggested in the last few years. For instance, Beer [4], Vallabhan [5] and Meek [6] are among the works where those techniques are proposed.

In the analysis of building structures it is usual to assume classical boundary conditions enforcing displacements equal to zero at all column ends in connection with the foundation. The problem is then simplified into two independent ones: the proper building structure and its foundation on which all loads coming from the building are applied. Then, structural analysis can be carried out by any well known micro-computer code, SAP-90 [7], SUPERSAP [8] and others.

Although this scheme is widely employed and accepted, taking into account the actual stiffness of the whole foundation could be a safer procedure for the problem under analysis. One possible way to consider foundation and building structure stiffnesses together is to adopt a boundary element formulation to model the infinite domain and join the corresponding equations into the finite element system defined for the structure.

In this work a formulation as described above is proposed introducing special rigid footing elements underneath each column to simplify the coupling model without significant loss of accuracy for the type of problem studied. The Three-dimensional solid is treated by the boundary element formulation accomplished with the Mindlin or Boussinesq-Cerruti fundamental solutions [9]. The stiffness matrix of the whole foundation is symmetrized by means of approximate procedures. Also in this case, the introduced errors are shown to be negligible regarding the nature of the problem.

DEFINITION OF THE PROBLEM.

The main objective of this work is to develop an element, i.e., to derive all its stiffness coefficients, for the whole foundation that supports the building structure under analysis. The basic characteristic of that element is seen to be its easy adaptation to be used in a classical finite element code, without influences on the other original system applications. For simplicity, but without loss of generality, a single elastic material zone combined with rigid rectangular footings has been chosen as the model of the whole foundation. Formulation extensions to a more complex foundation domain, layered soil media or soil of varying properties, are simple and do not require too much work.

The boundary element method was taken to derive the stiffness coefficients related to a single point defined on the rigid footing. Thus, six degrees of freedom will be associated with that node. Each new footing laid on or inserted in the infinite domain will give another six degrees of freedom. As all footings belong to the same media, stiffness coefficients relating them will appear.

In order to illustrate the algorithm to be developed its main steps are summarized as follows: a) the foundation element proposed consists of a semi-infinite domain plus one or more footings, each one connected to the main structure through one single node; b) the program automatically establishes the footing discretization, as well as computes all needed parameters to assemble the boundary element equation system; c) by definition, the stiffness matrix is achieved still related with footing centers. That is made by enforcing unit displacements corresponding to the six degrees of freedon and referred to the global system of coordinates; d) after computing all coefficients, $36N_f^2$ values where N_f is the number of footings, the element matrix is symmetrized by means of the minimum error concept and then joined into the main system matrix; e) the entire system of equations is analysed employing a classical solver; results of stresses throughout the infinite domain can be computed afterwards.

BASIC EQUATIONS AND APPROXIMATIONS

The displacement boundary integral equation for a three-dimensional solid, in absence of gravitational loads, is usually given in the following form [10],

$$C_{ik}(P)u_k(P) = \int_\Gamma p^*_{ik}(P,S)u_k(S)d\Gamma(S) + \int_\Gamma u^*_{ik}(P,S)p_k(S)d\Gamma(S) \quad (1)$$

where all values with the symbol "*" represent fundamental solutions; u_k and p_k stand for displacements and tractions respectively.

Taking into account that planes of distributed loads will be applied inside the domain due to footings inserted in the soil media, or laid on the ground surface, an extra integral term rises in the equation to give the corresponding influences. This new term can also be obtained by defining interfaces between rigid footing and soil media. In this case, due the small distance between the interfaces, all relating integrals can be properly studied. After carrying out this study the only remaining term is the one containing tractions, now assumed to be applied on single surfaces corresponding to all footings.

In order to make the problem simpler we are considering only the case of semi-infinite domains without any excavation, i.e., without boundaries. So, the displacement integral in (1) is eliminated. If ground loads were also not applied equation (1) is reduced to,

$$u_i(p) = \int_\Gamma u^*_{ik}(p,S)p_k(S)d\Gamma \quad (2)$$

where $p_k(S)$ now represents all footing reactions.

As no boundary is considered the constant $C_{ik}(P,S)$, given in this case by δ_{ik}, kronecker delta, is easily eliminated. Displacements at ground or internal points are, therefore, written only in terms of reactions, $p_k(S)$ as shown in (2).

Remembering N_f footings can be used for the whole

building, equation (2) becomes,

$$u_i(p) = \sum_{f=1}^{N_f} \int_{\Gamma_f} u^*_{ik}\,(p,S)p_k(S)d\Gamma(S) \tag{3}$$

Assuming rigid footing behaviour the reactions p_k could be approximate over cells defined throughout each region Γ_f(Fig.1). As the reactions underneath rigid footing have expected forms, the use of known shape function is more appropriate to model them. Initially the rectangular footing form was chosen together with a four node shape function approximation. In addition, a higher approximation function with eight nodes was also implemented into the code. In this case, the used shape functions are clearly more appropriate to model the expected reactions.

The local coordinate axes (Figure 2) adopted for this case are defined from the footing center with their directions parallel to the rectangle sides. The eight node shape functions used were borowed from classical finite element tests.

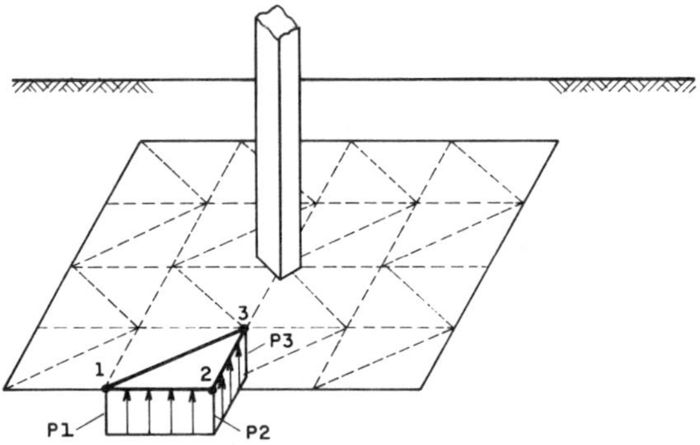

Figure 1 - Approximation of footing reactions

Using a four or eight node shape function the reaction $p_k(s)$ underneath any footing is approximated by,

$$p_k(S) = \phi^j(S)P^j_k \tag{4}$$

Replacing (4) in (3) gives,

$$u_i(p) = \sum_{f=1}^{N_f} \int_{\Gamma_f} u^*_{ik}(p,S)\phi^j(S)d\Gamma(S)P^j_k \tag{5}$$

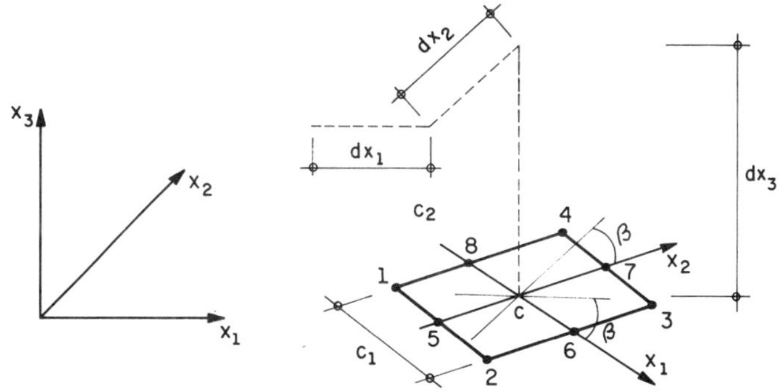

Figure 2 - Local coordinate axes and footing shape.

Integrating (5) analytically or numerically as convenient, the algebraic representation of displacements of any point P is achieved. Writing that algebraic representation for all nodes defined by the footing discretization, the following system of equations is found,

$$U = G\ P \tag{6}$$

where P gives the nodal reactions, U is the corresponding displacement vector and G is obtained after carrying out the integrals indicated in (5).

FOOTING STIFFNESS

After obtaining the simplified boundary element set of equations one can compute all reactions corresponding to the standard unit displacement, i.e. the vector P associated with six degrees of freedon of each footing (Fig. 3). For instance, assuming values equal to 1.0 for all nodal vertical displacements of a specific footing and enforcing all other displacements to be zero, vector P will be achieved by multiplying (6) by G^{-1}. For any other basic unit displacement vector U can be easily found and the corresponding vector P computed by solving (6) as shown above.

All those basic unit displacements can be associated, for simplicity, to the footing center. Translations and rotations can be provided afterwards if needed. Attention must be paid to the rotations. In this case, any unit displacement gives vector U written in terms of footing coordinates.

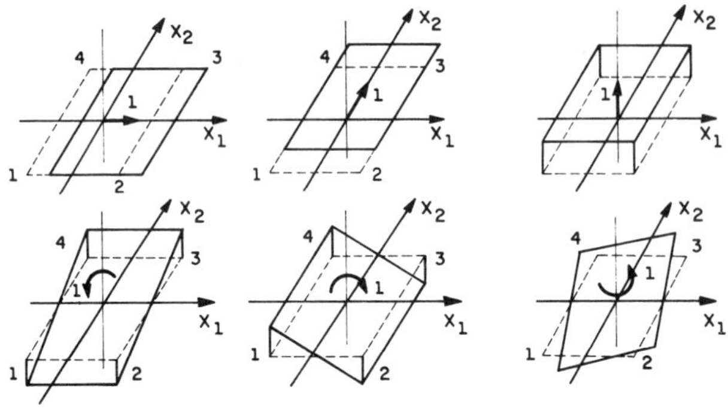

Figure 3 - Standard unit displacements of rectangular footing.

Integrating the reaction vectors P, computed by solving (6) over all footings, one can find the basic efforts, three forces and three moments for each footing. These values are, by definition, stiffness coefficients of the whole foundation.

As mentioned, coefficents referred to the footing center can be modified when they are associated with a different footing point using a convenient transformation as in [11].

$$\underset{\sim}{S} = \underset{\sim}{T}\,\underset{\sim}{\bar{S}}\,\underset{\sim}{T}^{T} \tag{7}$$

where $\bar{S}$ and S are the original and final matrices respectively, and T is the transformation matrix.

S obtained as proposed above is an unsymmetric matrix due to the nature of the adopted integral equations. Of course, the solution of the entire system of equations can be found taking into account both the banded symmetric part, coming from the finite element formulation, and the fully populated unsymmetric region just derived. However, in order to use well known and widely employed solvers symmetrizing the boundary matrix is an alternative recommended to be tried. As it is known the standard procedure to symmetrize matrices [12] is not entirely correct, but in this case, it is possible to verify that the amount of error introduced is acceptable taking into account the nature of the problems to be analysed. Thus, the matrix S will be modified by the expression,

$$S = \frac{1}{2}(S + S^{T}) \tag{8}$$

and then written into the system stiffness matrix as a simple element sub-matrix.

EXAMPLES

Following the formulation presented in the previons sections, is interesting now to show some applications to verify how the technique behaves. In the first place, a simple example to analyse the stiffness matrix and the order of all coefficients computed by the approximate process is discussed. A second example is dedicated to the analysis of a building structure assuming three distinct values for the foundation elastic modulus.

Two footing foundation element.
This example aims to clarify some important aspects of the type of foundation idealized here. In fact 6N spring-like coefficients associated with each degree of freedon have been derived. Now, it is important to know how great are the errors achieved when the symmetrization process is adopted. Another remaining question is concerned with the multiple interdependence of those springs. How far must one footing be from the others to be independent? The approximation reached by adopting a 4 or 8 node shape function is also important to give us the answer to how accurate the proposed method is.

The last question was earlier verified through examples where only one footing was defined. Previous works [9, 13], also using the Mindlin fundamental solution, show very well the type of reaction distributions that are obtained when constant displacements are applied to a particular rectangular region. Comparing those results it was clear that a very small amount of error is introduced by the 8 node shape function, while a considerable change in those values is verified for the 4 node shape function case. In spite of that, assembling the foundation element matrix using one or other approximation does not change significantly the final results for building structures.

In order to discuss properly the two footing problem we can start with the nature of the computed coefficients. Notice that, as they are stiffnesses, the errors we are looking for are given by themselves. Table 1, presented below, shows the coefficients for two $1m^2$ footings 2m distant from each other. One can see that diagonal terms are much greater than the unsymmetric values. Changes in small coefficients, like those made by equation (8), do not deteriorate the results. It is also important to notice that out of diagonal terms with signficant values are symmetric. Thus, even with an imprecise procedure such as that, no significant changes in the final results are expected for any building structure analysis. If one takes all coefficients like a group of springs, only the weak ones are being changed by equation (8) and it does not modifiy significantly the entire mechanical system.

Another important point to discuss with the proposed example is the real importance of the interaction between the two footings. Of course, many parameters are associated with

TABLE 1 - Stiffness Matrix for a two $1m^2$ footing 2,5 distant from each other.

5314	0	11	0	-501	0	-1169	0	185	0	129	0
0	5157	0	482	0	27	0	-771	0	-48	0	167
9	0	6559	0	-35	0	-172	0	-1191	0	-151	0
-164	0	-35	0	2226	0	60	0	161	0	37	0
0	7	0	1	0	3642	0	4	0	-19	0	13
-1169	0	-185	0	129	0	5314	0	-11	0	0501	0
0	1	0	-48	0	-167	0	7	0	482	0	-27
172	0	-1191	0	151	0	-9	0	6559	0	35	0
0	3	0	-26	0	-5	0	3	0	2220	0	0
60	0	-161	0	37	0	-164	0	35	0	2226	0
0	4	0	19	0	13	0	7	0	-1	0	3642

that interaction, but even so, some reference values can be achieved. The values shown in Table 1 refer to footings with centers 2.5m from each other, i.e. the separation is 2.5 times the footing side length. Those values show a very strong interaction between the two footings. Some coefficients relating different footings rise to the order of 20%. After studying several cases it was verified that 5 times the footing side length is the minimum value to diminish significantly the interaction. Even in this case the interacting coefficients are on average equal to 7.5%. This analysis is not so simple, other parameters not so strong, as footing position and the relation between their side lengths can also be important for this problem. Notice that the interaction is always considered for the present process. What has been discussed here is the amount of error introduced by processes that do not consider the interaction.

Building analysis under wind loads

This example shows some results taken from a complete analysis of a building structure. The greater interests are to know if the foundation stiffness has a strong influence on effort distributions in the building bars, including columns, and also on possible second order effects.

The building has 21 stories with 24 columns, as indicated in Figure 4, with definition of several flat plates. Three different values for the foundation stiffness have been assumed to carry out the proposed study: a) Rigid foundation with displacement and rotation equal to zero at all footings; b) Foundation with highly stiff soil, E = 10^5 kN/cm^2 and ν = 0.35; c) Foundation with soft soil, E = 10^4 kN/cm^2 and ν = 0.35.

Table 2 shows the absolute values of normal forces at column feet obtained only considering wind loads. Those values were computed with the 4 node shape function. Using the more precise approximation, the 8 node shape function, gives values with small changes. As it can be seen the distribution of loads

TABLE 2 - Column Normal Forces

Column	case a	case b	case c
c_1	2278	1667	719
c_3	1427	1251	715
c_4	7	36	335
c_7	2199	1552	246
c_9	804	514	184
c_{12}	185	199	130
c_{15}	185	199	130
c_{17}	637	770	1020
c_{18}	33	117	638

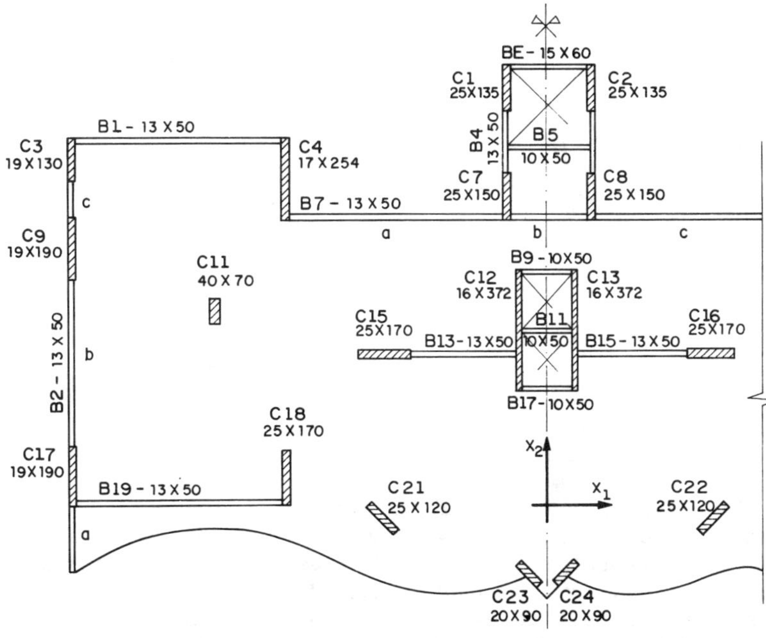

Figure 4 - Definition of columns and beams for all stores.

among columns is strongly influenced by the foundation stiffness. Analogous behaviour was also verified for the bending moment distribution. Using the classical model, case a), one obtains very big values of efforts that can never be reached when using real foundation stiffness. Spreading of loads, as shown in case b), followed by decreases of normal force and bending moment values will be found even for stiffer material foundations.

The second important aspect pointed out by this analysis is with regard to the horizontal displacement at all levels. As it can be seen in Figure 5, values obtained when highly stiff soil is assumed are twice as great as those computed when classical assumptions are made. This significant amount of displacements could give important second order effects.

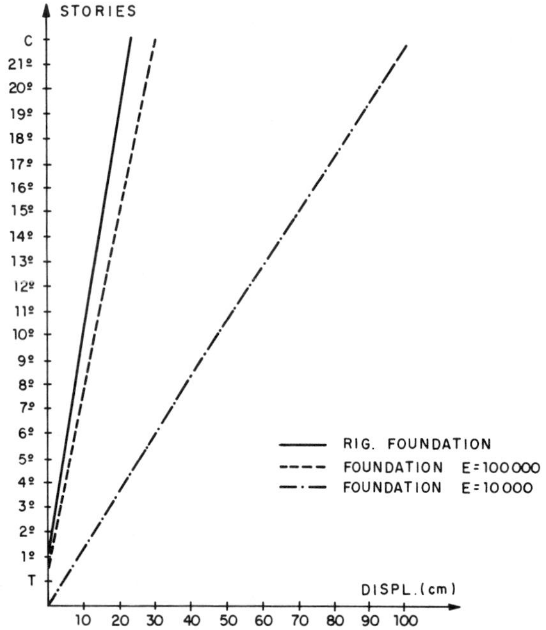

Figure 5 - Building displacements along all stories.

The reasons mentioned above are enough to justify the use of such an element. Although simple, the element is a first attempt to account for the real influences of the material on which buildings are constructed. Starting from this formulation it is not difficult to extend it to a more complete case with layered soil, accomplished with actual boundaries when needed.

CONCLUSIONS

The presented formulation is the first step treating building structures, taking into account the influence of the material foundation. The definition of a foundation element has been proposed in a very simple form. It can be easily inserted into any well established finite element code without loss of capability. The simplified symmetrization process required for that also does not disturb the expected solution. Improvements in the technique to consider either an accurate process to combine symmetric and unsymmetric matrices or more complete interactions are suggested.

REFERENCES

1. SHAW R.P. and FALBY W., "FEBIE: A Combined Finite Element-Boundary Integral Equation Method", in 1st Symposium on Innovative Numerical Analysis in Applied Engineering Science, Versailles, CETIN, 1977.

2. OSIAS J.R., WILSON R.B. and SEITELMAN L.A., "Combined Boundary Integral Equation - Finite Element Analysis of Solids", in 1st Symposium on Innorative Numerical Analysis in Applied Engineering Science, Versailles, CETIN, 1977.

3. ZIENKIEWICZ O.C., KELLY D.W. and BETTESS, P., The coupling of the Finite Element Method and Boundary Solution Procedures", Int. J. Num. Meth. Engng. 11, 355-375, 1977.

4. BEER G., "Implementation of Combined Boundary Element-Finite Analysis with Applications in Geomechanics", in Development in Boundary Elements-4, Banerjee, P.K. and Watson J.O. eds., appl. Science Publishers, UK, 1986.

5. VALLABHAN C.V.G., Coupling of BEM/FEM Technology: An Overview", In: Boundary Element Techniques: Applications in Stress Analysis and Heat Transfer, Brebbia, C.A. and Venturini, W.S. eds., Comp. Mech. Publications, UK., 1986.

6. MEEK, J.L.; ;BANDYOPADNYAY, B., "BECOUP - A Program for Coupled Boundary and Finite Element Analysis in 3-D Elasto-statics " In Boundary Elements X, Brebbia C.A. ed., Vol. 3, Computation Mechanics Publications, UK, 1988.

7. SAP90: User's Manual, Computer and Structures Inc. Berkeley, US, 1988.

8. SUPESAP: User's Manual, Algor Interactives systems, Inc., US, 1987.

9. NAKAGUMA, R.K., "Three - Dimensional elastostatics Using the Boundary Element Method", PhD. thesis, University of Southampton, UK, 1979.

10. BREBBIA, C.A., TELLES, J.C.F. and WROBEL L.C.,"Boundary Element Techniques: Theory and Applications in Engineering" Springer - Verlag, 1984.

11. MEEK, J.L., "Matrix Structural Analysis", McGraw-Hill, Tokyo, Japan, 1977.

12. BREBBIA, C.A. and GEORGIOU P., "Combination of Boundary and Finite Elements in Elastostatics", Appl. Math. Modelling, 3, 212-220, 1979.

13. De SÁ, P.A.C.O. and TELLES, J.C.F., "Analysis of 3D Linear Elastic Problems by the Boundary Element Method Using the Kelvin and Mindlin Fundamental Solutions", VII Latin-American Congress on Computational Methods for Engineering. São Carlos, BR., 1986. (in Portuguese)

On the Use of Geometrical Symmetry in the Boundary Element Methods for 3D Elasticity

M. Bonnet

Laboratoire de Mécanique des Solides (centre commun CNRS - X - Mines - Ponts), Ecole Polytechnique, 91128 PALAISEAU cedex, France

Summary . *A general approach for the use of geometrical symmetry in boundary element methods is proposed here . It relies upon an extension to the boundary element context of the algebraic theory developed by A. Bossavit for variational formulations of boundary-value problems . The geometrical domain under consideration is assumed to be invariant under a finite abelian (i.e. commutative) group, called symmetry group . The present paper shows how to decompose a given problem stated on a symmetrical domain into subproblems stated on a symmetry cell, ie a smallest subset of the original boundary which reproduces the entire domain under the action of the symmetry group ; in this sense, the BIE decomposition into subproblems presented here is optimal . The representation formula of the displacement for symmetrical domains is given . The steps allowing a complete BIE treatment using the symmetry of the domain are indicated . Finally a numerical application to an elastodynamic problem using our BEM code ASTRID is presented to demonstrate the efficiency of the present approach .*

I - Introduction .

A lot of problems arising in computational engineering deal with structures which have some geometrical symmetry, which can be described by a *symmetry group* of *n* isometries . In such cases, the problems under consideration can be reduced to *n* *subproblems* stated on a *symmetry cell*, that is, a part, *n* times smaller, of the original structure which generates the whole structure by the action of all the symmetry elements . Such a procedure allows substantial computational savings, since the computational cost increases much faster than the size of the problem considered .
It is easy to reduce a problem when the loading and the geometry share the same symmetry, but far less easy to obtain and implement the maximal reduction in the general case (ie on the smallest possible symmetry cell) when the loading has no, or weaker, symmetry . A systematic approach to the use of geometrical symmetry in solving boundary-value problems, based on the use of *group representation theory*, has been carried out by A. Bossavit [3] . As a result this work provides, in the framework of variational formulations of elliptic boundary-value problems, by means of algebraic formulas a systematic and automatic way (a "recipe") to *decompose* the initial problem into *n* subproblems and then *reconstruct* the global solution from the results of the *n* subproblems .
To our best knowledge, although symmetries are frequently used in this field, very few publications deal with a systematic treatment of the use of geometrical

symmetries in boundary integral equation (BIE) methods (see [6]) .
The framework and results of [3] are oriented towards variational formulations and finite-element methods . The present work shows how to adapt them to the systematic use of geometrical symmetries in the framework of boundary integral equation (BIE) methods . It is our hope that the present paper gives a clearer and more systematic view of a classical computational feature of the boundary element methods (and of other methods as well) which is of great practical importance .

For the sake of clarity of exposition we consider here elastostatic problems with vanishing body forces stated on a three-dimensional domain Ω, with boundary $\partial\Omega$. The following elastostatic boundary-value problem (P) is stated on Ω :

(P) $$\Delta^* \mathbf{u} \equiv \operatorname{div} \sigma(\mathbf{u}) = \mathbf{0}$$

$$\sigma.\mathbf{n} = \mathbf{t}^{(d)} \text{ on } S_t,\ \mathbf{u} = \mathbf{u}^{(d)} \text{ on } S_u\ ;\ S_t \cap S_u = \emptyset \text{ and } S_t \cup S_u = \partial\Omega \tag{1}$$

The problem (P) can equivalently be stated in a variational (or *weak*) form :

(P) Find $\mathbf{u} \in V = \{\mathbf{w} \in H^1(\Omega) \text{ and } \mathbf{w}|_{S_u} = \mathbf{u}^{(d)}\}$, $\forall\, \mathbf{v} \in V_0 = \{\mathbf{w} \in H^1(\Omega) \text{ and } \mathbf{w}|_{S_u} = \mathbf{0}\}$,

$$\int_\Omega \nabla\mathbf{u} : \mathbf{A} : \nabla\mathbf{v}\, dV = \int_{S_T} \mathbf{t}^{(d)}.\mathbf{u}\, dS \tag{2}$$

$\mathbf{A}$ denoting the fourth-order tensor of elastic coefficients in Hooke's law . The domain Ω is assumed to be invariant under a finite group G of isometries of R^3, that is :

$$\forall\, s \in G,\quad s(\Omega) = \Omega \tag{3}$$

Furthermore, the complementary components S_u and S_t of $\partial\Omega$ are also assumed to be *invariant under G* : symmetrical portions of the boundary are subjected to boundary conditions *of the same type* . On the contrary, no particular symmetry assumption is made regarding the boundary data ($\mathbf{u}^{(d)}$, $\mathbf{t}^{(d)}$), and one wishes to make the best use of the *geometrical* symmetry of the domain Ω .

II - A group-theoretical approach of geometrical symmetry in boundary-value problems [3] .

This section is devoted to a brief introduction of the essential concepts and notations used for the approach of geometrical symmetry in boundary-value problems in [3] .
For the geometrical symmetry to be of use, the differential operator governing the boundary-value problem (here $\operatorname{div}\sigma(\mathbf{u})$, which is understood as a second-order elliptic operator with respect to $\mathbf{u}$) must be invariant under the symmetry group G . This rather vague statement can be given a precise form using the variational formulation (2), stating that the bilinear form a($\mathbf{u}$, $\mathbf{v}$) which appears in the left-hand side of (2) must satisfy the following *equivariance property* :

$$a(\mathbf{Su}, \mathbf{Sv}) = a(\mathbf{u}, \mathbf{v}) \tag{4}$$

using the notation : $\mathbf{Su}(\mathbf{x}) = \mathbf{s}\,\mathbf{u}(\mathbf{s}^{-1}\mathbf{x})$

In the present context of linear elasticity, equation (4) is obviously true if the elastic coefficients are constant on Ω, but equivariance (4) could be violated in case of variable elastic coefficients . To put it another way, G must be the group of

geometrical and material symmetry of the structure considered . We restrict ourselves to *abelian* finite symmetry groups, ie for which $s_1 s_2 = s_2 s_1$, $s_1, s_2 \in G$; let n denote the *order* (ie the number of elements) of G .
The basic idea is well-known : replace the initial problem (P) on Ω by n *subproblems* on a *symmetry cell* C ; then the solutions of the subproblems (weighted with adequate coefficients, see later) add up to the global solution of (P), using the linearity of the boundary-value problem with respect to its data (superposition principle of linear elasticity) . A symmetry cell is a subdomain of Ω, n times smaller than Ω, such that $\Omega = s_1(C) \cup ... \cup s_n(C)$ (figure 1) . A more abstract and precise definition of a symmetry cell is given in [3], the "naïve" one being sufficient for the present purposes .

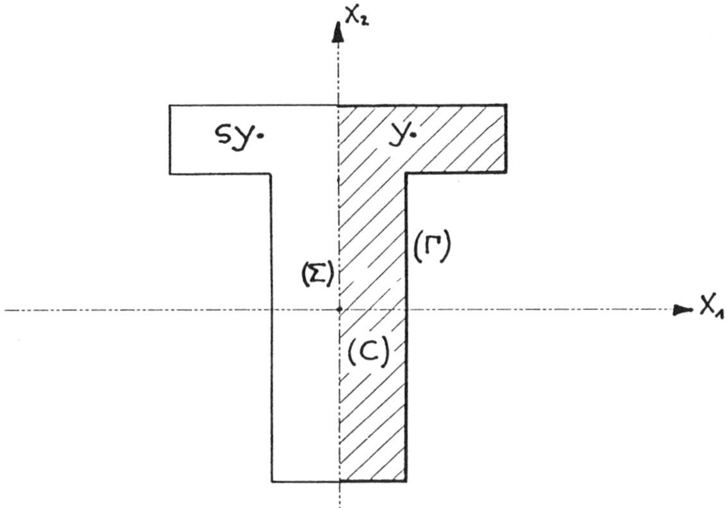

***Figure* 1** *Symmetrical domain Ω and symmetry cell C : case $G = \{ 1, s \}$, $\Omega = C \cup s(C)$.*

For example, if $G = \{1, s\}$ (s being a plane symmetry), one builds the even and odd parts of the boundary data :

$$(\mathbf{u}_E^{(d)}, \mathbf{t}_E^{(d)}) = (1/2(\mathbf{u}^{(d)} + \mathbf{S}\mathbf{u}^{(d)})(\mathbf{x}),\ 1/2(\mathbf{t}^{(d)} + \mathbf{S}\mathbf{t}^{(d)})(\mathbf{x})) \tag{5a}$$

$$(\mathbf{u}_O^{(d)}, \mathbf{t}_O^{(d)}) = (1/2(\mathbf{u}^{(d)} - \mathbf{S}\mathbf{u}^{(d)})(\mathbf{x}),\ 1/2(\mathbf{t}^{(d)} - \mathbf{S}\mathbf{t}^{(d)})(\mathbf{x})) \tag{5b}$$

Then $\mathbf{u}$ is the sum of its odd and even parts : $\mathbf{u} = \mathbf{u}_O + \mathbf{u}_E$, solutions of (P) with data given by (5a) and (5b) respectively . But, since $\mathbf{u}_O$ and $\mathbf{u}_E$ have known symmetry properties, the problem (P) with data (5a) and (5b) can be solved on the *symmetry cell* of figure 1, provided additional boundary conditions (which should be regarded as boundary *constraints* rather than *conditions*) are given on the additional boundary Σ, Σ being the intersection of Ω with the plane of symmetry :

$$\mathbf{n}.\mathbf{u}_E = 0 \ \text{ and } \ \boldsymbol{\tau}.(\boldsymbol{\sigma}.\mathbf{n})_E = 0 \ \text{ on } \Sigma \tag{6a}$$

$$\boldsymbol{\tau}.\mathbf{u}_O = 0 \ \text{ and } \ \mathbf{n}.(\boldsymbol{\sigma}.\mathbf{n})_O = 0 \ \text{ on } \Sigma \tag{6b}$$

Then one has : $\forall\, \mathbf{x} \in s(C)$, $\mathbf{u}_E(s\mathbf{x}) = s\mathbf{u}_E(\mathbf{x})$ and $\mathbf{u}_O(s\mathbf{x}) = -s\mathbf{u}_O(\mathbf{x})$ (7)

and finally $\forall\, x \in \Omega, \qquad u(x) = u_O(x) + u_E(x)$ (8)

Equations (5), (6) (decomposition) and (7), (8) (reconstruction) summarize the steps which are needed to make use of the symmetry properties of the geometrical domain Ω and of the material . Such an intuitive way of deriving these steps works nicely in the simplest case $G=\{1, s\}$, but becomes very cumbersome to manage for more complicated symmetry groups, to say nothing of *non-abelian* symmetry groups, see [3] . Thus the usefulness of a systematic treatment, which gives the algebraic decomposition and reconstruction formulas like (5), (6), (7) and (8) in the general case .

The systematic treatment in [3] uses as a key tool the concept of *irreducible representation* of a finite abelian group G . An irreducible representation is a complex-valued function ρ on G which preserves the group structure, and hence has the following properties :

$$\rho(1)=1 \quad \text{and} \quad \forall\, s_1,s_2 \in G,\ \rho(s_1 s_2)=\rho(s_1)\rho(s_2) \tag{9}$$

A finite group G of order n admits exactly n irreducible representations $\{\rho_\nu,\ \nu=0,\dots\ n-1\}$. Such irreducible representations are tabulated for many finite groups ; one has for example for a cyclic finite group $G=\{1, g, \dots, g^{n-1}\}$:

$$\rho_\nu(g)=\exp(2i\pi\nu/n)\ , \quad \nu=0, \dots, n-1 \tag{10}$$

The irreducible representations associated with plane symmetries (ie domains with one, two or three orthogonal planes of symmetry), which are the most often used, are given in Appendix .

The irreducible representations give the coefficients of the decomposition and reconstruction steps in the general case, as follows (we refer the reader to [3] for any further explanation about the theory and concentrate on the result and its subsequent use) .

(i) Choice of a symmetry cell . A symmetry cell C is chosen as an adequate subset of Ω such that :

$$\text{Meas}(C)=\frac{1}{n}\text{Meas}(\Omega) \quad \text{and} \quad \Omega=s_1(C)\cup\dots\cup s_n(C),\ s_1, \ , s_n \in G \tag{11}$$

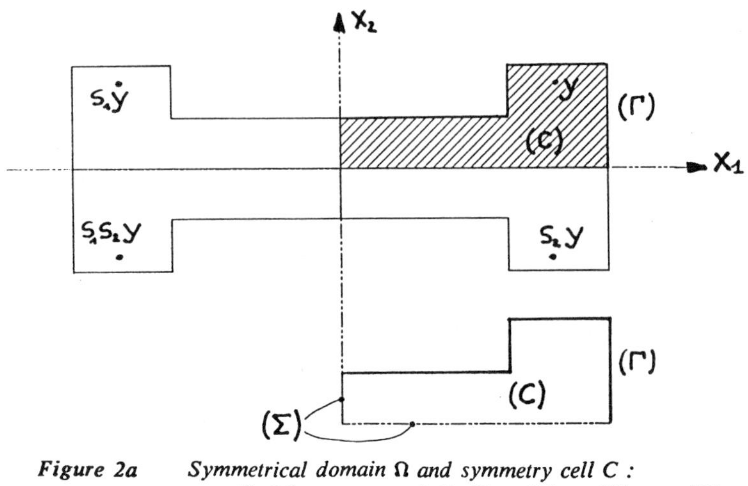

Figure 2a *Symmetrical domain Ω and symmetry cell C : case $G=\{1, s_1, s_2, s_1s_2\}$, $\Omega=C\cup s_1(C)\cup s_2(C)\cup s_1s_2(C)$.*

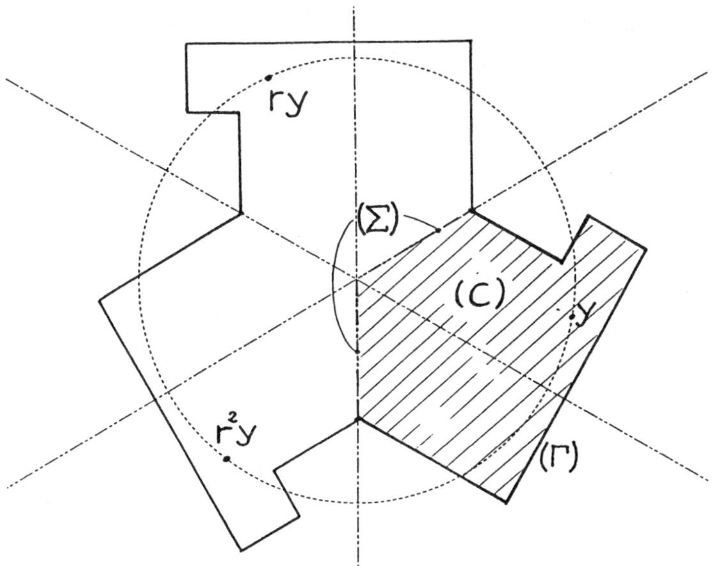

Figure 2b *Symmetrical domain Ω and symmetry cell C : case $G=\{1, r, r^2\}$, $\Omega = C \cup r(C) \cup r^2(C)$.*

We will denote by Γ_u and Γ_t the respective restrictions of S_u and S_t to C, and by Σ the additional boundary $\partial C - (\Gamma_u \cup \Gamma_t)$ (figure 2) ; S_u and S_t are generated by the action of G on Γ_u and Γ_t respectively .

(ii) Reduction to subproblems stated on C . The following sequence $(P_\nu)_{\nu=1,\dots,n}$ of *uncoupled* subproblems on Ω is introduced :

$$(P_\nu) \qquad \Delta^* \mathbf{u}^\nu = 0 \qquad (12a)$$

$$[\sigma(\mathbf{u}^\nu).\mathbf{n}](\mathbf{y}) = \mathbf{t}^{\nu(d)}(\mathbf{y}) = \frac{1}{n} \sum_{s \in G} \rho_\nu(s) \mathbf{S}\mathbf{t}^{(d)}(\mathbf{y}) \quad , \quad \forall\, \mathbf{y} \in S_t \qquad (12b)$$

$$\mathbf{u}^\nu(\mathbf{y}) = \mathbf{u}^{\nu(d)}(\mathbf{y}) = \frac{1}{n} \sum_{s \in G} \rho_\nu(s) \mathbf{S}\mathbf{u}^{(d)}(\mathbf{y}) \quad , \quad \forall\, \mathbf{y} \in S_u \qquad (12c)$$

which generalize the "odd" and "even" problems of eqns. (5) to (8), the subscript ν playing a role similar to the subscripts O, E . Subproblems (P_ν) are mixed symmetric/antisymmetric problems (e.g. symmetric with respect to the plane $x_1=0$ and antisymmetric w.r.t. the plane $x_2=0$ for a domain admitting plane symmetries) . Moreover, it may occur that the boundary data $(\mathbf{u}^{\nu(d)}, \mathbf{t}^{\nu(d)})$ appearing in (12b,c) vanish for certain values of ν ; this fact arises if the original boundary data $(\mathbf{u}^{(d)}, \mathbf{t}^{(d)})$ of (P) is itself invariant under some subgroup G′ of G . In particular :

- $G' = \{1\}$ (data with maximum symmetry) : the subproblem (P_1) only has nonzero data .
- $G' = G$ (no data symmetry) : the n subproblems are to be considered .

We will denote by n_0 the number of nontrivial (i.e. with nonzero boundary data $(\mathbf{u}^{\nu(d)}, \mathbf{t}^{\nu(d)})$) subproblems associated to a given problem (P) .

As the subproblems (P_ν) are either symmetric or antisymmetric with respect to any element of G, the following generalisation of (7) holds :

$$\mathbf{u}^\nu(\mathbf{s}\mathbf{y}) = \rho_\nu(\mathbf{s}^{-1}) \mathbf{s}\tilde{\mathbf{u}}^\nu(\mathbf{y}) \quad , \qquad \forall\, \mathbf{y} \in C, \ \forall\, \mathbf{s} \in G \qquad (13)$$

Therefore the (P_ν) can be restated, using (13), *on the symmetry cell* :

$$(P_\nu) \qquad \Delta^* \tilde{u}^\nu = 0 \tag{14a}$$

$$[\sigma(\tilde{u}^\nu).n](y) = \tilde{t}^{\nu(d)}(y) = \frac{1}{n} \sum_{s \in G} \rho_\nu(s) St^{(d)}(y) \quad , \forall y \in \Gamma_t \tag{14b}$$

$$\tilde{u}^\nu(y) = \tilde{u}^{\nu(d)}(y) = \frac{1}{n} \sum_{s \in G} \rho_\nu(s) Su^{(d)}(y) \quad , \forall y \in \Gamma_u \tag{14c}$$

$$[\tilde{u}^\nu - \rho_\nu(s^{-1}) S\tilde{u}^\nu](y) = 0 \qquad \forall y \in \Sigma, \; s^{-1}y \in \Sigma \tag{14d}$$

$$\Gamma_t \cap \Gamma_u = \emptyset \text{ and } \Gamma_t \cup \Gamma_u = \Gamma \tag{14e}$$

In (14) the notation $\tilde{u}^\nu$, $\tilde{t}^\nu$ is used instead of u^ν, t^ν in order to emphasize the fact that the solution of (P_ν) has its geometrical support restricted to the symmetry cell C . It is to be noticed that the restriction of (P_ν) to C leads to introduce additional boundary constraints (14d) *on* Σ which generalize (6a-b) and are a consequence of eqn. (13) ; these are of great importance . Eqns. (12b-c) and (14b-c) generalize the boundary constraints (5a-b) .

(iii) Reconstruction of the global solution . The subproblems (P_ν) are solved for $\tilde{u}$. Then the global solution of the problem of interest (P) is reconstructed using :

$$u(sy) = \sum_{\nu=0}^{n_0-1} \rho_\nu(s^{-1}) su^\nu(y) \quad , \qquad \forall y \in C, \forall s \in G \tag{15}$$

which generalizes (8) .

Case of complex-valued irreducible representations . An irreducible representation ρ of a finite abelian group G is basically a complex-valued function over G (see eqn. (10) for cyclic groups) . In some particular cases, including the most commonly used one (domains with orthogonal planes of symmetry), the irreducible representations ρ_ν are real-valued (see appendix) .
The trouble with complex-valued ρ_ν is that, when one considers a real-valued problem (as in elastostatics), the subproblems (P_ν) have complex-valued boundary data . However, it is shown that the complex-valued ρ_ν can be gathered into *conjugate pairs* $(\rho_{\nu_1}, \rho_{\nu_2})$, ie such that $\rho_{\nu_1}(s) = \rho_{\nu_2}(s)$ for every $s \in G$. Then, if a real-valued problem is considered, the pairs (u^{ν_1}, u^{ν_2}) and (t^{ν_1}, t^{ν_2}) are conjugate owing to (12b,c) . Let $u^{\nu_1} = v + iw$, $u^{\nu_2} = v - iw$. Then the boundary constraint (14d) becomes :

$$\begin{bmatrix} u(y) \\ w(y) \end{bmatrix} = \begin{bmatrix} \mathrm{Re}(\rho_\nu(s)) & -\mathrm{Im}(\rho_\nu(s)) \\ \mathrm{Im}(\rho_\nu(s)) & \mathrm{Re}(\rho_\nu(s)) \end{bmatrix} \begin{bmatrix} u(s^{-1}y) \\ w(s^{-1}y) \end{bmatrix} \tag{16}$$

Thus, complex-valued irreducible representations lead to pairs of real-valued subproblems *coupled by the boundary constraints (14d)* . This is to be taken into account when considering the implementation of cyclic symmetry with non-symmetric (in this sense) loadings .

III - Application to BEM elastic analysis .

III.1 - Regularized elastic BEM .

Let us now transpose the previous concepts to BEM elastic analysis, using the same assumptions and notations G, G', n, n_0 as in section II . The surface displacement and tractions **u**, **t** involved in (P) satisfy the well-known Somigliana singular boundary integral identity [4] :

$$c_{ki}(\mathbf{x})u_i(\mathbf{x}) + \int^{*}_{\partial\Omega} u_i(\mathbf{y})n_s(\mathbf{y})\Sigma^0_{is;k}(\mathbf{x},\mathbf{y})\, dS_y - \int_{\partial\Omega} t_i(\mathbf{y})U^0_{i;k}(\mathbf{x},\mathbf{y})\, dS_y = 0 \quad (17)$$

In (17) $\mathbf{U}^0(\mathbf{x},\mathbf{y})$ and $\mathbf{\Sigma}^0(\mathbf{x},\mathbf{y})$ are the classical Kelvin displacement and stress tensors ; the asterisk denotes a Cauchy principal value (CPV) integral and $c(\mathbf{x})$ the free term . In (17) and the sequel, the summation convention over repeated indices is used . The regularization problem has been studied in [2] (for time-domain elastodynamic BIE and cracks) and [1], [7] (steady-state elastodynamics), among others . These works state, as a partial result, the following regularized version of the elastostatic boundary integral equation (BIE) (17), free of CPVs, which will serve as a basis for the present purposes :

$$\int_{\partial\Omega} (u_i(\mathbf{y}) - u_i(\mathbf{x}))n_s(\mathbf{y})\Sigma^0_{is;k}(\mathbf{x},\mathbf{y})\, dS_y - \int_{\partial\Omega} t_i(\mathbf{y})U^0_{i;k}(\mathbf{x},\mathbf{y})\, dS_y = 0 \quad (18)$$

Eqn. (18) holds if $\mathbf{x}$ is a regular point as well as an edge or corner point on $\partial\Omega$. The surface $\partial\Omega$ is approximated by NE boundary elements, which are mapped, in a standard fashion, on a *reference element* E_0 , which is generally the square $\{\xi=(\xi_1,\xi_2) \in [-1, 1]^2\}$ or the triangle $\{0 \le \xi_1+\xi_2 \le 1\}$. The location of a point $\mathbf{y}$ on the approximate surface is written in terms of NU shape functions N_q and geometrical nodes $\mathbf{A}^q$, NU being the total number of geometrical nodes on C :

$$\mathbf{Oy} = N_q(\xi)\, \mathbf{OA}^q(\xi) \in E_0 \quad (19)$$

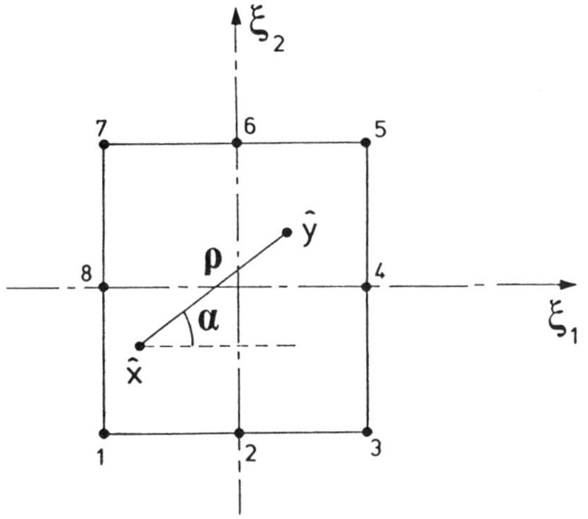

Figure 3 *Eight-noded quadrilateral element and reducd polar coordinate system .*

and we denote respectively by $\mathbf{a}_\alpha(\xi)$ $(\alpha=1,2)$ and $\mathbf{n}(\xi)$ the natural base of the tangent plane and the unit normal at $\mathbf{y}(\xi)$:

$$\mathbf{a}_\alpha(\xi) = \frac{\partial}{\partial \xi_\alpha}\mathbf{Oy}(\xi) \qquad (\alpha=1,2) \quad (20)$$

Furthermore the surface fields $\mathbf{u}$, $\mathbf{t}$ are interpolated isoparametrically, i.e. with the

same shape functions N_q as the surface $\partial\Omega$ itself .
During the assembly process, the surface integrals are evaluated element by element for each fixed collocation point $\mathbf{x}$. Singular integrals occur for the boundary elements containing $\mathbf{x}$. The CPV exclusion procedure apparent in the BIE (18) leads to element integrals of the form :

$$I^s = \int_E n_s(y)\Sigma^0_{is;k}(\mathbf{x},\mathbf{y})\,(N_q(\xi)-N_q(\eta))\,J(\xi)d\xi_1 d\xi_2 \tag{21}$$

where $(\eta_1,\eta_2)=\eta$ is the antecedent of $\mathbf{x}$ on the reference element E_0 . Let (ρ,α) be a polar coordinate system centered at $\hat{\mathbf{x}}$ on E_0 (figure 3) . Then for every standard polynomial shape function N_q a reduced shape function $\hat{N}^{\mathbf{x}}_q$ exists such that :

$$N_q(\xi)-N_q(\eta) = \rho\hat{N}^{\mathbf{x}}_q(\rho,\alpha) \tag{22}$$

The derivation of $\hat{N}^{\mathbf{x}}_q$ is straightforward, using for example a Taylor formula (which is exact for polynomials) . Furthermore $r = ||y-x||$ can be rewritten, using (31) and (22), as :

$$r = \rho r_0 = \rho\,||\,\hat{N}^{\mathbf{x}}_q(\rho,\alpha)\mathbf{OA}^q\,|| \qquad \text{(summation over the nodes of E)} \tag{23}$$

The factor $1/r_0$ is always nonsingular at $\rho = 0$. The kernel Σ^0 is transformed into :

$$\Sigma^0_{is;k}(\mathbf{x},\mathbf{y}) = \frac{1}{\rho^2}\,\hat{\Sigma}^0_{is;k}(\rho,\alpha) \qquad (\hat{\Sigma}^0 \text{ regular at } \rho = 0) \tag{24}$$

Finally the differential element $d\xi_1 d\xi_2$ becomes $\rho d\rho d\alpha$. This, combined with (22) and (23) allows us to recast the singular integral (21), in a completely regular form, as :

$$I^s = \int_E n_s(y)\,\hat{\Sigma}^0_{is;k}(\rho,\alpha)\,\hat{N}^{\mathbf{x}}_q(\rho,\alpha)\,J(\xi)\,d\rho d\alpha \tag{25}$$

The expression (25) takes full advantage of the CPV exclusion . The numerical evaluation of (25) can be performed with standard product Gaussian quadrature formulas, the complete procedure requiring a further coordinate change $(\rho,\alpha) \rightarrow (v_1,v_2)$ in order to recover an integral over the square $[-1,1]^2$. This involves some algebra but no major difficulty (see [7],[1]) .

III.2 - Making use of symmetry in BEM analysis : investigation .

To make use of geometrical symmetry in a BEM context one has to define a symmetry cell C, build the subproblems (P_ν) and apply the reconstruction formula (15) . A possible way of doing this is simply to translate into BEM language the subproblems (P_ν) defined by (14), that is, to apply (18) to the symmetry cell C of boundary $\partial C=\Sigma\cup\Gamma$ and to the boundary constraints (14b,c,d) . However, in a BEM context the computational efficiency of using geometrical symmetry is not related to the ratio Meas(C)/Meas(Ω) as in a FEM context but to the relative sizes of the *boundaries*, ie the ratio Meas(∂C)/Meas($\partial\Omega$) . If the additional boundary Σ of the symmetry cell is of similar size than the trace Γ of $\partial\Omega$, very few will be gained by using symmetry .
What we want is to define subproblems that use *integral equations stated on* Γ *only*, because the ratios Meas(Γ)/Meas($\partial\Omega$) and Meas(C)/Meas(Ω) are equal . In order to do so, let us revert to the definition (12) of (P_ν), where it is stated on the whole domain Ω . The boundary fields $(\mathbf{u}^\nu, \mathbf{t}^\nu)$ are then governed by the integral equation

(18) .

$$\int_{\partial\Omega} (u_i^\nu(y) - u_i^\nu(x))\, n_s(y)\, \Sigma^0_{is;k}(x,y)\, dS_y - \int_{\partial\Omega} t_i^\nu(y)\, U^0_{i;k}(x,y)\, dS_y = 0 \quad (26)$$

Now the integrals over Ω in (26) are reduced to a sum of integrals over the symmetry cell, which leads to the following equivalent integral equation :

$$\sum_{s\in G} \int_\Gamma (u_i^\nu(s\,y) - u_i^\nu(x))\, n_s(s\,y)\, \Sigma^0_{is;k}(x,s\,y)\, dS_y$$

$$- \sum_{s\in G} \int_\Gamma t_i^\nu(s\,y)\, U^0_{i;k}(x,s\,y)\, dS_y = 0 \quad (27)$$

Now comes the equivalent of the *equivariance property* : the Kelvin tensors satisfy the following symmetry properties, expressed in indicial tensor notation (a symmetry element s of G being represented by a matrix $[s_{ij}]$:

$$U^0_{i;k}(x,s\,y) = s_{ip}\, U^0_{p;j}(s^{-1}x,y)s_{jk} \qquad \forall s\in G \quad (28a)$$

$$n_s(s\,y)\Sigma^0_{is;k}(x,s\,y) = s_{ip}\, n_s(y)\Sigma^0_{ps;j}(s^{-1}x,y)s_{jk} \qquad \forall s\in G \quad (28b)$$

Property (28) can be established either by using a physical reasoning or by applying the rules of tensor calculus to the Kelvin tensors . These identities are inserted in (27), together with eqn. (13) . This transforms (27) into :

$$\sum_{s\in G} \rho_\nu(s^{-1}) \int_\Gamma \left(s_{im}\tilde{u}_m^\nu(y) - \tilde{u}_i^\nu(x) \right) s_{ip}\, n_s(y)\Sigma^0_{ps;j}(s^{-1}x,y)s_{jk}\, dS_y$$

$$- \sum_{s\in G} \rho_\nu(s^{-1}) \int_\Gamma s_{im}\tilde{t}_m^\nu(y)\, s_{ip}\, U^0_{p;j}(s^{-1}x,y)s_{jk}\, dS_y = 0 \quad (29)$$

but, as s is an isometry, one has $s^T s = s s^T = 1$, that is, $s_{im}s_{ip} = \delta_{mp}$, thus :

$$\sum_{s\in G} \rho_\nu(s^{-1}) \int_\Gamma \left(\tilde{u}_i^\nu(y) - \tilde{u}_p^\nu(x)\rho_\nu(s)s_{pi} \right) n_s(y)\Sigma^0_{is;j}(s^{-1}x,y)s_{jk}\, dS_y$$

$$- \sum_{s\in G} \rho_\nu(s^{-1}) \int_\Gamma \tilde{t}_i^\nu(y)\, U^0_{i;j}(s^{-1}x,y)s_{jk}\, dS_y = 0 \quad (30)$$

but (13) gives : $\tilde{u}_i^\nu(y) - \tilde{u}_p^\nu(x)\rho_\nu(s)s_{pi} = \tilde{u}_i^\nu(y) - \tilde{u}_p^\nu(s^{-1}x)$ which in turn transforms (30) into :

$$\sum_{s\in G} \rho_\nu(s^{-1}) \int_\Gamma \left(\tilde{u}_i^\nu(y) - \tilde{u}_p^\nu(s^{-1}x) \right) n_s(y)\Sigma^0_{is;j}(s^{-1}x,y)s_{jk}\, dS_y$$

$$- \sum_{s\in G} \rho_\nu(s^{-1}) \int_\Gamma \tilde{t}_i^\nu(y)\, U^0_{i;j}(s^{-1}x,y)s_{jk}\, dS_y = 0 \quad (31)$$

The boundary equation (31) is the desired result, which is the BIE statement of the subproblem (P_ν) using only $\Gamma = \partial\Omega \cap \partial C$ (and *not* the additional boundary Σ) .
Now we derive the discretized version of (31), using the concepts and notations recalled in §III.1 ; the discretized surface under consideration being Γ instead of $\partial\Omega$. In order to distinguish between the regular and singular integrals involving the

kernel $\Sigma(\mathbf{x},\mathbf{y})$, $\Gamma(\mathbf{x})$ denotes the union of the boundary elements containing the collocation point $\mathbf{x}$. Inserting nodal values for the surface fields $(\mathbf{u}^\nu, \mathbf{t}^\nu)$ and the shape functions $N_q(\xi)$ and $\hat{N}_q^{s^{-1}x}(\rho,\alpha)$, the discretized version of (31) reads :

$$\sum_{q=1}^{NN} \sum_{s \in G} \sum_{E \in \Gamma(s^{-1}x)} \rho_\nu(s^{-1}) s_{jk}\, \tilde{u}_p^{(q)\nu} \int_E \hat{N}_q^{s^{-1}x}(\rho,\alpha)\, n_s(y)\, \hat{\Sigma}^0_{is;j}(\rho,\alpha)\, J(\xi)\, d\rho d\alpha$$

$$+ \sum_{q=1}^{NN} \sum_{s \in G} \sum_{E \in \Gamma - \Gamma(s^{-1}x)} \rho_\nu(s^{-1}) s_{jk}\, \tilde{u}_p^{(q)\nu} \int_E N_q(\xi)\, n_s(y)\, \Sigma^0_{is;j}(s^{-1}x,y)\, J(\xi)\, d\xi_1 d\xi_2$$

$$- \sum_{s \in G} \sum_{E \in \Gamma - \Gamma(s^{-1}x)} \rho_\nu(s^{-1}) s_{jk}\, \tilde{u}_p^{\nu}(s^{-1}x) \int_E \Sigma^0_{is;j}(s^{-1}x,y)\, J(\xi)\, d\xi_1 d\xi_2$$

$$- \sum_{q=1}^{NN} \sum_{s \in G} \sum_{E \in \Gamma} \rho_\nu(s^{-1}) s_{jk}\, \tilde{t}_i^{(q)\nu} \int_E N_q(\xi)\, U^0_{i;j}(s^{-1}x,y)\, J(\xi)\, d\xi_1 d\xi_2 = 0 \qquad (32)$$

Equations (31), (32) above are our main result ; they allow the solution *on the external boundary* Γ *of the symmetry cell* C of the subproblem (P_ν) using a BEM formulation, and hence the optimal reduction of the initial problem (P) using symmetry . As another important feature, integral equation (32) has the following structure :

$$\sum_{s \in G} \sum_{q=1}^{NN} \rho_\nu(s^{-1}) s_{jk} \Big(\sum_{E \in \Gamma(s^{-1}x)} \tilde{u}_p^{(q)\nu}\, I^s(s^{-1}x,E) + \sum_{E \in \Gamma - \Gamma(s^{-1}x)} \tilde{u}_p^{(q)\nu}\, I^r(s^{-1}x,E) - \sum_{E \in \Gamma} \tilde{t}_i^{(q)\nu} J(s^{-1}x,E) \Big) \qquad (33)$$

in which the integrals $I^s(s^{-1}x,E)$, $I^r(s^{-1}x,E)$ and $J(s^{-1}x,E)$ *do not depend on the subproblem index* ν . This is logical since the dependence on ν of the subproblem (P_ν) appears only through the boundary conditions, the governing operator (or, equivalently, the bilinear form in a variational formulation) being the same for all subproblems . Hence this was bound to be true for a BIE formulation, but was far less apparent in the equations . As a practical computational consequence of this fact, each element integral $I^s(s^{-1}x,E)$, $I^r(s^{-1}x,E)$ or $J(s^{-1}x,E)$ is computed *once* and contributes to the building of *each* discretized subproblem, weighted by the adequate coefficients $\rho_\nu(s^{-1})$.

An examination of (32), (33) allows an approximate operation count . The boundary elements are supported on the symmetry cell Γ, n times smaller than the boundary $\partial\Omega$. Moreover, elementary integrals $I^s(s^{-1}x,E)$, $I^r(s^{-1}x,E)$ and $J(s^{-1}x,E)$ have to be evaluated for all elements of Γ and for all the images of the collocation points . In other words, the building loop scans the number of collocation points which would have been introduced *for the whole boundary* $\partial\Omega$, but a number of boundary elements n times smaller . As a result, the overall number of elementary integrals to be evaluated using symmetry is expected to be n times smaller than without the use of symmetry . However this is to be relaxed because in (32) the elementary integrals are dispatched to the subproblems using sums weighted by the coefficients $\rho_\nu(s)$, thus involving an additional amount of arithmetic operations which are not required if geometrical symmetry is not considered . Similarly, the solution step replaces one linear system of size (say) $n \times N$ by n (in the most general case, i.e. no symmetry assumption on the boundary data) linear systems of size N . As the matrices are fully-populated and nonsymmetric, the operation count for a linear system of size N is proportional to N^3 . As a consequence, the ratio of the respective operation counts is expected to be respectively :

$$\frac{\text{op. count without symmetry}}{\text{op. count using symmetry}} = \frac{(n\times N)^3}{[n(N^3)]} = n^2 \tag{34}$$

Finally the boundary integral equation (32) for subproblem (P_ν) is valid *whatever the symmetry group G considered*, provided it is abelian . This includes the common cases of *plane symmetry* and *cyclic symmetry* ; on the other hand the commutativity assumption excludes cases like combinations of cyclic and plane symmetries . Bossavit [3] treats non-abelian symmetry groups as well, but we will not go here into a BEM translation of such cases, leaving them for future work .

Integral representation for interior points . The boundary integral equation (32) allows to solve (P) for the surface fields, which is the first step of any boundary element method . Next comes the postprocessing step, in which one uses the well-known integral representations of $\mathbf{u}(\mathbf{x})$ and $\sigma(\mathbf{x})$ at user-specified interior points $\mathbf{x}$. Obviously much is to be gained by also using geometrical symmetry in this step .
In order to do so, let us first recall the integral representation of $\mathbf{u}(\mathbf{x})$:

$$u_k(\mathbf{x}) = -\int_{\partial\Omega} u_i(y)\,n_s(y)\,\Sigma^0_{is;k}(\mathbf{x},y)\,dS_\mathbf{y} + \int_{\partial\Omega} t_i(y)\,U^0_{i;k}(\mathbf{x},y)\,dS_\mathbf{y} \tag{35}$$

and reduce (35) to a sum of integrals over the symmetry cell, which leads to :

$$u_k(\mathbf{x}) = -\sum_{s\in G}\left[\int_\Gamma u_i(s\,y)\,n_s(s\,y)\,\Sigma^0_{is;k}(\mathbf{x},s\,y)\,dS_\mathbf{y} + \int_\Gamma t_i(s\,y)\,U^0_{i;k}(\mathbf{x},s\,y)\,dS_\mathbf{y}\right] \tag{36}$$

Then, insertion into (35) of identity (15) and use of the equivariance property (28a,b) yields :

$$\begin{aligned}
u_k(\mathbf{x}) &= -\sum_{s\in G}\sum_{\nu=0}^{n_0-1}\rho_\nu(s^{-1})\int_\Gamma s_{im}\,\tilde{u}^\nu_m(y)\,s_{ip}\,n_s(y)\Sigma^0_{ps;j}(s^{-1}\mathbf{x},y)s_{jk}\;dS_\mathbf{y}\\
&\quad+\sum_{s\in G}\sum_{\nu=0}^{n_0-1}\rho_\nu(s^{-1})\int_\Gamma s_{im}\,\tilde{t}^\nu_m(y)\,s_{ip}\,U^0_{p;j}(s^{-1}\mathbf{x},y)s_{jk}\;dS_\mathbf{y}\\
&= -\sum_{s\in G}\sum_{\nu=0}^{n_0-1}\rho_\nu(s^{-1})s_{jk}\int_\Gamma \tilde{u}^\nu_i(y)\,n_s(y)\Sigma^0_{is;j}(s^{-1}\mathbf{x},y)\;dS_\mathbf{y}\\
&\quad+\sum_{s\in G}\sum_{\nu=0}^{n_0-1}\rho_\nu(s^{-1})s_{jk}\int_\Gamma \tilde{t}^\nu_i(y)\,U^0_{i;j}(s^{-1}\mathbf{x},y)\;dS_\mathbf{y}
\end{aligned} \tag{37}$$

The discretized version of representation formula (37) reads :

$$\begin{aligned}
u_k(\mathbf{x}) = &-\sum_{q=1}^{NN}\sum_{s\in G}\rho_\nu(s^{-1})\sum_{E\in\Gamma}\sum_{\nu=0}^{n_0-1}s_{jk}\,u^{(q)\nu}_p\int_E N_q(\xi)\,n_s(y)\,\Sigma^0_{is;j}(s^{-1}\mathbf{x},y)\,J(\xi)\,d\xi_1 d\xi_2\\
&+\sum_{q=1}^{NN}\sum_{s\in G}\rho_\nu(s^{-1})\sum_{E\in\Gamma}\sum_{\nu=0}^{n_0-1}s_{jk}\,t^{(q)\nu}_i\int_E N_q(\xi)\,U^0_{i;j}(s^{-1}\mathbf{x},y)\,J(\xi)\,d\xi_1 d\xi_2
\end{aligned} \tag{38}$$

This result can be extended further by considering the representation at a symmetrical point $\mathbf{z}=s_0\mathbf{x}$, s_0 being a fixed element of G . Using (9), the representation formula (38) written for $s_0\mathbf{x}$ becomes :

$$u_k(s_0x) =$$

$$= - \sum_{q=1}^{NN} \sum_{s\in G} \rho_\nu(s^{-1}) \sum_{E\in\Gamma} \sum_{\nu=0}^{n_0-1} s_{jk}\, \tilde{u}_p^{(q)\nu} \int_E N_q(\xi)\, n_s(y)\, \Sigma^0_{is;j}(s^{-1}s_0x,y)\, J(\xi)\, d\xi_1 d\xi_2$$

$$\sum_{q=1}^{NN}+ \sum_{s\in G} \rho_\nu(s^{-1}) \sum_{E\in\Gamma} \sum_{\nu=0}^{n_0-1} s_{jk}\, \tilde{t}_i^{(q)\nu} \int_E N_q(\xi)\, U^0_{i;j}(s^{-1}s_0x,y)\, J(\xi)\, d\xi_1 d\xi_2$$

$$= - \sum_{q=1}^{NN} \sum_{s\in G} \rho_\nu(s^{-1})\rho_\nu(s_0) \sum_{E\in\Gamma} \sum_{\nu=0}^{n_0-1} s_{jk}\, \tilde{u}_p^{(q)\nu} \int_E N_q(\xi)\, n_s(y)\, \Sigma^0_{is;j}(s^{-1}x,y)\, J(\xi)\, d\xi_1 d\xi_2$$

$$\sum_{q=1}^{NN}+ \sum_{s\in G} \rho_\nu(s^{-1})\rho_\nu(s_0) \sum_{E\in\Gamma} \sum_{\nu=0}^{n_0-1} s_{jk}\, \tilde{t}_i^{(q)\nu} \int_E N_q(\xi)\, U^0_{i;j}(s^{-1}x,y)\, J(\xi)\, d\xi_1 d\xi_2 \qquad (39)$$

At this stage, one can remark that the representation formula (39) has the structure :

$$\sum_{s\in G} \sum_{q=1}^{NN} \rho_\nu(s^{-1}) s_{jk} \sum_{E\in\Gamma} \left(\tilde{u}_p^{(q)\nu}\, I(s^{-1}x,E) + \tilde{t}_i^{(q)\nu} J(s^{-1}x,E) \right) \qquad (40)$$

in which the integrals $I(s^{-1}x,E)$ and $J(s^{-1}x,E)$ *do not depend on the subproblem index* ν. This allows once again to compute the element integrals *once*, then they are used, weighted by the adequate coefficients $\rho_\nu(s^{-1})$, to the evaluation of $\mathbf{u}(s_0\mathbf{x})$ for all the images of $\mathbf{x}$.
A glance at (39) or (40) shows that the evaluation of $\mathbf{u}(s_0\mathbf{x})$ for a *fixed* symmetry s_0 requires as many integrations as the original representation formula (35) which makes no use of symmetry . However, once these integrals are computed, one is able to generate $\mathbf{u}(s_0\mathbf{x})$ for *all* the *n* images $s_0\mathbf{x}$ of $\mathbf{x}$ by an element of G instead of one single value $\mathbf{u}(\mathbf{x})$ using (35) .

Finally the representation formula for the stress tensor $\sigma(\mathbf{x})$ is obtained in a straightforward way by applying the stress-displacement relation (Hooke law) to (37) understood as a vector function of the interior point $\mathbf{x}$ (thus, application of Hooke's law leads to differentiate (37) under the integral sign with respect to the components of $\mathbf{x}$) .

III.3 - Making use of symmetry in BEM analysis : a practical recipe .

Now let us summarize the steps involved in making use of symmetry in BEM analysis using the present approach .
As a prerequisite for any symmetry group which is to be made available to the user, the BEM code must contain several informations :

- The *multiplication table of G*, ie a n×n integer array TABLE such that k=TABLE(i,j) is the index of the composition of the i-th and j-th symmetries of G .
- The table of *irreducible representations of G*, ie a n×n table containing the weighting coefficients $\rho_\nu(s)$.
- The action of a symmetry element s∈G on a point (or vector) $\mathbf{x}$, ie the n matrices $[s_{ij}]_{i,j=1,3}$ associated to the elements of G .

Then the code should perform the following solution procedure :

(i) Initialize . This preliminary step consists in :

- Read mesh data, ie nodes, connectivity table, ... This mesh is assumed to be the discrete *symmetry cell* Γ .

- Read information about the geometrical symmetry group G .
- Read boundary conditions data . Each point $\mathbf{y} \in \Gamma$ must be associated to *n* boundary data (n being the order of G) : if (say) $\mathbf{y}$ supports a given traction $\mathbf{t}^{(d)}(\mathbf{y})$, then the user must provide the value of $\mathbf{t}^{(d)}(s_1\mathbf{y})$, ..., $\mathbf{t}^{(d)}(s_n\mathbf{y})$, s_1, ...s_n being the elements of G .
- Using these informations, list up the subproblems needed for the case under consideration, and in particular list the *nontrivial subproblems*, ie those with nonzero boundary data .
- For each $s \in G$, identify the nodes $\mathbf{x} \in \Gamma$ such that $s\mathbf{x} \in \Gamma$ and work out the additional boundary constraints (14d) . Using these constraints ensure the adequate behaviour of the subproblems and allow a further reduction of the number of unknowns of each subproblem .

(ii) Build and solve the subproblems :

- Build the discretized boundary integral equation (32) is build *for each nontrivial subproblem* (P_ν) .
- Insert the boundary data $(\mathbf{u}^{\nu(d)}, \mathbf{t}^{\nu(d)})$.
- Take into account the boundary constraints (14d) (build during the initialization phase), either by eliminating unknowns and equations (no coupling between subproblems) or adding new equations (in case of coupled subproblems, e.g. using cyclic symmetry) .
- Solve these subproblems for the unknown components of $(\mathbf{u}^\nu, \mathbf{t}^\nu)|_\Gamma$.

(iii) Reconstruct the solution of (P) on $\partial\Omega$ by using the reconstruction identity (15) .

(iv) Compute $\mathbf{u}$ and σ at interior points if requested . For every specified interior point $\mathbf{x}$, $\mathbf{u}(s\mathbf{x})$, $\sigma(s\mathbf{x})$ are computed for all $s \in G$ using the discrete representation formula (39) .

IV - A numerical example of BEM analysis using symmetry .

The present approach has been applied to our BEM code ASTRID (which solves 3D elastodynamic and elastostatic problems) . We studied the following example in order to illustrate the computational efficiency of the use of geometrical symmetry .
A spherical cavity of radius unity is embedded into an infinite elastic space (figure 4) . The following traction is applied on the entire surface S of the cavity :

$$\mathbf{t}^d = (A + B\sin\theta - C\sin\theta\, e^{(i\phi)})\, e^{-i\omega t}\, \mathbf{e}_r \tag{41}$$

A, B, C being complex constants and θ, ϕ the spherical angles . This steady-state dynamic loading presents no particular symmetry . The Lamé constants and the pulsation ω are chosen such that the longitudinal and transverse wavenumbers k_L and k_T are equal to : $k_L = 0.5345$, $k_T = 1.$.
The spherical surface of the cavity is meshed using eight-noded conformal curved quadrilateral boundary elements . Four meshes M_3, M_{12}, M_{24}, M_{96} are considered :
• One-eighth of the sphere, with 3 elements and 16 nodes (M_3, coarse mesh using symmetry) .
In this case, the eight subproblems have respectively 33, 30, 30, 30, 26, 26, 26 and 21 scalar unknowns (observe that each subproblem would have 48 unknowns, had the boundary constraints (14d) not been taken into account) .
• The entire sphere, totalizing 24 elements, 74 nodes and 222 unknowns (M_{24}, coarse mesh without use of symmetry) .
• One-eighth of the sphere, with 12 elements and 49 nodes (M_{12}, fine mesh using

symmetry) .
In this case, the eight subproblems have respectively 120, 113, 113, 113, 105, 105, 105 and 96 scalar unknowns (observe that each subproblem would have 147 unknowns, had the boundary constraints (14d) not been taken into account) .
• The entire sphere, totalizing 96 elements, 290 nodes and 870 unknowns (M_{96}, fine mesh without use of symmetry) .

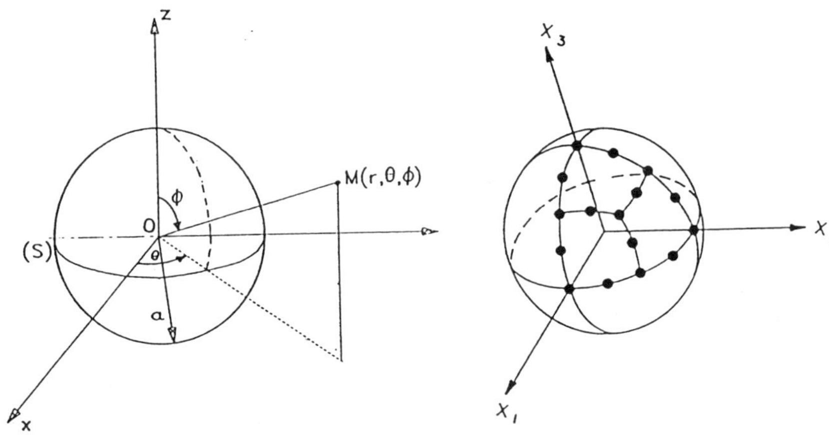

Figure 4 *(a) The pulsing spherical cavity, with spherical coordinate system ; (b) The mesh M_3 .*

The loading (41) is indeed a linear combination of spherical modes (0,0), (1,0) and (1,1) . Hence the elastic solution corresponding to the loading (41) is known analytically as a particular case of the general solution of elastic problems for spherical geometries obtained using spherical coordinates and separation of variables [5] . Thus we computed, for the present numerical example, the relative quadratic average error e for the surface displacement :

$$e = \left[\int_S || \mathbf{u}(\mathbf{y}) - \mathbf{u}_{exact}(\mathbf{y}) ||^2 \, dS_{\mathbf{y}} \right]^{1/2} \left[\int_S || \mathbf{u}_{exact}(\mathbf{y}) ||^2 \, dS_{\mathbf{y}} \right]^{-1/2} \quad (42)$$

which, using the meshes M_3, M_{24} or M_{12}, M_{96}, takes the value :

$$e = 6.5 \; 10^{-3} \; (M_3, M_{24}), \; 5.2 \; 10^{-4} \; (M_{12}, M_{96})$$

But the main interest of the present numerical study is obviously the computational accuracy of the method, expressed in terms of CPU time spent and of cumulated size of the arrays manipulated by ASTRID . Most of the CPU time of an execution is spent for building the linear system and solving it . The cost of these two main steps grows differently with the number of unknowns, and we display accordingly in table 1 below the CPU time needed for each step and for the four meshes (in the case of meshes M_3 and M_{12}, the solution time indicated is the cumulative time for solving the eight subproblems) . We also indicate in table 1 the "postprocessing time", which addresses the CPU time used to compute the integral representation for $\mathbf{u}(\mathbf{x})$ and $\sigma(\mathbf{x})$ for one point $\mathbf{x}$ (here we took $\mathbf{x}=(2., 2., 2.)$) .

	Building CPU time (s)	Solution CPU time (s)	Postprocessing CPU time (s)	Cumulative CPU time (s)	Cumulative array size (kb)
M_3	70.7	4.86	6.01	81.4	234.
M_{24}	210.	171.	3.49	383.	919.
M_{12}	488.	184.	23.3	698.	2362.
M_{96}	1786.	9430.	13.6	11171.	12600.

Table 1 : *CPU times and cumulated array sizes for the four meshes M_3, M_{12}, M_{24} and M_{96} .*

Some comments about the results displayed in table 1 .

• The ratio between the building CPU times using symmetry and without symmetry is roughly 1/4, instead of the ideal value of 1/8 . This is due to two reasons : the fact that, apart from the elementary integrals, the building process using full symmetry performs a large number of additions, as is apparent in (34), and also the increased complexity of the assembly loop .

• The ratio between the solution CPU times using symmetry and without symmetry is approximately equal to 1/35 (M_3 vs. M_{24}) and 1/51 (M_{12} vs M_{24}) . The finer meshes approach the "ideal" ratio of 1/64 .

• The "postprocessing time" is higher for M_3 than for M_{24} (the same for M_{12} and M_{96}) . This is due to similar reasons as above, keeping in mind that, as we pointed it out at the end of §III.2, the values $\mathbf{u}(s_0\mathbf{x})$, $\sigma(\mathbf{x})$ are generated for *all* the 8 images of $\mathbf{x}$ by G (instead of one single value $\mathbf{u}(\mathbf{x})$ in the cases M_{24} and M_{96}) .

These numerical computations have been made on a HP-Apollo DN400 workstation .

V - Conclusion .

In the present work, we extend to boundary element methods the algebraic approach of the use of geometrical symmetry for boundary-value problems which has been developed and applied to variational formulations in [1] . The present paper is restricted to abelian symmetry groups, which is by no means an essential limitation . In this context, we show how to formulate, in terms of boundary integral equations, the decomposition of a given elasticity problem stated on a symmetrical domain into a sum of subproblems . More precisely we show how to state these subproblems on a *symmetry cell*, i.e. a part of the original boundary which is smallest and on which the action of the symmetry group reproduces the entire original boundary . This is indeed the key point, as it allows an optimal use of the geometrical symmetry . The formalism encompasses all finite abelian symmetry groups in a unified manner, and no symmetry assumption is made on the boundary data . The computational efficiency of the present approach is demonstrated on a particular problem, on which both the solution error and the CPU times with or without use of symmetry are considered .

VI - References .

[1] BONNET M. - Méthode des équations intégrales régularisées en élastodynamique tridimensionnelle . *Thèse de doctorat, publiée dans le Bulletin EDF/DER série C, n° 1/2, 1987 .*

[2] BONNET M. - Regularized boundary integral equations for three-dimensional bounded or unbounded elastic bodies containing curved cracks of arbitrary shape under dynamic loading . *In "Boundary element techniques : applications in engineering (ed. C.A. Brebbia & N.G. Zamani), Computational Mechanics Publications (Southampton), 1989 .*

[3] BOSSAVIT A. - Symmetry, groups and boundary value problems : a progressive introduction to noncommutative harmonic analysis of partial differential equations in domains with geometrical symmetry . *Comp. Meth. in Appl. Mech. & Eng., vol. 56 pp 167-215, 1986 .*

[4] BREBBIA C.A., TELLES J.C.F., WROBEL L.C. - Boundary element techniques . Theory and application in engineering . *Springer - Verlag, 1984 .*

[5] ERINGEN A.C., SUHUBI E.S. - Elastodynamics (vol II - linear theory) . *Academic Press 1975 .*

[6] GUY J. & MANGEOT B. - Use of group theory in various integral equations . *SIAM J. Appl. Math. vol. 40, n° 3, pp 390-399, 1981 .*

[7] RIZZO F.J., SHIPPY D.J., REZAYAT M. - A boundary integral equation method for radiation and scattering of elastic waves in three-dimensions . *Int. J. Num. Meth. in Eng., vol. 21, 1985, p.115-129.*

Appendix - Irreducible representations of plane symmetry groups .

Let s_1, s_2 and s_3 denote the plane symmetries with respect to the coordinate planes $(x_1=0)$, $(x_2=0)$, $(x_3=0)$. The most commonly considered symmetry groups are those associatd to one, two or three orthogonal planes (which can then be chosen as the coordinate planes), denoted respectively D_1, D_2, D_3 and of respective order $n=2, 4, 8$. They admit the following tables of irreducible representations :

A	1	s_1
ρ_0	1	1
ρ_1	1	-1

Table 2A : $D_1=\{1,s_1\}$

B	1	s_1	s_2	q
ρ_0	1	1	1	1
ρ_1	1	-1	1	-1
ρ_2	1	1	-1	-1
ρ_3	1	-1	-1	1

Table 2B : $D_2=\{1,s_1,s_2,q=s_1s_2\}$

C	1	s_1	s_2	s_3	p_1	p_2	p_3	q
ρ_0	1	1	1	1	1	1	1	1
ρ_1	1	1	1	-1	-1	-1	1	-1
ρ_2	1	-1	1	1	1	-1	-1	-1
ρ_3	1	-1	1	-1	-1	1	-1	1
ρ_4	1	1	-1	1	-1	1	-1	-1
ρ_5	1	1	-1	-1	1	-1	-1	1
ρ_6	1	-1	-1	1	-1	-1	1	1
ρ_7	1	-1	-1	-1	1	1	1	-1

Table 2C : $D_3=\{1,s_1,s_2,s_3,p_1=s_2s_3,p_2=s_1s_3,p_3=s_1s_2,q=s_1s_2s_3\}$

The Multiple Reciprocity Method Applied to Thermoelastic Problems with Concentrated and Distributed Heat Sources

A.C. Neves, C.A. Brebbia

Wessex Institute of Technology, Computational Mechanics Institute, Ashurst Lodge, Ashurst, Southampton, U.K.

Abstract

In this paper, the multiple reciprocity method (MRM) is applied to transform domain integrals due to thermal loadings into equivalent boundary integrals for the case of point sources (i.e.'hot spots'), line and volume sources. The resulting formulations are restricted to two-dimensional elastic isotropic materials. Several numerical examples are presented to demonstrate the validity and accuracy of the method.

1 Introduction

Thermal stresses resulting from large temperature differences are frequently an important factor in determining the lifetime of mechanical components. To obtain accurate predictions, it is important to apply numerical methods such as the boundary element method (BEM). Since 1975 [2]-[6], the thermal stresses problem has been analysed numerically using BEM in conjunction with the Galerkin vector approach which transforms the domain integrals resulting from the body forces to the boundary.

In this paper, instead of using the classical Galerkin vector technique, the Multiple Reciprocity Method has been applied. The advantage of the latter is that it can model any thermal steady-state including those with an arbitrary internal source distribution. The MRM was originally developed by Nowak and Brebbia to solve Poisson [7] and Helmholtz equations [8], and has recently been extended by the present authors [9]-[10] to solve elastic problems governed by Navier's equations. The approach consists of a repeated application of the reciprocity theorem using a sequence of higher order fundamental solutions. It is essential that the resulting series of boundary integrals converges to the correct solution and this convergence has been studied here for a particular case.

Several examples of a circular disc subjected to different heat source distributions are presented and results are compared with analytical solutions [1],[6]. The numerical results demonstrate the accuracy of the method.

2 Boundary integral formulation

In the case of thermal loading acting on a body Ω with boundary Γ, the boundary integral equations for the displacements at any points and stresses at internal points are given respectively by [2]-[6]:

$$\begin{aligned} C_{ij}(\xi)u_j(\xi) &= \int_\Gamma u^*_{ij}(\xi,x)p_j(x)d\Gamma(x) - \int_\Gamma p^*_{ij}(\xi,x)u_j(x)d\Gamma(x) \\ &+2\mu(\frac{1+\nu}{1-2\nu})\alpha\int_\Omega u^*_{ij,j}(\xi,x)T(x)d\Omega(x) \end{aligned} \quad (1)$$

$$\begin{aligned} \sigma_{ij}(\xi) &= \int_\Gamma u^*_{ijk}(\xi,x)p_k(x)d\Gamma(x) - \int_\Gamma p^*_{ijk}(\xi,x)u_k(x)d\Gamma(x) \\ &-2\mu(\frac{1+\nu}{1-2\nu})\alpha[\int_\Omega \sigma^*_{ij}(\xi,x)T(x)d\Omega(x) + T(x)\delta_{ij}(\xi,x)] \end{aligned} \quad (2)$$

where
$C_{ij}(\xi)$ is the free coefficient;
$u_j(\xi)$ is the displacement at any point ξ in the j direction;
$\sigma_{ij}(\xi)$ is the stress tensor at an internal point ξ;
$u_j(x)$ is the displacement at a point x in the j direction;
$p_j(x)$ is the traction at x in the j direction;
$T(x)$ is the temperature distribution at x;
α is the coefficient of linear thermal expansion;
$\delta_{ij}(\xi,x)$ is the Kronecker delta;
ν is the Poisson's ratio;
μ is the shear modulus.

The functions $u^*_{ij}, p^*_{ij}, u^*_{ijk}$ and p^*_{ijk} appearing in the above boundary integrals are given by:

$$u^*_{ij} = \frac{1}{8\pi(1-\nu)\mu}\{(3-4\nu)ln(1/r)\delta_{ij} + r_{,i}r_{,j} - \frac{1}{2}\delta_{ij}\} \quad (3)$$

$$p^*_{ij} = \frac{-1}{4\pi(1-\nu)r}\{[(1-2\nu)\delta_{ij} + 2r_{,i}r_{,j}]\frac{\partial r}{\partial n} - (1-2\nu)(r_{,i}n_j - r_{,j}n_i)\} \quad (4)$$

$$u^*_{ijk} = \frac{1}{4\pi(1-\nu)r}\{(1-2\nu)(r_{,j}\delta_{ik} + r_{,i}\delta_{jk} - r_{,k}\delta_{ij}) + 2r_{,i}r_{,j}r_{,k}\} \quad (5)$$

$$\begin{aligned} p^*_{ijk} &= \frac{\mu}{2\pi(1-\nu)r^2}\{2\frac{\partial r}{\partial n}[(1-2\nu)\delta_{ij}r_{,k} + \nu(\delta_{ik}r_{,j} + \delta_{jk}r_{,i}) - 4r_{,i}r_{,j}r_{,k}] \\ &+2\nu(n_ir_{,j}r_{,k} + n_jr_{,i}r_{,k}) + (1-2\nu)(2n_kr_{,i}r_{,j} + n_j\delta_{ik} + n_i\delta_{jk}) - (1-4\nu)n_k\delta_{ij}\} \end{aligned} \quad (6)$$

where
r is the distance between the load point ξ and the field point x;
n_j are the direction cosines of the outward normal to the boundary.

The fundamental solutions in the domain integrals of equations (1) and (2) are given by the following expressions:

$$u^*_{ij,j} = \frac{-(1-2\nu)}{4\pi(1-\nu)\mu}\frac{r_{,i}}{r} \quad (7)$$

$$\begin{aligned} \sigma^*_{ij} &= -u^*_{ijk,k} - \frac{1}{2(1-\nu)}\delta_{ij}\Delta(\xi,x) = \\ &= \frac{(1-2\nu)}{\pi(1-\nu)}[\frac{1}{r^2}(r_{,i}r_{,j} - \frac{\delta_{ij}}{2}) - \frac{\pi}{2(1-2\nu)}\delta_{ij}\Delta(\xi,x)] \end{aligned} \quad (8)$$

where $\Delta(\xi,x)$ is the Dirac delta.

The boundary element method consists of the application of equation (1) at a series of points on the boundary using interpolation functions over each element for the displacements and tractions. This collocation process produces a system of algebraic equations once the necessary numerical integrations have been carried out. The system can be solved after taking into consideration the boundary conditions.

3 The Multiple Reciprocity Method

This method, proposed by Nowak and Brebbia [7], is a new and very powerful technique to transform domain integrals into boundary integrals.

The MRM consists in expressing the fundamental solution term in the domain integral by a set of new higher order functions while decreasing the order of the body force term. The technique is applied in what follows to transform those integrals in equations (1) and (2).

3.1 Domain Integral in Displacement Equations

The technique will be explained with reference to the domain integral of equation (1), *i.e.*:

$$D_i^{(0)} = \int_\Omega u_{ij,j}^{*(0)} T^{(0)} d\Omega \tag{9}$$

where the upper index (0) has been included to indicate that this is the original expression.

In this approach, the fundamental solution $u_{ij,j}^{*(0)}$ is transformed into a new higher order solution $u_{ij,j}^{*(1)}$ through the following recurrence formula:

$$\nabla^2 u_{ij,j}^{*(1)} = u_{ij,j}^{*(0)} \tag{10}$$

Hence, the domain integral term (9) can now be written as:

$$D_i^{(0)} = \int_\Omega \nabla^2 u_{ij,j}^{*(1)} T^{(0)} d\Omega \tag{11}$$

The next step consists of applying Green's reciprocity theorem, which gives:

$$D_i^{(0)} = \int_\Gamma \left(\frac{\partial u_{ij,j}^{*(1)}}{\partial n} T^{(0)} - u_{ij,j}^{*(1)} \frac{\partial T^{(0)}}{\partial n}\right) d\Gamma + \int_\Omega u_{ij,j}^{*(1)} \nabla^2 T^{(0)} d\Omega \tag{12}$$

Since the temperature distribution $T^{(0)}$ is known, it is now possible to define the Laplacian of the body force component as a new function $T^{(1)}$ through the use of another recurrence formula, *i.e.*:

$$T^{(1)} = \nabla^2 T^{(0)} \tag{13}$$

Application of equation (13) into (12) gives:

$$D_i^{(0)} = \int_\Gamma \left(\frac{\partial u_{ij,j}^{*(1)}}{\partial n} T^{(0)} - u_{ij,j}^{*(1)} \frac{\partial T^{(0)}}{\partial n}\right) d\Gamma + \int_\Omega u_{ij,j}^{*(1)} T^{(1)} d\Omega \tag{14}$$

Notice that the original domain integral has now been transformed into a boundary integral plus a new domain term similar to equation (9), *i.e.*:

$$D_i^{(1)} = \int_\Omega u_{ij,j}^{*(1)} T^{(1)} d\Omega \tag{15}$$

One can now proceed in the same way and reduce the integral $D_i^{(1)}$ to a boundary term and a domain term $D_i^{(2)}$. The operation can be repeated as many times as necessary. The recurrence law for the process in terms of a general index L can be defined in the following diagram:

$$D_i^{(L)} = \int_\Omega u_{ij,j}^{*(L)} T^{(L)} d\Omega$$

$$\nabla^2 u_{ij,j}^{*(L+1)} = u_{ij,j}^{*(L)}$$

$$D_i^{(L)} = \int_\Omega \nabla^2 u_{ij,j}^{*(L+1)} T^{(L)} d\Omega$$

2nd Green's Theorem

$$D_i^{(L)} = \int_\Gamma \left(\frac{\partial u_{ij,j}^{*(L+1)}}{\partial n} T^{(L)} - u_{ij,j}^{*(L+1)} \frac{\partial T^{(L)}}{\partial n}\right) d\Gamma + \int_\Omega u_{ij,j}^{*(L+1)} \nabla^2 T^{(L)} d\Omega$$

$$T^{(L+1)} = \nabla^2 T^{(L)}$$

$$D_i^{(L)} = \int_\Gamma \left(\frac{\partial u_{ij,j}^{*(L+1)}}{\partial n} T^{(L)} - u_{ij,j}^{*(L+1)} \frac{\partial T^{(L)}}{\partial n}\right) d\Gamma + \underbrace{\int_\Omega u_{ij,j}^{*(L+1)} T^{(L+1)} d\Omega}_{D_i^{(L+1)}}$$

In the limit, the domain integral $D_i^{(0)}$ is transformed into a series of equivalent boundary integrals, *i.e.*:

$$D_i^{(0)} = \sum_{L=0}^{\infty} \int_\Gamma \left(\frac{\partial u_{ij,j}^{*(L+1)}}{\partial n} T^{(L)} - u_{ij,j}^{*(L+1)} \frac{\partial T^{(L)}}{\partial n}\right) d\Gamma \tag{16}$$

The difficulty in applying the MRM for elasticity occurs in the calculation of the primitives in the recurrence formula (10). Since there are cross terms in the fundamental solution (7), the use of cylindrical coordinates in the Laplace's operator does not simplify the calculation of the primitives as was the case in Poisson's and Helmholtz's equations [7][8].

In spite of this dificulty, the higher order fundamental solutions in equation (16) and their normal derivatives were found by the present authors [10] and given by:

$$u_{ij,j}^{*(L+1)} = \frac{-(1-2\nu)}{4\pi(1-\nu)\mu} A^{(L+1)} r^{2L+1} r_{,i}(ln(r) - B^{(L+1)}) \tag{17}$$

$$\frac{\partial u_{ij,j}^{*(L+1)}}{\partial n} = \frac{-(1-2\nu)}{4\pi(1-\nu)\mu} A^{(L+1)} r^{2L}[(n_i + 2Lr_{,i}r_{,j}n_j)(ln(r) - B^{(L+1)}) + r_{,i}r_{,j}n_j] \tag{18}$$

where

$$A^{(L+1)} = \frac{A^{(L)}}{(2L+1)^2 - 1} \tag{19}$$

$$B^{(L+1)} = B^{(L)} + \frac{2(2L+1)}{(2L+1)^2 - 1} \tag{20}$$

For $L = 0$, $A^{(1)} = 1/2$ and $B^{(1)} = 0$.

3.2 Domain Integral in Stress Equations

Consider $S_{ij}^{(0)}$ the domain integral of equation (2),*i.e.*:

$$S_{ij}^{(0)} = \int_{\Omega} \sigma_{ij}^{(0)} T^{(0)} d\Omega(x) \tag{21}$$

The same procedure described in the preceding section can be applied to the above equation using the following recurrence formulas:

$$\begin{aligned} \nabla^2 \sigma_{ij}^{*(L+1)} &= \sigma_{ij}^{*(L)} \\ T^{*(L+1)} &= \nabla^2 T^{*(L)} \end{aligned} \tag{22}$$

After applying the MRM, $S_{ij}^{(0)}$ (equation (21)) will be represented by a series of boundary integrals, *i.e.*:

$$S_{ij}^{(0)} = \sum_{L=0}^{\infty} \int_{\Gamma} (\frac{\partial \sigma_{ij}^{*(L+1)}}{\partial n} T^{(L)} - \sigma_{ij}^{*(L+1)} \frac{\partial T^{(L)}}{\partial n}) d\Gamma \tag{23}$$

The higher order fundamental solution components in the above equation are given by:

$$\begin{aligned} \sigma_{ij}^{*(L+1)} = & -\frac{(1-2\nu)}{4\pi(1-\nu)} C^{(L+1)} [\frac{1}{1-2\nu}(ln(r) - D^{(L+1)})\delta_{ij} + (r_{,i}r_{,j} - \frac{\delta_{ij}}{2}) \\ & (E^{(L+1)} ln(r) - F^{(L+1)})] r^{2L} \end{aligned} \tag{24}$$

$$\begin{aligned} \frac{\partial \sigma_{ij}^{*(L+1)}}{\partial n} = & -\frac{(1-2\nu)}{4\pi(1-\nu)} C^{(L+1)} \{ \frac{1}{1-2\nu} [2L(ln(r) - D^{(L+1)}) + 1] r_{,k} n_k \delta_{ij} \\ & + [(r_{,i}r_{,j} - \frac{\delta_{ij}}{2})(2Lln(r) + 1) r_{,k}n_k + (n_i r_{,j} + n_j r_{,i} - 2r_{,i}r_{,j}r_{,k}n_k) ln(r)] E^{(L+1)} \\ & - [2Lr_{,k}n_k(r_{,i}r_{,j} - \frac{\delta_{ij}}{2}) + (n_i r_{,j} + n_j r_{,i} - 2r_{,i}r_{,j}r_{,k}n_k)] F^{(L+1)} \} r^{2L-1} \end{aligned} \tag{25}$$

where

$$C^{(L+1)} = \frac{C^{(L)}}{4L^2} \tag{26}$$

$$D^{(L+1)} = D^{(L)} + \frac{1}{L} \tag{27}$$

$$E^{(L+1)} = \frac{2L}{L+1} \tag{28}$$

$$F^{(L+1)} = (LF^{(L)} + E^{(L+1)})\frac{L}{L^2-1} \tag{29}$$

Notice that in the case of $L = 0$; $C^{(1)} = 1$ and $D^{(1)} = 0$ $E^{(1)} = 0$ and $F^{(1)} = -1$ and for $L = 1$; $F^{(2)} = 0$

It is interesting to point out that, mathematically, $\sigma_{ij}^{*(1)}$ has the same form as the fundamental solution for displacement $u_{ij}^{*(0)}$ (equation (3)).

3.3 Analysis of convergence

References [7] and [9] state that the convergence of the series (16) and (23) is guaranteed provided that the problem has been scaled, *i.e.*, all dimensions are divided by the maximum geometrical dimension of the problem. Those authors, however, only studied the convergence of the fundamental solution terms in the series. They did not take into consideration the possible effect of terms such as $T^{(L)}$ and $\frac{\partial T^L}{\partial n}$. If these terms diverge, a careful convergence study has to be undertaken. In the last example in what follows, the temperature distribution will generate infinite series which, under certain conditions, will converge. If the series are finite, the exact solution will be obtained.

4 Numerical Examples

For the case of steady-state without internal sources $\nabla^2 T^{(0)} = 0$ the domain integral on the right hand side of equation (12) disappears. If the medium has internal sources instead, one can write:

$$\nabla^2 T^{(0)} + \frac{\rho}{\kappa} = 0 \tag{30}$$

where

ρ is the source density;

κ is the thermal conductivity.

Substituting (30) into equation (12), one has:

$$D_i^{(0)} = \int_\Gamma (\frac{\partial u_{ij,j}^{*(1)}}{\partial n} T^{(0)} - u_{ij,j}^{*(1)} \frac{\partial T^{(0)}}{\partial n}) d\Gamma - \int_\Omega u_{ij,j}^{*(1)} \frac{\rho}{\kappa} d\Omega \tag{31}$$

In order to illustrate the accuracy of the MRM, a circular disc submitted to different heat source distributions is studied. In the following examples, the error norm is defined as:

$$Er = \frac{\sqrt{\sum_{i=1}^{p}(V_n - V_a)_i^2}}{\sqrt{\sum_{i=1}^{p}(V_a)_i^2}} \tag{32}$$

where

p is the number of points considered;

V_n is the numerical value;

V_a is the analytical value.

4.1 Case of Point Source (Hot-Spot)

For the case of hot-spot, the source density ρ can be represented by:
$\rho(\xi) = \rho_0$ at a given internal point, $\xi = \xi_p$
$\rho(\xi) = 0$ everywhere else $\xi \neq \xi_p$
Considering the above source density, equation (31) can be written as:

$$D_i^{(0)} = \int_\Gamma \left(\frac{\partial u_{ij,j}^{*(1)}}{\partial n} T^{(0)} - u_{ij,j}^{*(1)} \frac{\partial T^{(0)}}{\partial n}\right) d\Gamma - \frac{1}{\kappa} \sum_{p=1}^{P} \rho(\xi_p) u_{ij,j}^{*(1)}(\xi, \xi_p) \qquad (33)$$

where P is the number of point sources.

The same procedure can be applied to the domain integral of the stress equations.

As an example of point source, consider a thin circular disc of radius R_0 under a central hot-spot of intensity ρ as shown in figure 1 [6].

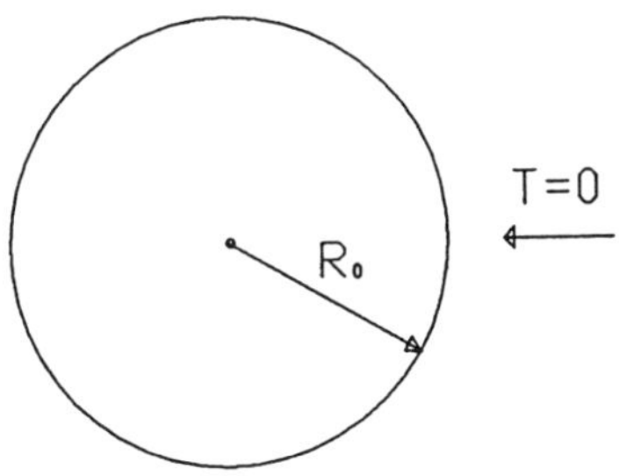

$$\begin{aligned} \nu &= 0.33 \\ E &= 2.1 * 10^5 N/cm^2 \\ \alpha &= 1.1 * 10^{-5} \\ \kappa &= 60 watts/(cm^0 K) \\ \rho &= 100 watts/cm^3 \\ R_0 &= 80cm \end{aligned}$$

Figure 1: Geometry of thin circular disc under a central hot-spot.

The analytical solution for this problem is given by:

$$\begin{aligned} u_R &= \frac{\rho\alpha}{4\pi\kappa}[1 - (1+\nu)\ln(\frac{R}{R_0})]R \\ \sigma_R &= \frac{\rho\alpha E}{4\pi\kappa}\ln(\frac{R}{R_0}) \\ \sigma_\theta &= \frac{\rho\alpha E}{4\pi\kappa}(1 + \ln(\frac{R}{R_0})) \end{aligned}$$

where u_R, σ_R and σ_θ are the radial displacement, radial stress and circumferential stress, respectively.

Figures 2 and 3 compare the analytical solution with the boundary element results obtained using a mesh of 8 and 16 linear elements and double precision.

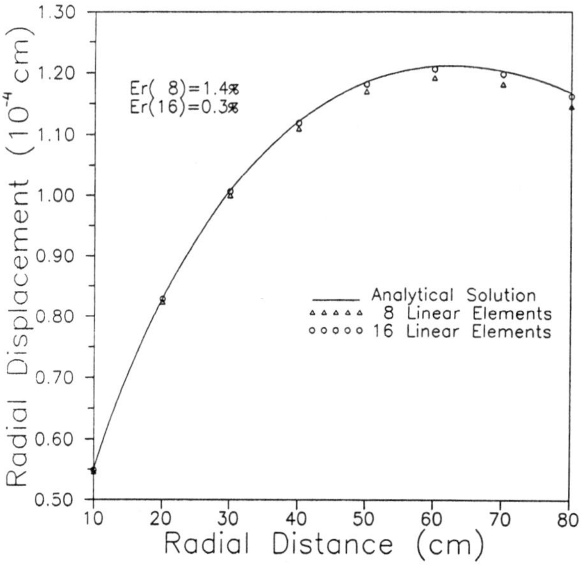

Figure 2: Radial displacements for the circular disc with central hot-spot.

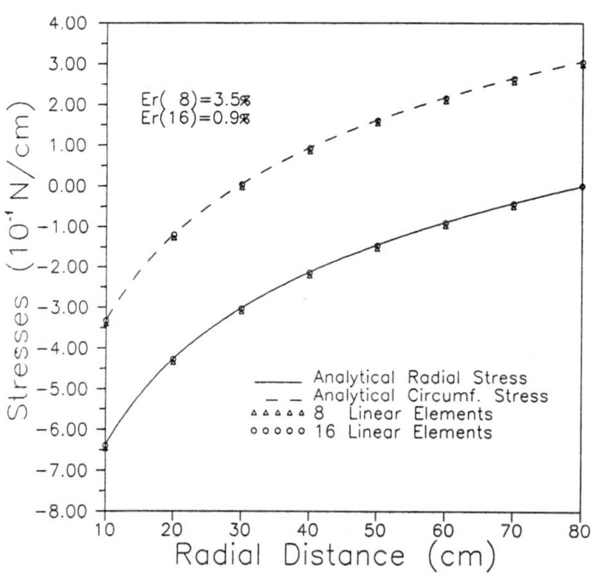

Figure 3: Radial and circumferential Stresses for the circular disc with central hot-spot.

4.2 Case of Line Source

The source density ρ for line sources can be represented by:

$\rho(\xi) = \rho_0$ for $\xi \in \xi_s$
$\rho(\xi) = 0$ for $\xi \ni \xi_s$ where ξ_s is the curve describing the line source.

For the case of a uniform line source ($\rho =$ const.) along ξ_s, equation (31) reduces to:

$$D_i^{(0)} = \int_\Gamma \left(\frac{\partial u_{ij,j}^{*(1)}}{\partial n} T^{(0)} - u_{ij,j}^{*(1)} \frac{\partial T^{(0)}}{\partial n}\right) d\Gamma - \frac{\rho}{\kappa} \int_s u_{ij,j}^{*(1)} ds \tag{34}$$

Consider a circular disc of the same geometry and material properties as the one in the previous example, but subject to a circular line thermal source of radius R_s as shown in figure 4.

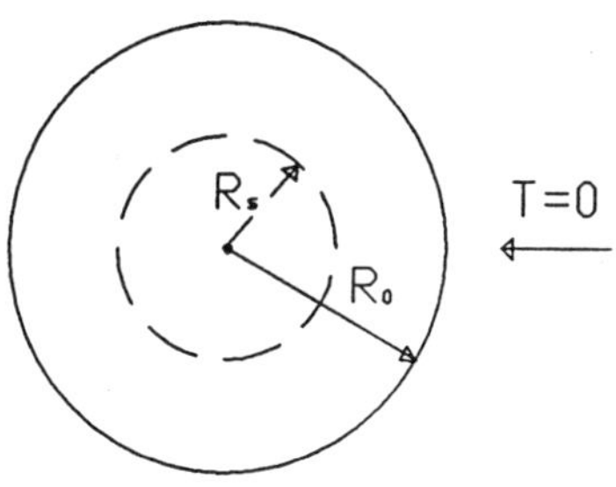

$$\begin{aligned} \nu &= 0.33 \\ E &= 2.1 * 10^5 N/cm^2 \\ \alpha &= 1.1 * 10^{-5} \\ \kappa &= 60 watts/(cm^0 K) \\ \rho &= 100 watts/cm^4 \\ R_0 &= 80cm \\ R_s &= 40cm \end{aligned}$$

Figure 4: Geometry of thin circular disc under a circular line source.

The analytical solution for this example in terms of displacements and stress components is given by:

$$R > R_s \begin{cases} u_R = -\frac{R_0 \rho \alpha}{8\kappa}\{2(1+\nu)\ln(\frac{R_s}{R_0}) - (1-\nu)(1-\frac{R_s^2}{R_0^2})\}R \\ \sigma_R = \sigma_\theta = \frac{E R_0 \rho \alpha}{8\kappa}\{1 - \frac{R_s^2}{R_0^2} + 2\ln(\frac{R_s}{R_0})\} \end{cases}$$

$$R \le R_s \begin{cases} u_R = -\frac{R_0 \rho \alpha}{8\kappa}\{(1+\nu)[2\ln(\frac{R}{R_0}) - (1-\frac{R_s^2}{R^2})] - (1-\nu)(1-\frac{R_s^2}{R_0^2})\}R \\ \sigma_R = \frac{E R_0 \rho \alpha}{8\kappa}\{2\ln(\frac{R}{R_0}) + R_s^2(\frac{1}{R^2} - \frac{1}{R_0^2})\} \\ \sigma_\theta = \frac{E R_0 \rho \alpha}{8\kappa}\{2(1+\ln(\frac{R}{R_0}) - R_0^2(\frac{1}{R^2} + \frac{1}{R_0^2})\} \end{cases}$$

The numerical results are plotted in figures (5) and (6). Notice that the boundary element solutions are in excellent agreement with the analytical results.

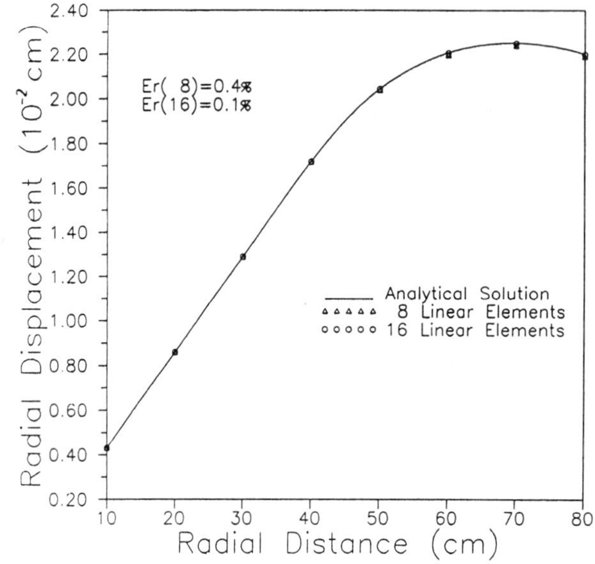

Figure 5: Radial displacements for the circular disc with circular line source.

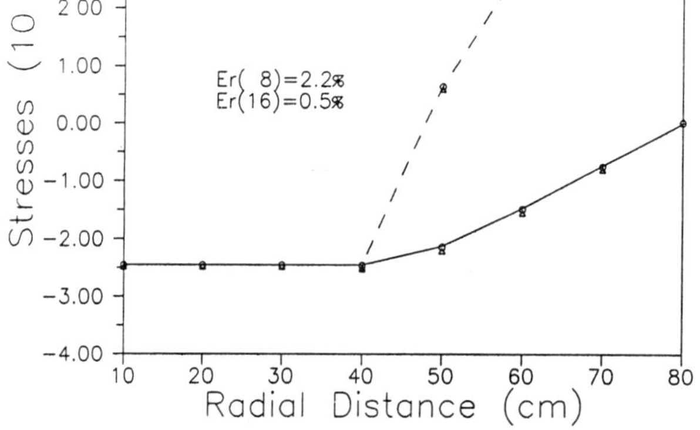

Figure 6: Radial and circumferential stresses for the circular disc with circular line source.

4.3 Case of distributed heat source function of R

Consider now the case of a disc with the temperature linearly varying with the distance from the centre, *i.e.*:

$$T^{(0)} = R \tag{35}$$

This problem is equivalent to a steady-state case with distributed heat source equal to:

$$T^{(1)} = \nabla^2 T^{(0)} = \frac{1}{R} \tag{36}$$

The above temperature distribution will generate an infinite series according to equations (16) and (23). In these series, the primitives of temperature and their normal derivatives are given respectively by:

$$T^{(L)} = \frac{K^{(L)}}{R^{2L-1}} \tag{37}$$

$$\frac{\partial T^{(L)}}{\partial n} = \frac{-(2L-1)K^{(L)}}{R^{2L}} \tag{38}$$

where

$$K^{(L)} = (2L-3)^2 K^{(L-1)} \tag{39}$$

and

$$K^{(0)} = 1 \tag{40}$$

Since the series are infinite, an analysis of convergence has to be undertaken. For this particular case, convergence is guaranteed only if $\frac{r}{R} < 1$, where R is the distance from the field point over the boundary element to the center of the disc, and r is the distance from the source point to the same field point. In practice, it has been found that convergence can also be achieved for values of r/R slightly higher than 1.

The discretization of the previous examples does not satisfy such condition at certain points; thus a new discretization is used according to figure 7.

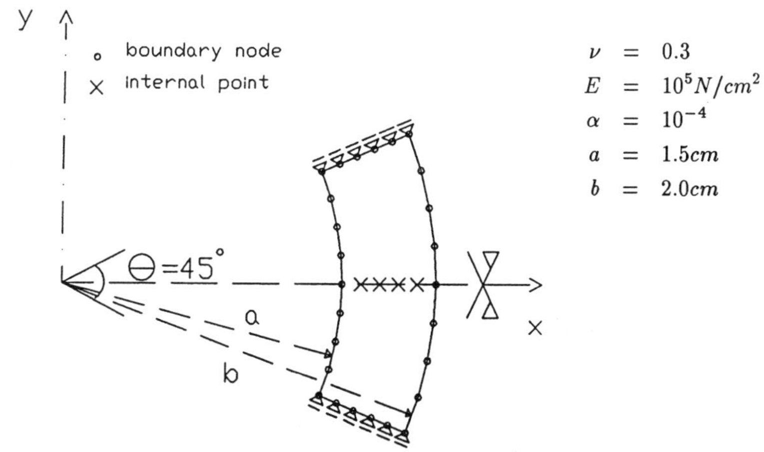

Figure 7: Discretization of the disc for example 3.

Notice that the boundary conditions have to be prescribed in a local coordinate system. Another way of satisfying the convergence request is to divide the domain into sub-regions in each of which the condition $\frac{r}{R} < 1$ has to be respected.

The analytical solutions for the displacements and stresses are given by [1]:

$$u_R = \frac{\alpha}{R}[(1+\nu)\int_a^R TRdR + \frac{(1-\nu)R^2+(1+\nu)a^2}{b^2-a^2}\int_a^b TRdR] \quad (41)$$

$$\sigma_\theta = \frac{\alpha E}{R^2}[\int_a^R TRdR + \frac{R^2+a^2}{b^2-a^2}\int_a^b TRdR - TR^2] \quad (42)$$

In order to show the convergence of the series, the problem was solved for different values of L. The results for the displacements and transverse stresses are plotted in figures 8 and 9, respectively.

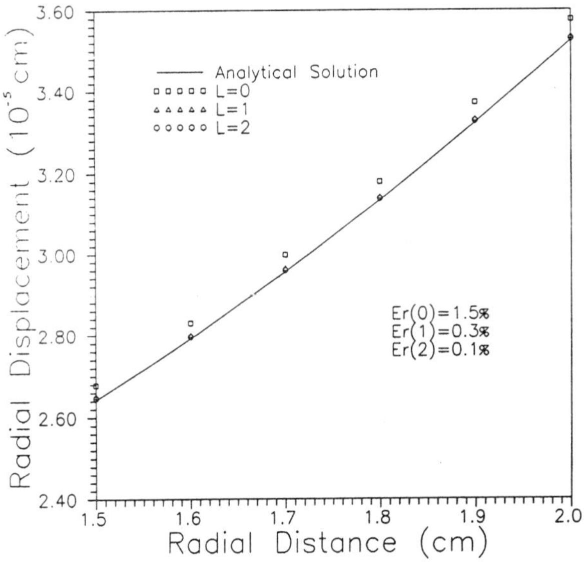

Figure 8: Radial displacements for example 3.

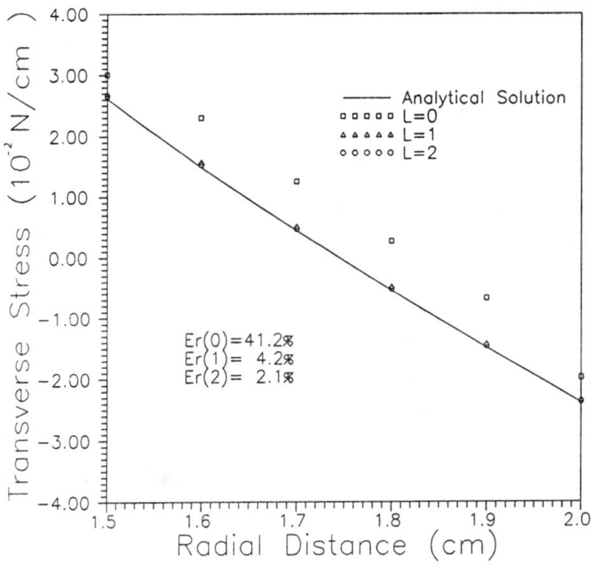

Figure 9: Angular stresses for example 3.

It can be seen that in this case the series converge to the analytical solution very rapidly and $L = 2$ produces accurate results.

By increasing the angle θ of the section analysed (figure 7), the error will reduce but more terms will be needed in the series and as θ continues to increase, the series start to diverge.

5 Conclusions

The paper demonstrates how the multiple reciprocity method (MRM) can be applied to solve any steady-state thermal problem including those with arbitrary internal source distribution.

The general formulation of the MRM has been presented and several examples have been studied to demonstrate the convergence and accuracy of the method. The approach gives the exact solution for the cases in which the internal effects can be represented in terms of a finite series. For the case of infinite series, the results may diverge and it is essential to determine the condition for convergence as it has been done for a particular example in the paper.

The numerical experiments validate the theory presented by the authors and demonstrate how the MRM can be applied in practice.

6 Acknowledgment

The first author would like to thank L. C. Wrobel, S. M. Niku and A. Nowak for many fruitful discussions, and CNPq (Brazilian Scientific Research Concil) for the financial support.

References

[1] Boley, B.A.; Weiner, J.H., "Theory of Thermal Stresses", John Wiley & Sons, New York, London (1960)

[2] Cruse, T.A., "Boundary Integral Equation Method for Three Dimensional Elastic Fracture Mechanics Analysis", AFOSR-TR-75 0813 Report (1975).

[3] Danson, D.J., "A Boundary Element Formulation of Problems in Linear Isotropic Elasticity with Body Forces", Boundary Element Methods, Ed. C.A. Brebbia, Springer Verlag (1981).

[4] Brebbia, C.A.; Telles, J.C.F.; Wrobel, L.C., "Boundary Element Techniques: Theory and Applications in Engineering", Springer Verlag, (1984).

[5] Telles, J.C.F.,"Two Dimensional Elastostatic Analysis Using Boundary Elements", Computational Mechanics Publication (1986).

[6] Niku, S.M.; Brebbia, C.A.," The Treatment of Thermoelastic Problems with Concentrated and Distributed Heat Sources", 10th International Conference on B.E.M., Southampton University, Computational Mechanics Publication and Springer-Verlag, Vol 3, pp 591-600 (1988).

[7] Nowak, A.J.; Brebbia, C.A.,"The Multiple-Reciprocity Method. A New Approach for Transforming B.E.M. Domain Integrals to the Boundary", Enginnering Analysis with Boundary Elements, No.3, Vol.6, pp 164-168, (1989).

[8] Nowak, A.J.; Brebbia, C.A.,"Solving Helmholtz Equation by Multiple Reciprocity Method", Computer and Experiments in Fluid Flow, Ed. Carlomagno; G.M. and Brebbia; C.A., Computational Mechanics Publication, Southampton, pp 265-270 (1989).

[9] Neves, A.C.; Brebbia, C.A.,"The Multiple Reciprocity Boundary Element Method in Elasticity: A New Approach for Transforming Domain Integrals to the Boundary", International Journal for Numerical Methods in Engineering, No.31, Vol.4, pp 709-727 (1991).

[10] Neves, A.C.; Brebbia, C.A.,"The Multiple Reciprocity Method Applied to Thermal Stress Problems", Submitted for publication in International Journal for Numerical Methods in Engineering.

Solving Plate Bending Problems using Boundary Elements Method

L.M. Castro, J. Pinho, P. Parreira

Department of Civil Engineering, Instituto Superior Técnico, Av. Rovisco Pais, 1096 Lisboa, Portugal

SUMMARY

The direct boundary element method is used to solve Kirchhoff plate bending problems. The algorithm uses linear elements to model the geometry. Hermite polynomials to interpolate the displacements. and linear interpolation of slopes. bending moments and shear forces. All integrations are performed numerically. Numerical examples are used to test the implementation.

INTRODUCTION

Finite element discretizations are usually used in the solution of Kirchhoff plate bending problems[1,2]. More recently. boundary element formulations have been developed. Hartmann and his co-workers[3,4] used the direct boundary element method based on the Rayleigh-Green identity to perform the static analysis of Kirchhoff plates. Other authors[5,6,7,8,9] have been working in the same subject.

The aim of this paper is to present some results in solving Kirchhoff plates using boundary element techniques. Linear elements are used to model the geometry. Hermite polynomials are adopted in the interpolation of the displacements on the boundary. Slopes. bending moments and shear forces are linearly interpolated.

All integrations are performed numerically. using Telles[10] transformation for weak singularities. and the work developed by Guiggiani and his co-workers[11,12,13] to solve Cauchy principal value integrals and hypersingular integrals.

KIRCHHOFF PLATES THEORY REVIEW

Let w denote the deflection in a thin plate. If the curvatures χ and the bending moments $\mathbf{M}$ are collected in arrays,

$$\chi = \begin{bmatrix} \chi_{11} \\ \chi_{22} \\ 2\,\chi_{12} \end{bmatrix}, \quad \text{and} \quad \mathbf{M} = \begin{bmatrix} M_{11} \\ M_{22} \\ M_{12} \end{bmatrix},$$

the fundamental relationships can be summarized as follows:

$$\chi = \mathbf{L}\, w; \tag{1}$$

$$\mathbf{L}^T\, \mathbf{M} + p = 0; \tag{2}$$

$$\mathbf{M} = \mathbf{D}_f\, \chi. \tag{3}$$

where,

$$\mathbf{L} = \begin{bmatrix} -\partial^2/\partial\, x_1^2 \\ -\partial^2/\partial\, x_2^2 \\ -2\partial^2/\partial\, x_1\, \partial\, x_2 \end{bmatrix}$$

and:

$$\mathbf{D}_f = \frac{E\, t^3}{12\,(1-\nu^2)} \begin{bmatrix} 1 & \nu & 0 \\ \nu & 1 & 0 \\ 0 & 0 & (1-\nu)/2 \end{bmatrix},$$

In the above definitions, p is the transversal distributed load, t the thickness of the plate, E the Young modulus and ν the Poisson's coefficient.

Equations (1),(2) and (3) are equivalent to the following fourth-order differential equation on the deflection w.

$$\mathbf{L}^T\, \mathbf{D}_f\, \mathbf{L}\, w + p = 0. \tag{4}$$

The static variables in the local coordinates of the boundary element are defined thus in terms of the bending moments.

$$Q_\alpha = \frac{\partial\, M_{i\alpha}}{\partial\, x_i}; \tag{5}$$

$$M_n = M_{\alpha\beta}\, n_\alpha\, n_\beta; \tag{6}$$

$$M_{nt} = M_{\alpha\beta}\, n_\alpha\, t_\beta, \tag{7}$$

The kinematic variables on the boundary are the deflection (w) and the slope ($\partial\, w/\partial\, n$), which are related through the following condition:

$$\frac{\partial\, w}{\partial\, n} = \frac{\partial\, w}{\partial\, x_i}\, n_i. \tag{8}$$

On the same boundary the static variables involved are the bending moment (M_n) and the Kirchhoff shear force (V_n), defined by

$$V_n = Q_n + \frac{\partial M_{nt}}{\partial s}. \tag{9}$$

BOUNDARY ELEMENT DISCRETIZATION

The boundary element method is based on Betti's principle. In a plate bending problem, two of four Betti data are prescribed and the other two are unknown. Then, for a boundary element formulation to be applied, two integral equations are required, for instance the integral equation of the deflection and the integral equation of the normal slope.

Following Hartmann[3], these equations can be written in the form:

$$\mathbf{c}(\mathbf{x})\, w(\mathbf{x}) = \int_\Gamma \Big[g_0(\mathbf{y},\mathbf{x})\, V_\nu(w)(\mathbf{y}) - \frac{\partial}{\partial \nu} g_0(\mathbf{y},\mathbf{x})\, M_\nu\,(w)(\mathbf{y}) -$$
$$- V_\nu(g_0(\mathbf{y},\mathbf{x}))\, w(\mathbf{y}) + M_\nu(g_0(\mathbf{y},\mathbf{x}))\, \frac{\partial w}{\partial \nu}(\mathbf{y}) \Big]\, ds_{\mathbf{y}} +$$
$$+ \int_\Omega g_0(\mathbf{y},\mathbf{x})\, p(\mathbf{y})\, d\Omega_{\mathbf{y}} + \sum_c [\, g_0(\mathbf{y}^c,\mathbf{x})\, F(w)(\mathbf{y}^c) - w(\mathbf{y}^c)\, F(g_0)(\mathbf{y}^c,\mathbf{x}) \,], \tag{10}$$

and

$$\mathbf{c}_1(\mathbf{x}) w_{,1}(\mathbf{x}) + \mathbf{c}_2(\mathbf{x}) w_{,2}(\mathbf{x}) = \int_\Gamma \Big[g_1(\mathbf{y},\mathbf{x}) V_\nu(w)(\mathbf{y}) - \frac{\partial}{\partial \nu} g_1(\mathbf{y},\mathbf{x}) M_\nu(w)(\mathbf{y}) -$$
$$- V_\nu(g_1(\mathbf{y},\mathbf{x}))\, [w(\mathbf{y}) - w(\mathbf{x})] + M_\nu(g_1(\mathbf{y},\mathbf{x})) \frac{\partial w}{\partial \nu}(\mathbf{y}) \Big] ds_{\mathbf{y}} + \int_\Omega g_1(\mathbf{y},\mathbf{x}) p(\mathbf{y}) d\Omega_{\mathbf{y}} +$$
$$+ \sum_c \{ g_1(\mathbf{y}^c,\mathbf{x}) F(w)(\mathbf{y}^c) - [w(\mathbf{y}^c) - w(\mathbf{x})]\, F(g_1)(\mathbf{y}^c,\mathbf{x}) \}, \tag{11}$$

respectively, wherein $w_{,i} = \partial w / \partial x_i$. In the above equations, $\mathbf{x}$ is the source point and $\mathbf{y}$ is the field point and F denotes the corner forces, which can be obtained through the following definition:

$$F(\mathbf{x}^c) = M_{nt}(\mathbf{x}^c_+) - M_{nt}(\mathbf{x}^c_-). \tag{12}$$

The summation index includes all the corner points of the boundary, and if the source point coincides with one of the corners this contribution is neglected.

In equation (10), $g_0(\mathbf{y},\mathbf{x})$, $\partial g_0(\mathbf{y},\mathbf{x})/\partial \nu$, $M_\nu(g_0(\mathbf{y},\mathbf{x}))$, $V_\nu(g_0(\mathbf{y},\mathbf{x}))$ and $M_{\nu\tau}(g_0(\mathbf{y},\mathbf{x}))$ are, respectively, the deflection, the slope, the bending moment, the Kirchhoff shear force and the twisting moment that are obtained in the field point y of an infinite plate, when a concentrated unit force is acting at the source point x. The following are their form expressions[3]:

$$g_0(\mathbf{y},\mathbf{x}) = \frac{1}{8\,\pi\,K}\, r^2 \ln r; \tag{13}$$

$$\frac{\partial\, g_0(\mathbf{y},\mathbf{x})}{\partial\, \nu} = \frac{1}{8\,\pi\,K}\, r\, r_\nu\, (1 + 2 \ln r); \tag{14}$$

$$M_\nu(g_0(\mathbf{y},\mathbf{x})) = -\frac{1}{8\,\pi} \left[2(1+\nu)\ln r + (3+\nu)\, r_\nu^2 + (1+3\nu)\, r_\tau^2\right]; \tag{15}$$

$$V_\nu(g_0(\mathbf{y},\mathbf{x})) = -\frac{2}{8\,\pi\, r} \left[2\, r_\nu + (1-\nu)\, r_\nu\, (r_\nu^2 - r_\tau^2)\right]; \tag{16}$$

$$M_{\nu\tau}(g_0(\mathbf{y},\mathbf{x})) = -\frac{1}{4\,\pi}(1-\nu)\, r_\nu\, r_\tau, \tag{17}$$

where $K = E\,t^3/12(1-\nu^2)$, and ν and τ are, respectively, the normal and the tangential vectors to the boundary at field point $\mathbf{y}$.

The directional derivatives are given by:

$$r_\nu = r_{,1}\,\nu_1 + r_{,2}\,\nu_2 \tag{18}$$

$$r_\tau = -r_{,1}\,\nu_2 + r_{,2}\,\nu_1 \tag{19}$$

where,

$$r_{,i} = \frac{y_i - x_i}{r}. \tag{20}$$

In equation (11), $g_1(\mathbf{y},\mathbf{x})$, $\partial\, g_1(\mathbf{y},\mathbf{x})/\partial\,\nu$, $M_\nu(g_1(\mathbf{y},\mathbf{x}))$, $V_\nu(g_1(\mathbf{y},\mathbf{x}))$ and $M_{\nu\tau}(g_1(\mathbf{y},\mathbf{x}))$ denote the same variables, due to a concentrated unit bending moment acting at x in direction n. They are defined thus:

$$g_1(\mathbf{y},\mathbf{x}) = \frac{1}{8\,\pi\,K}\, r\,(1 + 2\ln r)\, r_n; \tag{21}$$

$$\frac{\partial\, g_1(\mathbf{y},\mathbf{x})}{\partial\,\nu} = \frac{1}{8\,\pi\,K}\left[2(r_\nu\, r_n + r_\tau\, r_t)\ln r + 3\, r_\nu\, r_n + r_t\, r_\tau\right]; \tag{22}$$

$$M_\nu(g_1(\mathbf{y},\mathbf{x})) = -\frac{1}{4\,\pi\, r}\left[(1+\nu)\, r_n + 2(1-\nu)\, r_\nu\, r_\tau\, r_t\right]; \tag{23}$$

$$\begin{aligned} V_\nu(g_1(\mathbf{y},\mathbf{x})) &= -\frac{1}{4\,\pi\, r^2}\Big\{\left[3 - \nu - 2(1-\nu)\, r_\tau^2\right](r_\tau\, r_t - r_\nu\, r_n) + \\ &\quad + 4\,(1-\nu)\, r_\nu^2\, r_\tau\, r_t\Big\}; \end{aligned} \tag{24}$$

$$M_{\nu\tau}(g_1(\mathbf{y},\mathbf{x})) = -\frac{1-\nu}{4\,\pi\, r}\,(r_\tau^2 - r_\nu^2)\, r_t. \tag{25}$$

The directional derivatives evaluated at x are:

$$r_n = -r_{,1}\, n_1 - r_{,2}\, n_2; \tag{26}$$

$$r_t = r_{,1}\, n_2 - r_{,2}\, n_1. \tag{27}$$

The characteristic functions in (10) and (11) are

$$\begin{cases} \text{if } \mathbf{x} \in \Omega & c(\mathbf{x}) = 1; \quad c_1(\mathbf{x}) = n_1 \quad \text{and} \quad c_2(\mathbf{x}) = n_2, \\ \text{if } \mathbf{x} \in \Gamma & c(\mathbf{x}) = \Delta\,\rho/2\,\pi; \quad c_1(\mathbf{x}) = c_1'(\mathbf{x}) \quad \text{and} \quad c_2(\mathbf{x}) = c_2'(\mathbf{x}), \end{cases} \tag{28}$$

where,

$$c_1'(\mathbf{x}) = \frac{\Delta\,\rho}{2\,\pi}\, n_1 + \frac{\nu}{2\,\pi} \left[\frac{1}{2} \sin(2\rho)\, n_1 + \sin^2(\rho)\, n_2 \right]_{\rho_2}^{\rho_1}, \tag{29}$$

$$c_2'(\mathbf{x}) = \frac{\Delta\,\rho}{2\,\pi}\, n_2 + \frac{\nu}{2\,\pi} \left[\sin^2(\rho)\, n_1 - \frac{1}{2} \sin(2\rho)\, n_2 \right]_{\rho_2}^{\rho_1}. \tag{30}$$

and ρ_1 and ρ_2 represent the angles which the two tangents at a boundary point form with x_1.

To obtain a numerical solution, at each smooth node on the boundary two equations, (10) and (11), must be considered. At the corners, due to the discontinuity of the normal, equation (11) must be used twice. Then, the total number of equations is two times the number of nodes plus the number of corners. Hence, these equations should be arranged to form the matrix equation

$$[\mathbf{H}] \begin{bmatrix} \mathbf{w} \\ \mathbf{w}_{,n} \end{bmatrix} = [\mathbf{G}] \begin{bmatrix} \mathbf{M}_\nu \\ \mathbf{V}_\nu \end{bmatrix} + [\mathbf{d}], \tag{31}$$

which, after the known boundary conditions enforced, leads to a determined system of linear equations.

NUMERICAL IMPLEMENTATION

As the fundamental solution for the Kirchhoff shear force due to unit applied bending moment, equation (24), presents a hypersingularity of second order when the source and the field points are coincidents, a C^1 continuity in the nodes for the interpolation functions used to approximate the deflection is required. Hermite polynomials are adopted for that purpose. For the slopes, bending moments and Kirchhoff shear forces, linear interpolations are sufficient.

All the integrations required to form equation (31) are performed numerically. When the integrand presents a weak singularity, Telles[10] transformation is used. For Cauchy Principal Value integrals and hypersingular integrals Guiggiani[11,12,13] methods were applied.

The unsymmetric system of equations is solved using direct Gauss elimination with pivot searching.

NUMERICAL EXAMPLES

Two fast examples are presented to illustrate the performance of the formulation presented here.

First, the square plate with clamped edges, with $E = 10 \times 6$ kN/m^2, $\nu = 0.15$ and $t = 0.1$ m, was analysed (figure 1). A uniformly distributed load $p = 10$ kN was considered. To perform the analysis of that plate, three different boundary element discretizations were used, respectively with 2,4 and 20 uniform elements on each edge of the plate. In figure 2, the corresponding distributions of the bending moments along edge AB,are presented. In figure 3, the deflection along line 1-2 is plotted. In figure 4, the distribution of the bending moments obtained using the 20 element boundary mesh is compared with the one obtained with a 100 uniform finite element mesh using the COSMOS/M code. As it can be seen from figure 5, the boundary and finite elements solutions for the deflection along line 1-2 are in good agreement. In table 1 the minimum value of the bending moments along AB and the maximum value of the deflection in line 1-2 are presented. Those values were obtained using the developed code (BEM solution), the finite element code COSMOS/M (FEM solution) and the Bareš tables[14].

The second test example is an L-shaped plate is (figure 6) with $E = 10 \times 6$ kN/m^2, $\nu = 0.2$ and t 0.2 m. A uniformly distributed load with $p = 10$ kN/m^2 was also considered. In figure 7 the distribution of bending moments along AB is plotted. In figure 8 the deflection along line 1-2 is presented. The boundary element solution was obtained using a 10-element mesh for each edge of the plate. The solution thus obtained is compared with a finite element solution. For this purpose a 48-element finite element mesh was considered.

CONCLUSIONS

A boundary element method formulation for the solution of Kirchhoff plates was presented and a numerical implementation procedure was discussed. Two classic examples have been studied, where the boundary element method reveals to be numerically more efficient than the finite element method, regarding both CPU time and memory allocation requirements. The schemes of numerical integration implemented also proved to be efficient and well suited for this kind of problem.

ACKNOWLEDGEMENTS

The autors acknowledge Dr. Guiggiani for his contribution on the implementation of his numerical integration schemes. This work has been sponsored by Junta Nacional de Investigação Científica e Tecnológica as part of grant BM/1/90-RM and research project PMCT/C/MPF/638/90.

REFERENCES

1. Timoshenko, S. and Woinowsky-Krieger (1959), Theory of Plates and Shells, McGraw-Hill Book Company.

2. Hrabok, M.M. and Hrudey, T.M. (1984), A review and catalogue of plate bending finite elements, Computers & Structures, Vol. 19, n° 3, 479–495.

3. Hartmann, F. (1989), Introduction to Boundary Elements, Springer Verlag.

4. Hartmann, F. and Zotemantel, R. (1986), The direct boundary element method in plate bending, Int. Jou. Num. Meths. Engng., Vol. 23, 2049–2069

5. Stern, M. (1979), A general boundary integral formulation for the numerical solution of plate bending problems, Int. Jou. Solids Struct. Vol. 15.

6. Bézine, G. and Gamby, D. (1978), A new integral equation formulation for plate bending problems, C.A. Brebbia (ed.) Recent Advances in Boundary Element Menthods, Pentech Press, London.

7. Costa, Jr. and Brebbia, C.A. (1984), Plate bending problems usig BEM, Bondary Element VI, C.A. Brebbia (ed.), Springer Verlag.

8. Forbes, D.J. and Robinson, A.R. (1969), Numerical analysis of elastic plates and shallow shells by an integral equation method, Univ. of Illinois, Structural Research Series, Report 345.

9. Tanaka, M. and Miyazaki, K. (1985), A direct boundary element method for elastic bending analysis of plates, Trans. Japan Soc. Mech. Engng. 51, serie A, n° 466

10. Telles, J.C.F. (1987), A self-adaptive co-ordinate transformation for efficient numerical evaluation of general boundary element integrals, Int. Jou. Num. Meth. in Engng., **24**, 959–973.

11. Guiggiani, M. and Casalini, P. (1987), Direct computation of Cauchy principal value integrals in advanced boundary elements, Int. Jou. Num. Meth. in Engng. **24**, 1711–1720.

12. Guiggiani, M. and Gigante, A. (1990). A general algorithm for multi-dimensional Cauchy principal value integrals in the boundary element method, ASME Jou. Appl. Mechs.

13. Guiggiani, M., Krishnasamy, G. Rizzo, F.J. and Rudolphi, T.J. (1990). Hypersingular boundary integral equations: a new approach for their numerical treatment, IABEM 90, Rome.

14. Bareš, R. (1981). Tablas para el Calculo de Placas y Vigas Pared, Editorial Gustavo Gili, S.A., Barcelona.

	Bareš	BEM	FEM
M_n^{min}	-51.50	-51.46	-49.55
w^{max}	1.490	1.484	1.504

Table 1: Boundary minimum bending moment and maximum deflection for square clamped plate.

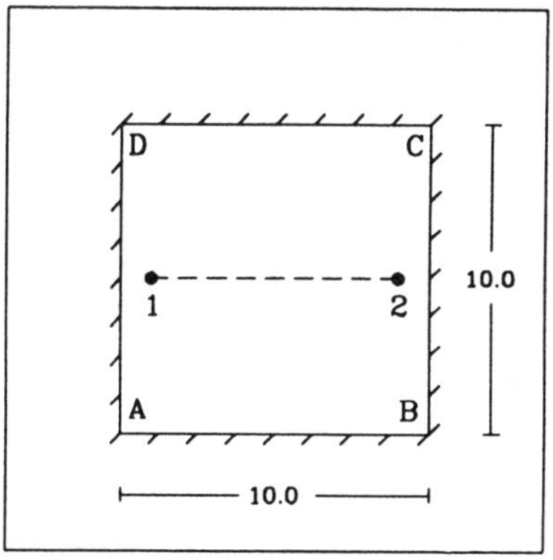

Figure 1: Square clamped plate example.

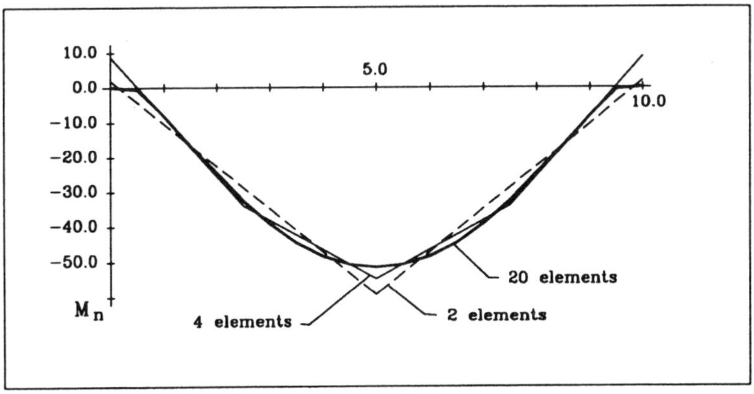

Figure 2: Bending moments distribution along edge AB. Square plate example.

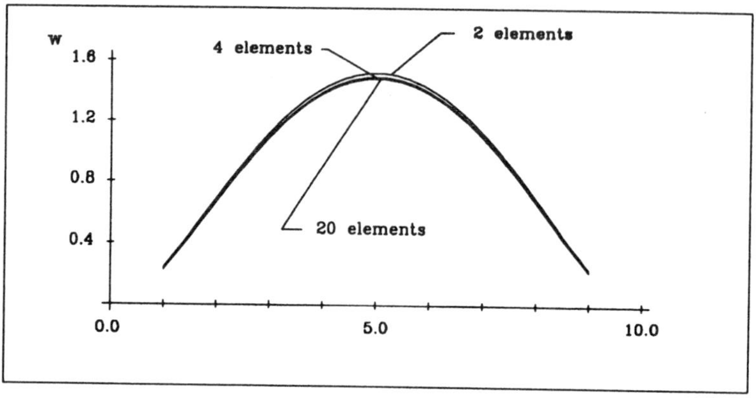

Figure 3: Deflection along 1-2. Square plate example.

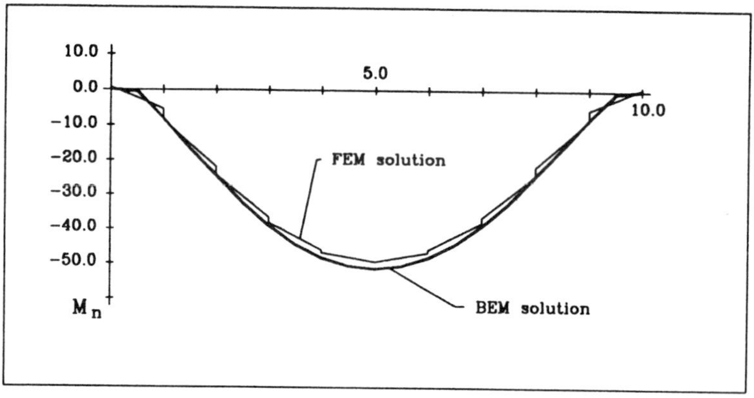

Figure 4: BEM and FEM bending moment solutions. Square plate example.

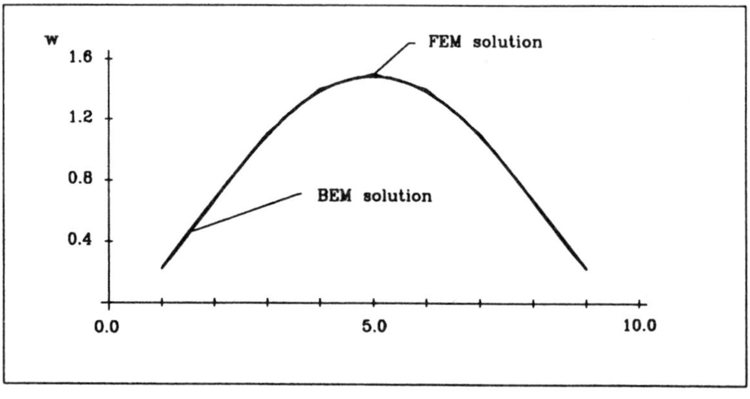

Figure 5: BEM and FEM deflection solutions. Square plate example.

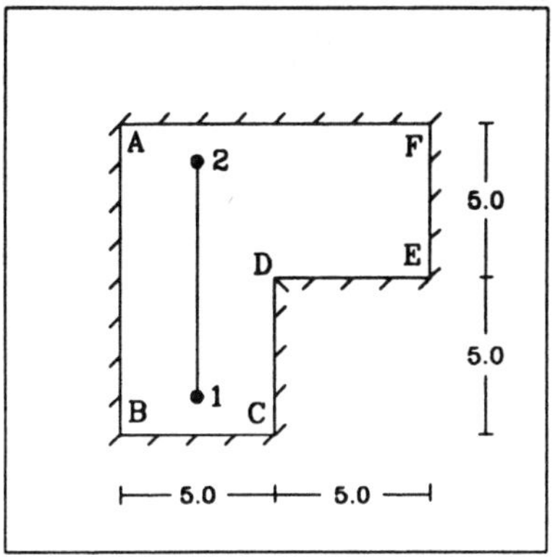

Figure 6: L-shaped plate example.

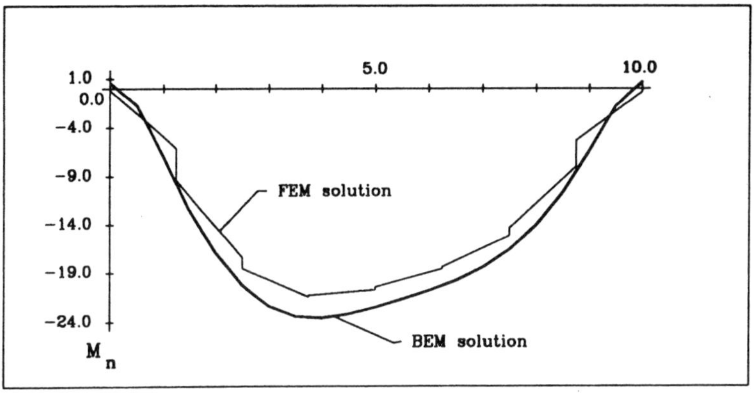

Figure 7: Boundary bending moments distribution. L-shaped plate example

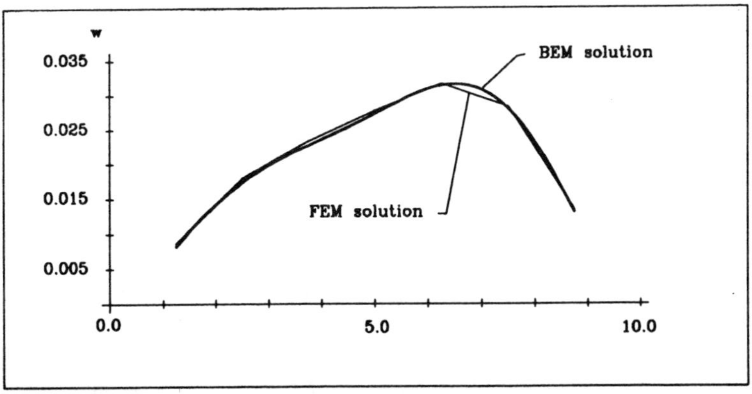

Figure 8: Deflection along 1-2. L-shaped plate example

Kernel Functions of Mindlin's Solution for Evaluation of Internal Stresses

Y.S. Jiang

Service of Machine Elements & Tribology, Department of Mechanical Engineering, State University of Liège, 4000 Liège, Belgium

Abstract

Mindlin's solution of point forces acting in elastic half-space has been used as the fundamental solution in boundary element implementations since 1979. However, the kernel functions of Mindlin's solution which are necessary for calculation of the internal stresses have not been presented in any published literature. The present author has deduced these kernel functions. All the derivatives involved in the kernel functions are checked with an algebraic programming system, REDUCE. Numerical results of the evaluation of the internal stresses are compared with analytical solutions.

Introduction

In boundary element solutions of half-space problems in solid mechanics, the fundamental solutions of half-space such as Melan's solution (2D) and Mindlin's solution (3D) play an important role. Number of authors have studied solutions of two dimensional problems in half-plane (Telles and Brebbia [2], Jiang [3]). However, little attention has been given to the equivalent three-dimensional problems. To the best of author's knowledge, only Rakaguma [1] has used Mindlin's solution in 3D boundary element implementations. But he did not derive the kernel functions, often represented by $\mathbf{D}^*_{kij}$ and $\mathbf{S}^*_{kij}$, from Mindlin's solution for the evaluation of the internal stresses because "they are very complicated", as he noted ([1], p.134). Instead, he used the kernel functions derived from Kelvin's solution. Unfortunately, this makes it necessary to discretize the free ground surface if the internal stresses are required.

In a recent development of a 3D boundary element program, the author has deduced the kernel functions $\mathbf{D}^*_{kij}$ and $\mathbf{S}^*_{kij}$ from Mindlin's solution. As a lot of differentiations and algebraic operations are involved, the author used an algebraic programming system, REDUCE, to verify all the final expressions of derivatives. The kernel functions have been implemented into a BEM program. Numerical results of the internal stresses are compared with analytical solutions.

In the author's opinion, although this work would not be of great theoretical importance, it may save uncountable time for an interested BEM program developer as he will not need to deduce all the formulas again himself.

Theory and Development

In boundary element formulations, the kernel functions $\mathbf{D}^{*}_{kij}$ and $\mathbf{S}^{*}_{kij}$ are related to the internal stresses by the following integral equation [4]:

$$\sigma_{ij} = - \int_{\Gamma} S^{*}_{kij} u_k \, d\Gamma + \int_{\Gamma} D^{*}_{kij} p_k \, d\Gamma + \int_{\Omega} D^{*}_{kij} b_k \, d\Omega \tag{1}$$

in which u_k and p_k are displacements and surface tractions on boundary Γ, b_k are components of body forces distributed within domain Ω. The kernel functions $\mathbf{D}^{*}_{kij}$ and $\mathbf{S}^{*}_{kij}$ may be split into two parts: a part corresponding to Kelvin's solution represented by $\mathbf{D}^{k}_{kij}$ and $\mathbf{S}^{c}_{kij}$, and a complementary part represented by $\mathbf{D}^{c}_{kij}$ and $\mathbf{S}^{c}_{kij}$:

$$\mathbf{D}^{*}_{kij} = \mathbf{D}^{k}_{kij} + \mathbf{D}^{c}_{kij} \tag{2}$$

$$\mathbf{S}^{*}_{kij} = \mathbf{S}^{k}_{kij} + \mathbf{S}^{c}_{kij} \tag{3}$$

As expressions of $\mathbf{D}^{k}_{kij}$ and $\mathbf{S}^{k}_{kij}$ may be found in [4], only the complementary parts $\mathbf{D}^{c}_{kij}$ and $\mathbf{S}^{c}_{kij}$ will be developed in what follows. Using an expression in [2], these two tensors may be calculated by

$$\mathbf{D}^{c}_{kij} = G\cdot\left(\frac{\partial u^{c}_{ik}}{\partial x_j} + \frac{\partial u^{c}_{jk}}{\partial x_i} \right) + \frac{2G\upsilon}{1-2\upsilon}\cdot\frac{\partial u^{c}_{lk}}{\partial x_l}\cdot\delta_{ij} \tag{4}$$

$$\mathbf{S}^{c}_{kij} = G\cdot\left[\frac{\partial \sigma^{c}_{kmi}}{\partial x_j} + \frac{\partial \sigma^{c}_{kmj}}{\partial x_i} + \frac{2\upsilon}{1-2\upsilon}\cdot\frac{\partial \sigma^{c}_{kml}}{\partial x_l}\cdot\delta_{ij}\right]\cdot n_m \tag{5}$$

where G is elastic shear module, υ is Poisson's ratio, δ_{ij} is Kronecker Delta, n_m is the normal direction cosines, expressions for displacements u^{c}_{ij} and stresses σ^{c}_{ijk} may be found in [4]. All the derivatives in Equation (4) and (5) are taken with reference to the coordinates of the load point. There are in total 81 different derivatives to be developed and they are listed in Appendix. All of these derivatives have been verified with an algebraic programming system, REDUCE [5], in order to exclude any possible human errors.

Numerical example

As a numerical example, the internal stresses due to a uniform vertical loading over a square area on the surface of a half space, figure 1, are solved with the boundary element solution. Mindlin's fundamental solution together with the kernel functions given in Appendix are used in boundary element calculations. Numerical results are compared with Boussinesq's solution. Although the latter corresponds to the point loading, stresses at points 3 times far from the load point should be precise enough to be used as the reference of comparison.

In table 1 the coordinates of points and the stresses calculated at these points are given. The analytical solution is represented by *italic* characters while boundary element solution by normal ones.

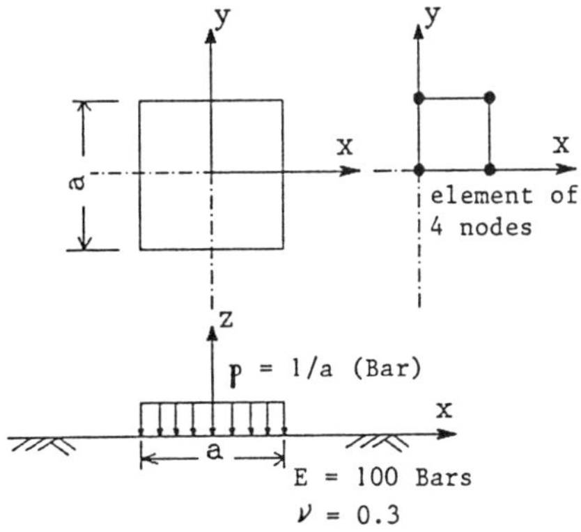

Figure 1. Uniform vertical loading

Table 1 Stresses at points within a half space (x 10^{-3} bars)

x/a	y/a	z/a	σ_x/p	σ_y/p	σ_z/p	σ_{xy}/p	σ_{yz}/p	σ_{xz}/p
4	0	0	3.978 *3.979*	-3.978 *-3.979*	0.000 *0.000*	0.000 *0.000*	0.000 *0.000*	0.000 *0.000*
6	0	0	1.768 *1.768*	-1.768 *-1.768*	0.000 *0.000*	0.000 *0.000*	0.000 *0.000*	0.000 *0.000*
0	0	-4	1.809 *1.989*	1.809 *1.989*	-29.08 *-29.84*	0.000 *0.000*	0.000 *0.000*	0.000 *0.000*
0	0	-6	0.848 *0.884*	0.848 *0.884*	-13.11 *-13.26*	0.000 *0.000*	0.000 *0.000*	0.000 *0.000*
6	6	-6	-0.680 *-0.681*	-0.680 *-0.681*	-0.855 *-0.851*	-0.646 *-0.647*	0.852 *0.851*	0.852 *0.851*

One can observe from the above table that the errors of the boundary element solution compared with the analytical solution (Boussinesq's solution) are generally small, especially on the ground surface (z = 0). However, the relative error reaches 10% at a point on z axis (x=0, y=0, z=-4). Improvements on numerical integrations over boundary elements would be necessary.

Conclusions

The aim of this work is to offer the interested BEM program developer a complete list of the kernel functions derived from Mindlin's solution for the evaluation of the internal stresses within the half-space. All the 81 derivatives involved in these functions are verified with an algebraic programming system, REDUCE. They are given in Appendix. Numerical results of the evaluation of the internal stresses are compared with an analytical solution and good precision is obtained.

Acknowledgements

The financial support of *La Région Wallonne* of Belgium is sincerely appreciated. I wish also to express my gratitude to MSM (*Service de Mécanique des matériaux - Stabilité des constructions - Mécanique des structure*) of University of Liège in Belgium for their offer of computer facilities which allowed me to use REDUCE.

References

[1] Rakaguma, R.K. (1979)
Three Dimensional Elastics Using The Boundary Element Method
Ph.D. Thesis. University of Southampton, England

[2] Telles, J.C.F. and Brebbia, C.A. (1981)
Boundary Element Solution for Half-Plane Problems
Int.J. Solid Structures, Vol.17, No.12, pp.1149-1158.

[3] Jiang, Y.S. (1986)
Half Plane With Body Force Problem and Its Uses in Geomechanics by BEM. BETECH 86. Proceedings of 2nd Boundary Element Technology Conference. July, 1986. MIT Boston. Computational Mechanics Publications.

[4] Brebbia, C.A., Telles, J.C.F. and Wrobel, L.C.(1984)
Boundary Element Techniques, Theory and Applications in Engineering Springer-Verlag. Berlin, Heidelberg, New York 1984

[5] Hearn, C.A. (1987)
Reduce, User's Manual, Version 3.3
The Rand Corporation, Santa Monica, CA 90406-2138.

Appendix

In this appendix, all the 81 derivatives involved in Equation (4) and (5) are listed. They are deduced from the complementary expressions for the displacements and stresses due to unit loads applied within the half-space given in Brebbia *et al* ([4], Appendix B.1, page 455-458). Symbols used in formulas are referenced to figure 2.

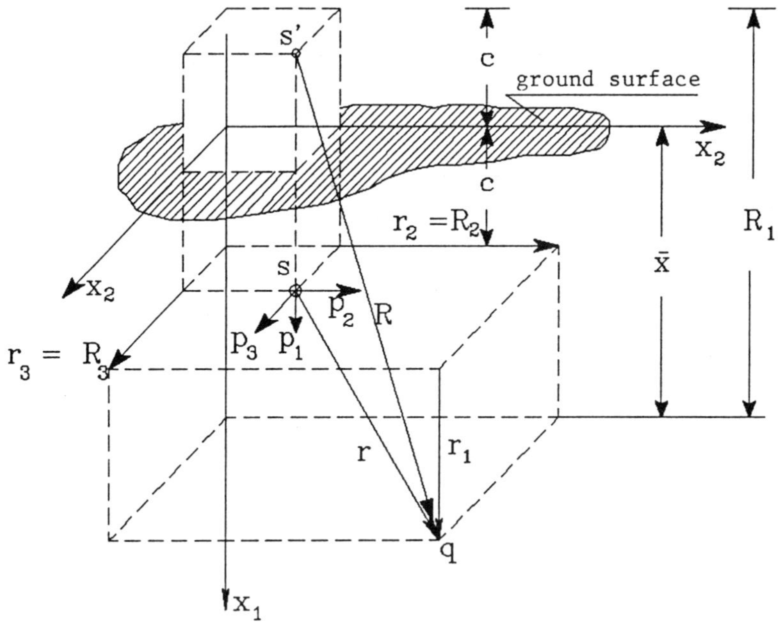

$$K_d = \frac{1}{16\pi(1-\upsilon)G}, \qquad K_s = \frac{1}{8\pi(1-\upsilon)},$$

$$R = (R_i R_i)^{1/2}, \qquad r_i = x_i(q) - x_i(s),$$

$$R_i = x_i(q) - x_i(s'), \qquad c = x_1(s) \geq 0,$$

$$\bar{x} = x_1(q) \geq 0.$$

$$(i = 1,2,3)$$

Figure 2. Unit point loads applied within the half-space ($|p_1|=1$)

$$\frac{\partial u^c_{11}}{\partial x_1} = K_d \cdot \left\{ -\frac{[8(1-\upsilon)^2-3(3-4\upsilon)]R_1 + 2\bar{x}}{R^3} + \frac{R_1^2[-3R_1(3-4\upsilon)+6\bar{x}] + 18c\bar{x}R_1}{R^5} - \frac{30c\bar{x}R_1^3}{R^7} \right\}$$

$$\frac{\partial u^c_{11}}{\partial x_2} = K_d \cdot r_2 \cdot \left\{ \frac{8(1-\upsilon)^2-(3-4\upsilon)}{R^3} + \frac{3(3-4\upsilon)R_1^2 - 6c\bar{x}}{R^5} + \frac{30c\bar{x}R_1^2}{R^7} \right\}$$

$$\frac{\partial u^c_{11}}{\partial x_3} = K_d \cdot \frac{r_3}{r_2} \cdot \frac{\partial u^c_{11}}{\partial x_2}$$

$$\frac{\partial u^c_{12}}{\partial x_1} = K_d \cdot r_2 \cdot \left\{ \frac{[4(1-\upsilon)(1-2\upsilon)-(3-4\upsilon)]}{R^3} + \frac{6\bar{x}(R_1+c) - 3(3-4\upsilon)r_1R_1}{R^5} - \frac{30c\bar{x}R_1^2}{R^7} \right\}$$

$$\frac{\partial u^c_{12}}{\partial x_2} = K_d \cdot \left\{ -\frac{(3-4\upsilon)r_1}{R^3} + \frac{3(3-4\upsilon)r_1r_2^2 - 6c\bar{x}R_1}{R^5} + \frac{30c\bar{x}R_1r_2^2}{R^7} + \frac{4(1-\upsilon)(1-2\upsilon)}{R(R+R_1)}\left[1 - \frac{r_2^2(2R+R_1)}{R^2(R+R\ )}\right] \right\}$$

$$\frac{\partial u^c_{12}}{\partial x_3} = K_d \cdot r_2 \cdot r_3 \cdot \left\{ \frac{3(3-4\upsilon)r_1}{R^5} + \frac{30c\bar{x}R_1}{R^7} - \frac{4(1-\upsilon)(1-2\upsilon)(2R+R_1)}{R^3(R+R_1)^2} \right\}$$

$$\frac{\partial u^c_{13}}{\partial x_1} = \frac{r_3}{r_2} \cdot \frac{\partial u^c_{12}}{\partial x_1}, \qquad \frac{\partial u^c_{13}}{\partial x_2} = \frac{\partial u^c_{12}}{\partial x_3}$$

$$\frac{\partial u^c_{13}}{\partial x_3} = K_d \cdot \left\{ -\frac{(3-4\upsilon)r_1}{R^3} + \frac{3(3-4\upsilon)r_1r_3^2 - 6c\bar{x}R_1}{R^5} + \frac{30c\bar{x}R_1r_3^2}{R^7} + \frac{4(1-\upsilon)(1-2\upsilon)}{R(R+R_1)}\left[1 - \frac{r_3^2(2R+R_1)}{R^2(R+R\ )}\right] \right\}$$

$$\frac{\partial u^c_{21}}{\partial x_1} = K_d \cdot r_2 \cdot \left\{ -\frac{4(1-\upsilon)(1-2\upsilon)+(3-4\upsilon)}{R^3} - \frac{3(3-4\upsilon)r_1R_1 + 6\bar{x}(R_1+c)}{R^5} + \frac{30c\bar{x}R_1^2}{R^7} \right\}$$

$$\frac{\partial u^c_{21}}{\partial x_2} = K_d \cdot \left\{ -\frac{(3-4\upsilon)r_1}{R^3} + \frac{3(3-4\upsilon)r_1r_2^2 + 6c\bar{x}R_1}{R^5} - \frac{30c\bar{x}R_1r_2^2}{R^7} - \frac{4(1-\upsilon)(1-2\upsilon)}{R(R+R_1)}\left[1 - \frac{r_2^2(2R+R_1)}{R^2(R+R\ )}\right] \right\}$$

$$\frac{\partial u^c_{21}}{\partial x_3} = K_d \cdot r_2 \cdot r_3 \cdot \left\{ \frac{3(3-4\upsilon)r_1}{R^5} - \frac{30c\bar{x}R_1}{R^7} + \frac{4(1-\upsilon)(1-2\upsilon)(2R+R_1)}{R^3(R+R_1)^2} \right\}$$

$$\frac{\partial u^c_{22}}{\partial x_1} = K_d \cdot \left\{ \frac{2\bar{x}-R_1}{R^3} - \frac{3(3-4\upsilon)R_1r_2^2 + 6\bar{x}(R_1c+r_2^2)}{R^5} + \frac{30c\bar{x}R_1r_2^2}{R^7} - \frac{4(1-\upsilon)(1-2\upsilon)}{R(R+R_1)}\left[1 - \frac{r_2^2(2R+R_1)}{R^2(R+R\ )}\right] \right\}$$

$$\frac{\partial u^c_{22}}{\partial x_2} = K_d \cdot r_2 \cdot \left\{ \frac{(-5+8\upsilon)}{R^3} + \frac{3(3-4\upsilon)r_2^2 + 18c\bar{x}}{R^5} - \frac{30c\bar{x}r_2^2}{R^7} + \frac{4(1-\upsilon)(1-2\upsilon)}{R(R+R_1)^2}\left[3 - \frac{r_2^2(3R+R_1)}{R^2(R+R_1)}\right] \right\}$$

$$\frac{\partial u^c_{22}}{\partial x_3} = K_d \cdot r_3 \cdot \left\{ \frac{1}{R^3} + \frac{3(3-4\upsilon)r_2^2 + 6c\bar{x}}{R^5} - \frac{30c\bar{x}r_2^2}{R^7} + \frac{4(1-\upsilon)(1-2\upsilon)}{R(R+R_1)^2}\left[1 - \frac{r_2^2(3R+R_1)}{R^2(R+R_1)}\right] \right\}$$

$$\frac{\partial u^c_{23}}{\partial x_1} = K_d \cdot r_2 \cdot r_3 \cdot \left\{ -\frac{3(3-4\upsilon)R_1 + 6\bar{x}}{R^5} + \frac{30c\bar{x}R_1}{R^7} + \frac{4(1-\upsilon)(1-2\upsilon)(2R+R_1)}{R^3(R+R_1)^2} \right\}$$

$$\frac{\partial u^c_{23}}{\partial x_2} = K_d \cdot r_3 \cdot \left\{ -\frac{(3-4\upsilon)}{R^3} + \frac{3(3-4\upsilon)r_2^2 + 6c\bar{x}}{R^5} - \frac{30c\bar{x}r_2^2}{R^7} + \frac{4(1-\upsilon)(1-2\upsilon)}{R(R+R_1)^2}\left[1 - \frac{r_2^2(3R+R_1)}{R^2(R+R_1)}\right] \right\}$$

$$\frac{\partial u^c_{23}}{\partial x_3} = K_d \cdot r_2 \cdot \left\{ -\frac{(3-4\upsilon)}{R^3} + \frac{3(3-4\upsilon)r_3^2 + 6c\bar{x}}{R^5} - \frac{30c\bar{x}r_3^2}{R^7} + \frac{4(1-\upsilon)(1-2\upsilon)}{R(R+R_1)^2}\left[1 - \frac{r_3^2(3R+R_1)}{R^2(R+R_1)}\right] \right\}$$

$$\frac{\partial u^c_{31}}{\partial x_1} = K_d \cdot \frac{r_3}{r_2} \cdot \frac{\partial u^c_{21}}{\partial x_1}, \qquad \frac{\partial u^c_{31}}{\partial x_2} = \frac{\partial u^c_{21}}{\partial x_3}$$

$$\frac{\partial u^c_{31}}{\partial x_3} = K_d \cdot \left\{ -\frac{(3-4\upsilon)r_1}{R^3} + \frac{3(3-4\upsilon)r_1 r_3^2 + 6c\bar{x}R_1}{R^5} - \frac{30c\bar{x}R_1 r_3^2}{R^7} - \frac{4(1-\upsilon)(1-2\upsilon)}{R(R+R_1)}\left[1 - \frac{r_3^2(2R+R_1)}{R^2(R+R\ )} \right] \right\}$$

$$\frac{\partial u^c_{32}}{\partial x_1} = \frac{\partial u^c_{23}}{\partial x_1}, \qquad \frac{\partial u^c_{32}}{\partial x_2} = \frac{\partial u^c_{23}}{\partial x_2} \qquad \frac{\partial u^c_{32}}{\partial x_3} = \frac{\partial u^c_{23}}{\partial x_3}$$

$$\frac{\partial u^c_{33}}{\partial x_1} = K_d \cdot \left\{ \frac{2\bar{x}-R_1}{R^3} - \frac{3(3-4\upsilon)R_1 r_3^2 + 6\bar{x}(R_1 c + r_3^2)}{R^5} + \frac{30c\bar{x}R_1 r_3^2}{R^7} - \frac{4(1-\upsilon)(1-2\upsilon)}{R(R+R_1)}\left[1 - \frac{r_3^2(2R+R_1)}{R^2(R+R\ )}\right] \right\}$$

$$\frac{\partial u^c_{33}}{\partial x_2} = K_d \cdot r_2 \cdot \left\{ \frac{1}{R^3} + \frac{3(3-4\upsilon)r_3^2 + 6c\bar{x}}{R^5} - \frac{30c\bar{x}r_3^2}{R^7} + \frac{4(1-\upsilon)(1-2\upsilon)}{R(R+R_1)^2}\left[1 - \frac{r_3^2(3R+R_1)}{R^2(R+R_1)} \right] \right\}$$

$$\frac{\partial u^c_{33}}{\partial x_3} = K_d\, r_3 \cdot \left\{ \frac{(-5+8\upsilon)}{R^3} + \frac{3(3-4\upsilon)r_3^2 + 18c\bar{x}}{R^5} - \frac{30c\bar{x}r_3^2}{R^7} + \frac{4(1-\upsilon)(1-2\upsilon)}{R(R+R_1)^2}\left[3 - \frac{r_3^2(3R+R_1)}{R^2(R+R_1)} \right] \right\}$$

$$\frac{\partial \sigma^c_{111}}{\partial x_1} = K_s \cdot \left\{ -\frac{1-2\upsilon}{R^3} - \frac{3R_1[(1-2\upsilon)r_1 + (1-8\upsilon)\bar{x} + 2c] - 3c\cdot(5\bar{x}-c)}{R^5} + \frac{15R_1^2[(1-4\upsilon)\bar{x}R_1 - c\cdot(11\bar{x}-c)]}{R^7} + \frac{210c\bar{x}R_1^4}{R^9} \right\}$$

$$\frac{\partial \sigma^c_{111}}{\partial x_2} = K_s \cdot r_2 \left\{ \frac{3(1-2\upsilon)r_1}{R^5} - \frac{15(3-4\upsilon)\bar{x}R_1^2 - 15cR_1(5\bar{x}-c)}{R^7} - \frac{210c\bar{x}R_1^3}{R^9} \right\}$$

$$\frac{\partial \sigma^c_{111}}{\partial x_3} = K_s \frac{r_3}{r_2} \frac{\partial \sigma^c_{111}}{\partial x_2}$$

$$\frac{\partial \sigma^c_{121}}{\partial x_1} = K_s\, r_2 \cdot \left\{ -\frac{3(1-2\upsilon)R_1 - 12\upsilon\bar{x} - 6c}{R^5} + \frac{15(1-4\upsilon)\bar{x}R_1^2 - 15cR_1(7\bar{x}+c)}{R^7} + \frac{210c\bar{x}R_1^3}{R^9} \right\}$$

$$\frac{\partial \sigma^c_{121}}{\partial x_2} = K_s \cdot \left\{ -\frac{1-2\upsilon}{R^3} + \frac{3(3-4\upsilon)\bar{x}R_1 - 3c(3\bar{x}+c) + 3(1-2\upsilon)r_2^2}{R^5} + \frac{30c\bar{x}R_1^2 - 5r_2^2[3(3-4\upsilon)\bar{x}R_1 - 3c(3\bar{x}+c)]}{R^7} - \frac{210c\bar{x}R_1^2 r_2^2}{R^9} \right\}$$

$$\frac{\partial \sigma^c_{121}}{\partial x_3} = K_s\, r_2\, r_3 \cdot \left\{ \frac{3(1-2\upsilon)}{R^5} - \frac{15(3-4\upsilon)\bar{x}R_1 - 15c(3\bar{x}+c)}{R^7} - \frac{210c\bar{x}R_1^2}{R^9} \right\}$$

$$\frac{\partial \sigma^c_{131}}{\partial x_1} = K_s \frac{r_3}{r_2} \frac{\partial \sigma^c_{121}}{\partial x_1}, \qquad \frac{\partial \sigma^c_{131}}{\partial x_2} = \frac{\partial \sigma^c_{121}}{\partial x_3}$$

$$\frac{\partial \sigma^c_{131}}{\partial x_3} = K_s \cdot \left\{ -\frac{1-2\upsilon}{R^3} + \frac{3(3-4\upsilon)\bar{x}R_1 - 3c(3\bar{x}+c) + 3(1-2\upsilon)r_3^2}{R^5} + \frac{30c\bar{x}R_1^2 - 5r_3^2[3(3-4\upsilon)\bar{x}R_1 - 3c(3\bar{x}+c)]}{R^7} - \frac{210c\bar{x}R_1^2 r_3^2}{R^9} \right\}$$

$$\frac{\partial \sigma^c_{221}}{\partial x_1} = K_s \cdot \left\{ \frac{(1-2\upsilon)(1-8\upsilon)}{R^3} + \frac{[24\upsilon(1-\upsilon)-3]r_2^2 + 6(c+R_1)[(1-2\upsilon)\bar{x} - 2\upsilon c] - 12\upsilon cR_1 - 3(1-2\upsilon)(3r_1 - 4\upsilon R_1)R_1}{R^5} \right.$$

$$+\frac{15r_2^2[(3-4\nu)r_1R_1-2\bar{x}(c+R_1)] -30cR_1^2[(1-2\nu)\bar{x}-2\nu c]}{R^7}+\frac{210c\bar{x}R_1^2r_2^2}{R^9}\Bigg\}$$

$$\frac{\partial\sigma^c_{221}}{\partial x_2} = K_s\cdot r_2\cdot\Bigg\{\frac{(27-42\nu)r_1 - 12(1-2\nu)\nu R_1}{R^5} - \frac{15(3-4\nu)r_1r_2^2 - 30cR_1[(3-2\nu)\bar{x} - 2\nu c]}{R^7} - \frac{210c\bar{x}R_1r_2^2}{R^9}$$
$$- \frac{4(1-\nu)(1-2\nu)}{R^3(R+R_1)^2}\Big[6R + 3R_1 - \frac{(8R^2+9RR_1+3R_1^2)r_2^2}{R^2(R+R_1)}\Big]\Bigg\}$$

$$\frac{\partial\sigma^c_{221}}{\partial x_3} = K_s\cdot r_3\cdot\Bigg\{\frac{3(1-2\nu)(3r_1-4\nu R_1)}{R^5} - \frac{15(3-4\nu)r_1r_2^2 - 30cR_1[(1-2\nu)\bar{x}-2\nu c]}{R^7} - \frac{210c\bar{x}R_1r_2^2}{R^9}$$
$$- \frac{4(1-\nu)(1-2\nu)}{R^3(R+R_1)^2}\Big[2R + R_1 - \frac{(8R^2+9RR_1+3R_1^2)r_2^2}{R^2(R+R_1)}\Big]\Bigg\}$$

$$\frac{\partial\sigma^c_{231}}{\partial x_1} = K_sr_2r_3\Bigg\{\frac{3(3-4\nu)}{R^5} + \frac{15(3-4\nu)R_1r_1 - 30\bar{x}(c+R_1)}{R^7} + \frac{210c\bar{x}R_1^2}{R^9} - \frac{4(1-\nu)(1-2\nu)}{R^4(R+R_1)}\Big(2 + \frac{R + 3R_1}{R}\Big)\Bigg\}$$

$$\frac{\partial\sigma^c_{231}}{\partial x_2} = K_sr_3\Bigg\{\frac{3(3-4\nu)r_1}{R^5} + \frac{30c\bar{x}R_1 - 15(3-4\nu)r_1r_2^2}{R^7} - \frac{210c\bar{x}R_1r_2^2}{R^9}$$
$$- \frac{4(1-\nu)(1-2\nu)}{R^3(R+R_1)^2}\Big[2R + R_1 - \frac{(8R^2+9RR_1+3R_1^2)r_2^2}{R^2(R+R_1)}\Big]\Bigg\}$$

$$\frac{\partial\sigma^c_{231}}{\partial x_3} = K_sr_2\Bigg\{\frac{3(3-4\nu)r_1}{R^5} + \frac{30c\bar{x}R_1 - 15(3-4\nu)r_1r_3^2}{R^7} - \frac{210c\bar{x}R_1r_3^2}{R^9}$$

$$- \frac{4(1-\nu)(1-2\nu)}{R^3(R+R_1)^2}\Big[2R + R_1 - \frac{(8R^2+9RR_1+3R_1^2)r_3^2}{R^2(R+R_1)}\Big]\Bigg\}$$

$$\frac{\partial\sigma^c_{331}}{\partial x_1} = K_s\cdot\Bigg\{\frac{(1-2\nu)(1-8\nu)}{R^3} + \frac{[24\nu(1-\nu)-3]r_3^2 + 6(c+R_1)[(1-2\nu)\bar{x}-2\nu c] -12\nu cR_1 -3(1-2\nu)(3r_1-4\nu R_1)R_1}{R^5}$$
$$+ \frac{15r_3^2[(3-4\nu)r_1R_1-2\bar{x}(c+R_1)] -30cR_1^2[(1-2\nu)\bar{x}-2\nu c]}{R^7} + \frac{210c\bar{x}R_1^2r_3^2}{R^9}\Bigg\}$$

$$\frac{\partial\sigma^c_{331}}{\partial x_2} = K_s\cdot r_2\cdot\Bigg\{\frac{3(1-2\nu)(3r_1-4\nu R_1)}{R^5} - \frac{15(3-4\nu)r_1r_3^2 - 30cR_1[(1-2\nu)\bar{x}-2\nu c]}{R^7} - \frac{210c\bar{x}R_1r_3^2}{R^9}$$
$$- \frac{4(1-\nu)(1-2\nu)}{R^3(R+R_1)^2}\Big[2R + R_1 - \frac{(8R^2+9RR_1+3R_1^2)r_3^2}{R^2(R+R_1)}\Big]\Bigg\}$$

$$\frac{\partial\sigma^c_{331}}{\partial x_3} = K_s\cdot r_3\cdot\Bigg\{\frac{(27-42\nu)r_1 - 12(1-2\nu)\nu R_1}{R^5} - \frac{15(3-4\nu)r_1r_3^2 - 30cR_1[(3-2\nu)\bar{x} - 2\nu c]}{R^7} - \frac{210c\bar{x}R_1r_3^2}{R^9}$$
$$- \frac{4(1-\nu)(1-2\nu)}{R^3(R+R_1)^2}\Big[6R + 3R_1 - \frac{(8R^2+9RR_1+3R_1^2)r_3^2}{R^2(R+R_1)}\Big]\Bigg\}$$

$$\frac{\partial\sigma^c_{112}}{\partial x_1} = K_s\cdot r_2\Bigg\{\frac{(-9+6\nu)R_1 + 6(3-2\nu)c}{R^5} + \frac{15R_1^2[(3-4\nu)R_1+2\bar{x}] - 30cR_1[c+(1-2\nu)R_1 - 2\bar{x}]}{R^7} - \frac{210c\bar{x}R_1^3}{R^9}\Bigg\}$$

$$\frac{\partial\sigma^c_{112}}{\partial x_2} = K_s\cdot\Bigg\{\frac{1-2\nu}{R^3} + \frac{3(3-4\nu)R_1^2 - 6c[c+(1-2\nu)R_1] - 3(1-2\nu)r_2^2}{R^5}$$

$$- \frac{30c\bar{x}R_1^2 + 15(3-4\nu)R_1^2 r_2^2 - 30[c+(1-2\nu)R_1]cr_2^2}{R^7} + \frac{210c\bar{x}R_1^2 r_2^2}{R^9} \Big\}$$

$$\frac{\partial\sigma^c_{112}}{\partial x_3} = K_s\, r_2 r_3 \Big\{ - \frac{3(1-2\nu)}{R^5} - \frac{15(3-4\nu)R_1^2 - 30c[c+(1-2\nu)R_1]}{R^7} + \frac{210c\bar{x}R_1^2}{R^9} \Big\}$$

$$\frac{\partial\sigma^c_{122}}{\partial x_1} = K_s \cdot \Big\{ - \frac{1-2\nu}{R^3} - \frac{3(1-2\nu)R_1 r_1 + 3r_2^2 + 6\bar{x}(c+R_1)}{R^5}$$

$$+ \frac{15(3-4\nu)R_1^2 r_2^2 + 30c\bar{x}R_1^2 - 30(1-2\nu)cR_1 r_2^2 + 30\bar{x}R_1 r_2^2 + 30c\bar{x}r_2^2}{R^7} - \frac{210c\bar{x}R_1^2 r_2^2}{R^9} \Big\}$$

$$\frac{\partial\sigma^c_{122}}{\partial x_2} = K_s\, r_2 \Big\{ \frac{3(1-2\nu)r_1 + 6(3-4\nu)R_1 - 12(1-2\nu)c}{R^5} - \frac{15(3-4\nu)R_1 r_2^2 + 90c\bar{x}R_1 - 30(1-2\nu)cr_2^2}{R^7}$$

$$+ \frac{210c\bar{x}R_1 r_2^2}{R^9} \Big\}$$

$$\frac{\partial\sigma^c_{122}}{\partial x_3} = K_s\, r_3 \Big\{ \frac{3(1-2\nu)r_1}{R^5} - \frac{15(3-4\nu)R_1 r_2^2 + 30c\bar{x}R_1 - 30(1-2\nu)cr_2^2}{R^7} + \frac{210c\bar{x}R_1 r_2^2}{R^9} \Big\}$$

$$\frac{\partial\sigma^c_{132}}{\partial x_1} = K_s\, r_2 r_3 \Big\{ - \frac{3}{R^5} + \frac{15(3-4\nu)R_1^2 - 30(1-2\nu)cR_1 + 30\bar{x}(c+R_1)}{R^7} - \frac{210c\bar{x}R_1^2}{R^9} \Big\}$$

$$\frac{\partial\sigma^c_{132}}{\partial x_2} = K_s\, r_3 \Big\{ \frac{3(3-4\nu)R_1 - 6c(1-2\nu)}{R^5} - \frac{30c\bar{x}R_1 + 15(3-4\nu)R_1 r_2^2 - 30c(1-2\nu)r_2^2}{R^7} + \frac{210c\bar{x}R_1 r_2^2}{R^9} \Big\}$$

$$\frac{\partial\sigma^c_{132}}{\partial x_3} = K_s\, r_2 \Big\{ \frac{3(3-4\nu)R_1 - 6c(1-2\nu)}{R^5} - \frac{30c\bar{x}R_1 + 15(3-4\nu)R_1 r_3^2 - 30c(1-2\nu)r_3^2}{R^7} + \frac{210c\bar{x}R_1 r_3^2}{R^9} \Big\}$$

$$\frac{\partial\sigma^c_{222}}{\partial x_1} = K_s\, r_2 \Big\{ - \frac{3(11-18\nu+8\nu^2)R_1 - 6(3+2\nu)c}{R^5} + \frac{15(3-4\nu)R_1 r_2^2 - 90c^2 R_1 + 30(3-2\nu)cR_1^2 + 30\bar{x}r_2^2}{R^7}$$

$$- \frac{210c\bar{x}R_1 r_2^2}{R^9} + \frac{4(1-\nu)(1-2\nu)}{R^3(R+R_1)^2}\Big[3(2R+R_1) - \frac{(8R^2+9RR_1+3R_1^2)r_2^2}{R^2(R+R_1)} \Big] \Big\}$$

$$\frac{\partial\sigma^c_{222}}{\partial x_2} = K_s \Big\{ - \frac{(1-2\nu)(5-4\nu)}{R^3} + \frac{3(14-26\nu+8\nu^2)r_2^2 - 18c^2 + 6c(3-2\nu)R_1}{R^5} + \frac{210c\bar{x}r_2^4}{R^9}$$

$$- \frac{15r_2^2[6c\bar{x} + (3-4\nu)r_2^2 - 6c^2 + 2(3-2\nu)cR_1]}{R^7} + \frac{210c\bar{x}r_2^4}{R^9}$$

$$+ \frac{4(1-\nu)(1-2\nu)}{R(R+R_1)^2}\Big[3 - \frac{6r_2^2(3R+R_1)}{R^2(R+R_1)} + \frac{3r_2^4(5R^2+4RR_1+R_1^2)}{R^4(R+R_1)^2} \Big] \Big\}$$

$$\frac{\partial\sigma^c_{222}}{\partial x_3} = K_s\, r_2 r_3 \Big\{ \frac{3(1-2\nu)(5-4\nu)}{R^5} - \frac{15(3-4\nu)r_2^2 - 90c^2 + 30(3-2\nu)cR_1}{R^7} + \frac{210c\bar{x}r_2^2}{R^9}$$

$$- \frac{4(1-\nu)(1-2\nu)}{R^3(R+R_1)^3}\Big[3(3R+R_1) - \frac{3r_2^2(5R^2+4RR_1+R_1^2)}{R^2(R+R_1)} \Big] \Big\}$$

$$\frac{\partial\sigma^c_{232}}{\partial x_1} = K_s\, r_3 \Big\{ - \frac{3(1-2\nu)R_1 + 6\bar{x}}{R^5} + \frac{15(3-4\nu)R_1 r_2^2 + 30c\bar{x}R_1 + 30\bar{x}r_2^2}{R^7} - \frac{210c\bar{x}R_1 r_2^2}{R^9}$$

$$+ \frac{4(1-\nu)(1-2\nu)}{R^3(R+R_1)^2}\left[2R+R_1 - \frac{(8R^2+9RR_1+3R_1^2)r_2^2}{R^2(R+R_1)} \right] \Bigg\}$$

$$\frac{\partial\sigma^c_{232}}{\partial x_2} = K_s\, r_2 r_3 \Bigg\{ \frac{3(7-10\nu)}{R^5} - \frac{15(3-4\nu)r_2^2 + 90c\bar{x}}{R^7} + \frac{210c\bar{x}r_2^2}{R^9}$$
$$- \frac{4(1-\nu)(1-2\nu)}{R^3(R+R_1)^3}\left[3(3R+R_1) - \frac{3r_2^2(5R^2+4RR_1+R_1^2)}{R^2(R+R_1)} \right] \Bigg\}$$

$$\frac{\partial\sigma^c_{232}}{\partial x_3} = K_s \Bigg\{ -\frac{1-2\nu}{R^3} + \frac{3(3-4\nu)r_2^2 + 6c\bar{x} + 3(1-2\nu)r_3^2}{R^5} - \frac{30c\bar{x}r_2^2 + 15(3-4\nu)r_2^2r_3^2 + 30c\bar{x}r_3^2}{R^7}$$
$$+ \frac{210c\bar{x}r_2^2r_3^2}{R^9} + \frac{4(1-\nu)(1-2\nu)}{R(R+R_1)^2}\left[1 - \frac{(3R+R_1)(R-R_1)}{R^2} + \frac{3r_2^2r_3^2(5R^2+4RR_1+R_1^2)}{R^4(R+R_1)^2} \right] \Bigg\}$$

$$\frac{\partial\sigma^c_{332}}{\partial x_1} = K_s\, r_2 \Bigg\{ \frac{6(1+2\nu)c - 3(1-2\nu)(5-4\nu)R_1}{R^5} + \frac{15(3-4\nu)R_1r_3^2 - 30c^2R_1 + 30(1-2\nu)cR_1^2 + 30\bar{x}r_3^2}{R^7}$$
$$- \frac{210c\bar{x}R_1r_3^2}{R^9} + \frac{4(1-\nu)(1-2\nu)}{R^3(R+R_1)^2}\left[2R+R_1 - \frac{(8R^2+9RR_1+3R_1^2)r_3^2}{R^2(R+R_1)} \right] \Bigg\}$$

$$\frac{\partial\sigma^c_{332}}{\partial x_2} = K_s \Bigg\{ -\frac{(1-2\nu)(3-4\nu)}{R^3} + \frac{3(3-4\nu)r_3^2 - 6c^2 + 6(1-2\nu)cR_1 + 3(1-2\nu)(3-4\nu)r_2^2}{R^5}$$
$$- \frac{30c\bar{x}r_3^2 + 15(3-4\nu)r_2^2r_3^2 - 30c^2r_2^2 + 30(1-2\nu)cR_1r_2^2}{R^7} + \frac{210c\bar{x}r_2^2r_3^2}{R^9}$$
$$+ \frac{4(1-\nu)(1-2\nu)}{R(R+R_1)^2}\left[1 - \frac{(3R+R_1)(R-R_1)}{R^2} + \frac{3r_2^2r_3^2(5R^2+4RR_1+R_1^2)}{R^4(R+R_1)^2} \right] \Bigg\}$$

$$\frac{\partial\sigma^c_{332}}{\partial x_3} = K_s\, r_2 r_3 \Bigg\{ \frac{3(3-4\nu)(3-2\nu)}{R^5} - \frac{15(3-4\nu)r_3^2 - 30c^2 + 30(1-2\nu)cR_1 + 60c\bar{x}}{R^7} + \frac{210c\bar{x}r_3^2}{R^9}$$
$$- \frac{4(1-\nu)(1-2\nu)}{R^3(R+R_1)^3}\left[3(3R+R_1) - \frac{3r_3^2(5R^2+4RR_1+R_1^2)}{R^2(R+R_1)} \right] \Bigg\}$$

$$\frac{\partial\sigma^c_{113}}{\partial x_1} = \frac{r_3}{r_2}\frac{\partial\sigma^c_{112}}{\partial x_1}, \qquad \frac{\partial\sigma^c_{113}}{\partial x_2} = \frac{\partial\sigma^c_{112}}{\partial x_3}$$

$$\frac{\partial\sigma^c_{113}}{\partial x_3} = K_s \cdot \Bigg\{ \frac{1-2\nu}{R^3} + \frac{3(3-4\nu)R_1^2 - 6c[c+(1-2\nu)R_1] - 3(1-2\nu)r_3^2}{R^5}$$
$$- \frac{30c\bar{x}R_1^2 + 15(3-4\nu)R_1^2r_3^2 - 30[c+(1-2\nu)R_1]cr_3^2}{R^7} + \frac{210c\bar{x}R_1^2r_3^2}{R^9} \Bigg\}$$

$$\frac{\partial\sigma^c_{123}}{\partial x_1} = \frac{\partial\sigma^c_{132}}{\partial x_1}, \qquad \frac{\partial\sigma^c_{123}}{\partial x_2} = \frac{\partial\sigma^c_{132}}{\partial x_2}, \qquad \frac{\partial\sigma^c_{123}}{\partial x_3} = \frac{\partial\sigma^c_{132}}{\partial x_3},$$

$$\frac{\partial\sigma^c_{133}}{\partial x_1} = K_s \cdot \Bigg\{ -\frac{1-2\nu}{R^3} - \frac{3(1-2\nu)R_1r_1 + 3r_3^2 + 6\bar{x}(c+R_1)}{R^5}$$

$$+ \frac{15(3-4\upsilon)R_1^2 r_3^2 + 30c\bar{x}R_1^2 - 30(1-2\upsilon)cR_1 r_3^2 + 30\bar{x}R_1 r_3^2 + 30c\bar{x}r_3^2}{R^7} - \frac{210c\bar{x}R_1^2 r_3^2}{R^9}\Bigg\}$$

$$\frac{\partial\sigma^c_{133}}{\partial x_2} = K_s\, r_2 \left\{\frac{3(1-2\upsilon)r_1}{R^5} - \frac{15(3-4\upsilon)R_1 r_3^2 + 30c\bar{x}R_1 - 30(1-2\upsilon)cr_3^2}{R^7} + \frac{210c\bar{x}R_1 r_3^2}{R^9}\right\}$$

$$\frac{\partial\sigma^c_{133}}{\partial x_3} = K_s\, r_3 \left\{\frac{3(1-2\upsilon)r_1 + 6(3-4\upsilon)R_1 - 12(1-2\upsilon)c}{R^5} - \frac{15(3-4\upsilon)R_1 r_3^2 + 90c\bar{x}R_1 - 30(1-2\upsilon)cr_3^2}{R^7} + \frac{210c\bar{x}R_1 r_3^2}{R^9}\right\}$$

$$\frac{\partial\sigma^c_{223}}{\partial x_1} = K_s\, r_3 \left\{\frac{6(1+2\upsilon)c - 3(1-2\upsilon)(5-4\upsilon)R_1}{R^5} + \frac{15(3-4\upsilon)R_1 r_2^2 - 30c^2R_1 + 30(1-2\upsilon)cR_1^2 + 30\bar{x}r_2^2}{R^7} - \frac{210c\bar{x}R_1 r_2^2}{R^9} + \frac{4(1-\upsilon)(1-2\upsilon)}{R^3(R+R_1)^2}\left[2R+R_1 - \frac{(8R^2+9RR_1+3R_1^2)r_2^2}{R^2(R+R_1)}\right]\right\}$$

$$\frac{\partial\sigma^c_{223}}{\partial x_2} = K_s\, r_2 r_3 \left\{\frac{3(3-4\upsilon)(3-2\upsilon)}{R^5} - \frac{15(3-4\upsilon)r_2^2 - 30c^2 + 30(1-2\upsilon)cR_1 + 60c\bar{x}}{R^7} + \frac{210c\bar{x}r_2^2}{R^9} - \frac{4(1-\upsilon)(1-2\upsilon)}{R^3(R+R_1)^3}\left[3(3R+R_1) - \frac{3r_2^2(5R^2+4RR_1+R_1^2)}{R^2(R+R_1)}\right]\right\}$$

$$\frac{\partial\sigma^c_{223}}{\partial x_3} = K_s \left\{-\frac{(1-2\upsilon)(3-4\upsilon)}{R^3} + \frac{3(3-4\upsilon)r_2^2 - 6c^2 + 6(1-2\upsilon)cR_1 + 3(1-2\upsilon)(3-4\upsilon)r_3^2}{R^5} - \frac{30c\bar{x}r_2^2 + 15(3-4\upsilon)r_2^2r_3^2 - 30c^2r_3^2 + 30(1-2\upsilon)cR_1r_3^2}{R^7} + \frac{210c\bar{x}r_2^2r_3^2}{R^9} + \frac{4(1-\upsilon)(1-2\upsilon)}{R(R+R_1)^2}\left[1 - \frac{(3R+R_1)(R-R_1)}{R^2} + \frac{3r_2^2r_3^2(5R^2+4RR_1+R_1^2)}{R^4(R+R_1)^2}\right]\right\}$$

$$\frac{\partial\sigma^c_{233}}{\partial x_1} = K_s\, r_2 \left\{-\frac{3(1-2\upsilon)R_1 + 6\bar{x}}{R^5} + \frac{15(3-4\upsilon)R_1 r_3^2 + 30c\bar{x}R_1 + 30\bar{x}r_3^2}{R^7} - \frac{210c\bar{x}R_1 r_3^2}{R^9} + \frac{4(1-\upsilon)(1-2\upsilon)}{R^3(R+R_1)^2}\left[2R+R_1 - \frac{(8R^2+9RR_1+3R_1^2)r_3^2}{R^2(R+R_1)}\right]\right\}$$

$$\frac{\partial\sigma^c_{233}}{\partial x_2} = K_s \left\{-\frac{1-2\upsilon}{R^3} + \frac{3(3-4\upsilon)r_3^2 + 6c\bar{x} + 3(1-2\upsilon)r_2^2}{R^5} - \frac{30c\bar{x}r_3^2 + 15(3-4\upsilon)r_2^2r_3^2 + 30c\bar{x}r_2^2}{R^7} + \frac{210c\bar{x}r_2^2r_3^2}{R^9} + \frac{4(1-\upsilon)(1-2\upsilon)}{R(R+R_1)^2}\left[1 - \frac{(3R+R_1)(R-R_1)}{R^2} + \frac{3r_2^2r_3^2(5R^2+4RR_1+R_1^2)}{R^4(R+R_1)^2}\right]\right\}$$

$$\frac{\partial\sigma^c_{233}}{\partial x_3} = K_s\, r_2 r_3 \left\{\frac{3(7-10\upsilon)}{R^5} - \frac{15(3-4\upsilon)r_3^2 + 90c\bar{x}}{R^7} + \frac{210c\bar{x}r_3^2}{R^9} - \frac{4(1-\upsilon)(1-2\upsilon)}{R^3(R+R_1)^3}\left[3(3R+R_1) - \frac{3r_3^2(5R^2+4RR_1+R_1^2)}{R^2(R+R_1)}\right]\right\}$$

$$\frac{\partial \sigma^{c}_{333}}{\partial x_1} = K_s\, r_3 \left\{ - \frac{3(11-18\upsilon+8\upsilon^2)R_1 - 6(3+2\upsilon)c}{R^5} + \frac{15(3-4\upsilon)R_1 r_3^2 - 90c^2R_1 + 30(3-2\upsilon)cR_1^2 + 30\bar{x}r_3^2}{R^7} \right.$$

$$\left. - \frac{210c\bar{x}R_1 r_3^2}{R^9} + \frac{4(1-\upsilon)(1-2\upsilon)}{R^3(R+R_1)^2}\left[3(2R+R_1) - \frac{(8R^2+9RR_1+3R_1^2)r_3^2}{R^2(R+R_1)} \right] \right\}$$

$$\frac{\partial \sigma^{c}_{333}}{\partial x_2} = K_s\, r_2 r_3 \left\{ \frac{3(1-2\upsilon)(5-4\upsilon)}{R^5} - \frac{15(3-4\upsilon)r_3^2 - 90c^2 + 30(3-2\upsilon)cR_1}{R^7} + \frac{210c\bar{x}r_3^2}{R^9} \right.$$

$$\left. - \frac{4(1-\upsilon)(1-2\upsilon)}{R^3(R+R_1)^3}\left[3(3R+R_1) - \frac{3r_3^2(5R^2+4RR_1+R_1^2)}{R^2(R+R_1)} \right] \right\}$$

$$\frac{\partial \sigma^{c}_{333}}{\partial x_3} = K_s \left\{ - \frac{(1-2\upsilon)(5-4\upsilon)}{R^3} + \frac{3(14-26\upsilon+8\upsilon^2)r_3^2 - 18c^2 + 6c(3-2\upsilon)R_1}{R^5} \right.$$

$$- \frac{15r_3^2[6c\bar{x} + (3-4\upsilon)r_3^2 - 6c^2 + 2(3-2\upsilon)cR_1]}{R^7} + \frac{210c\bar{x}r_3^4}{R^9}$$

$$\left. + \frac{4(1-\upsilon)(1-2\upsilon)}{R(R+R_1)^2}\left[3 - \frac{6r_3^2(3R+R_1)}{R^2(R+R_1)} + \frac{3r_3^4(5R^2+4RR_1+R_1^2)}{R^4(R+R_1)^2} \right] \right\}$$

Boundary Element Formulation for Non-linear Problems using the Reissner's Theory

G.O. Ribeiro and W.S. Venturini

Federal University of Minas Gerais, Br. and São Carlos School of Engineering, University of São Paulo, Br.

ABSTRACT

Many works dedicated to analysis of plate bending using the boundary element method have already been published so far. The majority of them is based on the Kirchhoff's approximations and represents improvements in the context of linear analysis using BEM. In this paper a nonlinear approach to model important plate behaviour usually found in engineering applications is proposed together with the Reissner's model.

INTRODUCTION

The use of boundary elements to analyse plates in bending was proposed in 1968, based on the so-called indirect methods [1]. Since this first work many others have already been elaborated using both indirect and direct approaches to establish the proper basis of the method within the contex of Kirchhoff's theory. Among those works it is important to point out those conducted by BEZINE [2,3], STERN [4], TOTTENHAM [5] which represent important contributions to the boundary methods. The simplifications given by that theory lead to a simple boundary formulation appropriate for many applications in engineering. Nevertheless, several other engineering problems require more precise models. Analysis of thick plate, for instance, can only be carried out if the shear effects on the displacements are taken into account. The Reissner's model [6,7], therefore, appears to have the right assumptions to derive all needed boundary integral equations. On the other hand, all singularity troubles of the boundary elements based on the Kirchhoff's hypotheses are eliminated. Thus,for some particular applications in engineering, even those related to thin plates, the Reissner's model could be more appropriate.

In this work the boundary element formulation for plate bending considering the Reissner's model is reviewed. The formulation proposed by WEEËN [8] is extended to incorporate terms suitable to model initial moment fields applied over the domain. This complement permits temperature and shrinkage effects modelling, as well as to approach nonlinear behaviours modelled by an initial stress-like procedure. As often used in boundary elements, the collocation points needed to write the integral representations are defined out of the domain, avoiding therefore dealing with singular boundary integrals.

REISSNER'S THEORY FOR PLATES. BASIC EQUATIONS

For a finite plate defined for a domain Ω with boundary Γ and thickness h, the following assumptions related to stresses are made

$$\sigma_{x3} = 0 \qquad \text{and} \qquad \sigma_{33} = q/2 \tag{1}$$

x_3 is defined as an axis perpendicular to plate plane while x_1 and x_2 are implane axes. Greek letters are adopted to represent indices varying from 1 to 2 and latin letters indicate indices in the 1 to 3 range.

It is also assumed, according to the Reissner's hypotheses, the following effort/stress relation,

$$\begin{aligned} \sigma_{\alpha\beta} &= (12x_3/h^3)\, M_{\alpha\beta} \\ \sigma_{\alpha 3} &= (3/2h)[1-(2x_3/h)^2]\, Q_\alpha \\ \sigma_{33} &= 0.25(2x_3/h)[3-(2x_3/h)^2]q \end{aligned} \tag{2}$$

where $M_{\alpha\beta}$ represents bending and twisting moments and Q_α stands for the shear forces.

The displacement/effort relations, also following the Reissner's theory, are given by,

$$\begin{aligned} M_{\alpha\beta} &= \frac{(1-\nu)}{2} D\left(\phi_{\alpha,\beta} + \phi_{\beta,\alpha} + \frac{2\nu}{1-\nu}\phi_{\gamma,\gamma}\,\delta_{\alpha\beta}\right) + \frac{\nu}{1-\nu}\frac{q}{\lambda^2}\delta_{\alpha\beta} \\ Q_\alpha &= D\frac{1-\nu}{2}\lambda^2(\phi_\alpha + w_{,\alpha}) \end{aligned} \tag{3}$$

where

$$D = Eh^3/[12(1-\nu^2)] \qquad\qquad \lambda = \sqrt{10}/h$$

and ϕ_α and w are rotations and middle surface transversal displacement respectively assumed to be evaluated by the following formulas,

$$\phi_\alpha = \int_{-h/2}^{h/2} \frac{12x_3}{h^3} v_\alpha dx_3$$

$$w = \int_{-h/2}^{h/2} \frac{3}{2h} \left[1 - (2 \frac{x_3}{h})\right] v_3 dx_3 \tag{4}$$

Therefore, ϕ_α and w are values averaged through the plate thickness using the actual displacements v_1, v_2 and v_3.

For simplicity and whenever possible, rotations and displacements will be denoted by their generalized forms, u_1, u_2 and u_3; the two first components are rotations and the last one is the transversal displacement. Thus, the two classical conditions to be enforced at any boundary point are given by,

$$u_i(S) = \bar{u}_i(S) \qquad S \text{ on } \Gamma_1$$

$$p_i(S) = \bar{p}_i(S) \qquad S \text{ on } \Gamma_2 \tag{5}$$

where $p_i(S)$, generalized tractions, are obtained from

$$p_\alpha(S) = M_{\alpha\beta}(S)\eta_\beta(S)$$

$$p_3(S) = Q_\alpha(S)\eta_\alpha(S) \tag{6}$$

and η_α stands for the outward normal cosines as indicated in Figure 1.

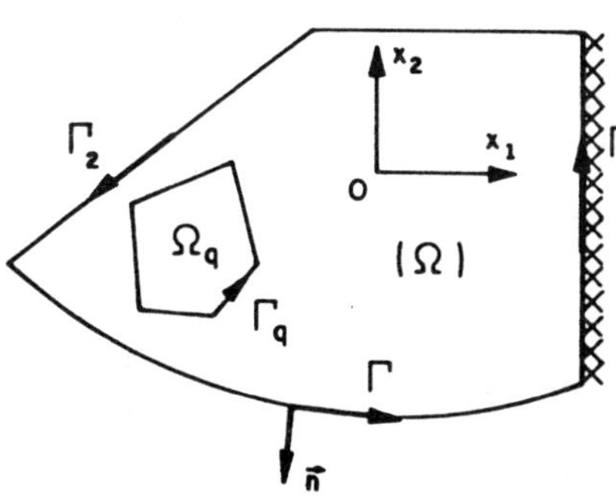

Figure 1 - Plate definition

Substituting equation (3) into the equilibrium relations gives a six order differential equation system that makes it possible to take into account the three boundary conditions in terms of displacements or the corresponding efforts. It substantially differs from the Kirchhoff's hypotheses for which it is only possible to prescribe two conditions.

As for any nonlinear problem, one has to consider an initial moment field applied to the plate domain. Thus, the actual moment values at any plate point is given by its elastic part less the applied moment, i.e.,

$$M_{\alpha\beta}(s) = {}^{e}M_{\alpha\beta}(s) - M^{0}_{\alpha\beta}(s) \tag{7}$$

where

$$M^{e}_{\alpha\beta}(s) = \frac{1-\nu}{2} D\left[u_{\alpha,\beta}(s) + u_{\beta,\alpha}(s) + \frac{2\nu}{1-\nu} u_{\gamma,\gamma} \delta_{\alpha\beta})\right] \tag{8}$$

or

$$M_{\alpha\beta}(s) = C_{\alpha\beta\gamma\theta} \chi_{\gamma\theta} (s) \tag{9}$$

in which $\chi_{\gamma\theta}$ is the curvature tensor.

In the case of applying an initial curvature field, as used to model temperature variation effects, the corresponding initial moments can be obtained by,

$$M^{0}_{\alpha\beta}(s) = C_{\alpha\beta\gamma\theta} \chi^{0}_{\gamma\theta}(s) \tag{10}$$

For a classical nonlinear problem the initial moment field is obtained directly from the application of the constitutive relation that governs the problem.

INTEGRAL EQUATIONS

First of all, one can obtain the integral representation of displacements using either a reciprocal identity or the weighted residual method. Using one of those procedures the integral displacement representation for a point P, given in terms of generalized tractions and displacements, is expressed in the following form [8],

$$K(P)u_i(P) = \int\left[u^*_{ij}(P,S)p_i(S) - p^*_{ij}(P,S)u_j(S)\right]d\Gamma(S) +$$

$$+ \int_{\Omega_g}\left[u^*_{i3}(P,s) - \frac{\nu}{(1-\nu)\lambda^2} u^*_{i\alpha,\alpha}(P,s)\right]g(s)d\Omega(s) +$$

$$+ \int_{\Omega} \hat{M}_{\alpha\beta}(s)u^*_{i\alpha,\beta}(P,s)d\Omega(s) \tag{11}$$

where $u^*_{ij}(P,S)$ and $p^*_{ij}(P,S)$ are fundamental values of displacements and tractions respectively; K(P) is 1.0 for points inside the domain, zero when the point does not belong to the domain and given in terms of boundary geometry if defined along the contour; it is equal to 0.5 for any smooth boundary point.

For simplicity, the generation of all integral displacement representations needed to define the system of equations was conducted by selecting only outside collocation points, therefore K(P) vanishes.

Differentiating equation (1), given for an internal point and taking into account the proper concept of derivatives of singular integrals, and substituting these values into equations (3), one obtains the integral representations for bending and twisting moments as well as for shear forces, as follows,

$$M_{\alpha\beta}(p) = \int_{\Gamma} \left[u^*_{\alpha\beta k}(p,S)p_k(S) - p^*_{\alpha\beta k}(p,S)\, u_k(S) \right] d\Gamma(S) +$$

$$+ q \int_{\Gamma_q} w^*_{\alpha\beta}(p,S)d\Gamma(S) + \int_{\Omega} e^*_{\alpha\beta\gamma\theta}(p,s)\, M^o_{\gamma\theta}(s)d\Omega(s) +$$

$$+ M^0_{\alpha\beta}(p)\; g_{\alpha\beta\gamma\theta}(p) - M^0_{\alpha\beta}(p)$$

$$Q_{\beta}(p) = \int_{\Gamma} \left[u^*_{3\beta k}(p,S)p_k(S) - p^*_{3\beta k}(p,S)u_k(S) \right] d\Gamma(S)$$

$$+ q \int_{\Gamma_q} w^*_{3\beta}(p,S)d\Gamma(S) + \int_{\Omega} e^*_{3\beta\gamma\theta}(p,s)M^0_{\gamma\theta}(s)d\Omega(s) \qquad (12)$$

All the fundamental values used above to write either bending and twisting moments or shear forces are given in the appendix.

NUMERICAL SOLUTION TECHNIQUE

The boundary was assumed to be divided into straight elements over which quadratic shape functions were adopted to approach displacements and tractions. The nodes were chosen at the element ends and also at element middle points. For sharp boundary points, the double node concept was adopted enforcing discontinuities of all variables. For each boundary node an external point is defined following the element outward normal direction. Its position is fixed by averaging the adjacent element lengths, with an appropriate reduction.

The domain integrals are approached by the definition of internal cells over which all initial moment values are assumed to be linear with nodal values defined at the cell corners. Taking into account these approximations, equation (11) can be rewritten in its algebraic form, for an appropriate number of external collocation points, establishing a system of equations given in the following matrix form,

$$\underset{\sim}{H}\,\underset{\sim}{U} = \underset{\sim}{G}\,\underset{\sim}{P} + \underset{\sim}{B} + \underset{\sim}{E}\,\underset{\sim}{M}_0 \tag{13}$$

Vectors U and P contain displacement and traction nodal values respectively while M_o is the vector that gives initial moment values obtained at all cell corners defined by the internal discretization.

After rearranging equation (13), keeping all unknowns at the left hand side, one obtains,

$$\underset{\sim}{A}\,\underset{\sim}{X} = \underset{\sim}{F} + \underset{\sim}{E}\,\underset{\sim}{M}_0 \tag{14}$$

where vector F considers all known values. E contains the influences of the applied initial moment field. It is computed by numerical integration over all cells defined over the plate domain. For this integral, no strong singularity appears then computing those values does not require special technique.

Following a similar way equations (12) can be written in their discretized form, as follows,

$$\underset{\sim}{M} + -\underset{\sim}{H}'\underset{\sim}{U} + \underset{\sim}{G}'\underset{\sim}{P} + B' + \underset{\sim}{E}'\underset{\sim}{M}^0$$

and (15)

$$\underset{\sim}{Q} = -\underset{\sim}{H}''\underset{\sim}{U} + \underset{\sim}{G}''\underset{\sim}{P} + \underset{\sim}{B}'' + \underset{\sim}{E}''\underset{\sim}{M}^0$$

The vectors M and Q contain efforts at internal points only. For boundary nodes, one has to write moments and shear forces using the components already known after finding out the vector X in equation (14) and applying a finite difference scheme to express the unknown components in terms of boundary displacements and rotations. Employing such a scheme all efforts can be written in the form shown in equation (15).

All the matrices in equation (15), can be computed by numerical schemes, even when computing E' and E" that lead one to deal with strong singular integrals. These matrices, E' and E", contain influence coefficients due to the application of initial moments fields. To integrate cells that do not contain the singular point a numerical procedure may be more appropriate. However, if the singular point is a cell corner, strong singular integrals must be carried out what might indicate an analytical scheme.

In order to compute analytically those singular terms the cell can be divided into three small triangular regions, as indicated in Figure 2. The triangle adjacent to the singular point P must have the two adjacent sides not longer than 2/λ to allow the use of appropriate Bessel expantions. Then, the three sub-cells are defined as follows: the adjacent one, $K_1K_4K_5$, with two sides smaller than 2/λ, and the non singular ones, $K_4K_3K_5$ and $K_2K_3K_4$, with no side length restriction.

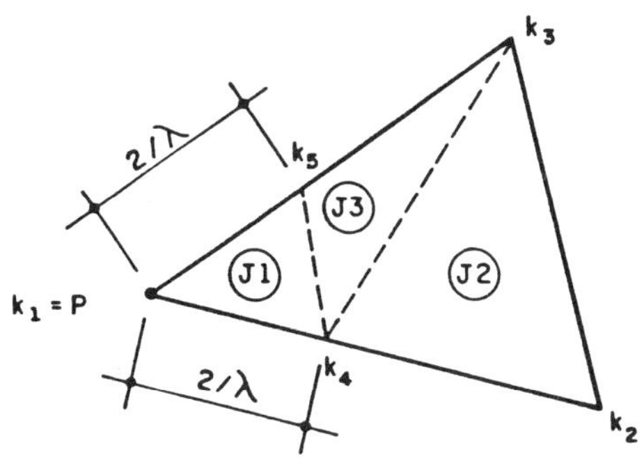

Figure 2 - Cell J sub-division.

By applying Bessel expantions we can clearly see the type of singularity we have. After that an integration over one dimension, the distance r(P,s) considering a cylindrical system of coordinates with origin at P, can be performed. Special attention must be paid to integrate the adjacent triangle for which principal values must be properly taken into account. One must also note that different expantions have to be used for the adjacent triangular area and the other two triangles (see Abramovitz [9]).

After integrating all cells numerically, or analytically for the case of small regions around singularities, matrices E' and E" are found and all initial moment field influences algebraically represented in equations (15).

EXAMPLES

In order to show the accuracy of the presented technique two simple examples are analysed. In both cases a square plate is studied when subjected to a known initial moment field. The obtained results can illustrate very well the quality of the responses that the technique can achieve.

Example 1 - A unit moment field applied to a square plate.

Initial moments, $M^0_{11} = 1.0$, are applied all over a square plate simply supported along only two sides, as in Figure 3. Each side has been divided into four quadratic elements giving the total amount of 36 boundary nodes. 128 triangular cells defined by 49 internal points were also employed to perform all domain integrals needed take into account the initial moment influences.

The results computed for this case are given in Table 1. As one can see, the errors are at the fourth figure. Note that such a loading is equivalent to a uniform temperature gradient applied all over the plate domain.

Table 1 - Central line displacements

x_1	$\frac{w}{K\ell^2}$		$\frac{\sigma x_1}{K\ell}$	
	EXACT	B.E.M.	EXACT	B.E.M.
0.	0.	0.	0.50000	0.49976
$\ell/8$	0.05469	0.05467	0.37500	0.37489
$\ell/4$	0.09375	0.09372	0.25000	0.24991
$3\ell/8$	0.11719	0.11715	0.12500	0.12495
$\ell/2$	0.12500	0.12496	0.	0.

Example 2 - Clamped plate

The same geometry adopted for the example 1 has been used again. As boundary conditions, displacements and rotations are enforced to be zero along all plate sides. Unit initial moments are applied over the whole domain, i.e.

$$M^0_{11} = M^0_{22} = D(1+\nu)\,\frac{\alpha t}{h} = 1.0$$

where t is an equivalent temperature gradient.

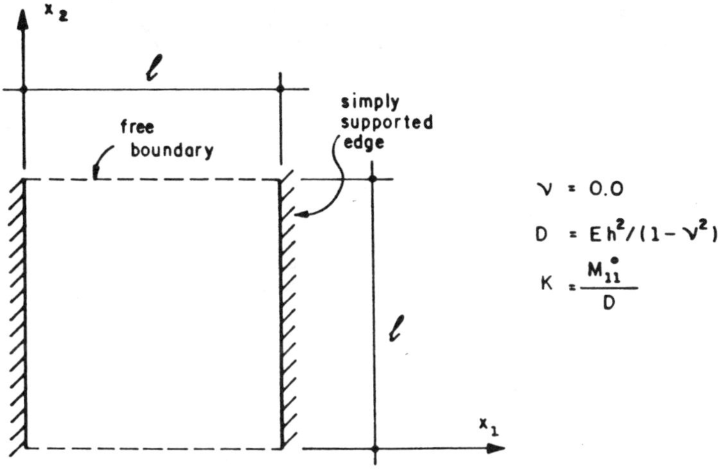

Figure 3 - Plate under analysis

For this field of initial moments, exactly the same values should be obtained by solving the proper system of equations. The numerical solution found by applying the proposed scheme and using the same discretizations described for the example 1, shows the accuracy of the technique. Once again the errors appear only at the fourth figure. At all points where moments should vanish values around 10^{-7} were computed.

CONCLUSIONS

The paper presents the boundary element formulation for plate bending based on the Reissner's theory. It has been extended to include the possibility of applying, to the plate, initial moment or curvature loadings such as those used to analyse temperature gradient effects. The formulation accuracy has been verified with simple numerical examples. How it can be adapted to model nonlinear behaviours was also discussed throughout the paper.

REFERENCES

1. JASWON, M.A. and MAITI, M. An Integral Equation Formulation of Plate Bending Problems, J. Eng.Math., 2, 33-93, 1968.

2. BEZINE, G.P. and GAMBY, D.A. A New Integral Equation Formulation for Plate Bending Problems. Proc. of the First Int. Sem. on Recent Advances in Boundary Element Methods, Pentech Press, p. 327-342, 1978.

3. BEZINE, G.P., Boundary Integral Formulation for Plate Flexure with Arbitrary Boundary Conditions, Mech. Res. Commun., 5, p.197-206, 1978.

4. STERN, M. A General Integral Formulation for the Numerical Solution of Plate Bending Problems. Int. J. Solids and Structures, 15, p.769-782, 1979.

5. TOTTENHAM, H. The Boundary Element Method for Plates and Shells. In: Developments in Boundary Element Methods, ed. by Banerjee, P.K. and Butterfield, R., Appl. Sciences Publ. 1979.

6. REISSNER, "On the Theory of Bending of Elastic Plates". J.Math. Ph., 23, p.184-191, 1944.

7. REISSNER, E. The Effect of Transverse Shear Deformation on the Bending of Elastic Plates. J.Appl.Mech., 12, p. A69-A77, 1945.

8. WEEËN, V.F. Application of the Boundary Integral Equation Method to Reissner's Plate Model. Int. J. Num. Meth. Eng., 18, p.1-10, 1982.

9. ABRAMOVITZ, M. and STEGNN, I.A. Handbook of Mathematical Functions. Dover Publications, New York, 1965.

APPENDIX

Fundamental Solutions

In this complementary section fundamental solutions for the plate bending problem formulated from the Reissner's approximations, as derived by WEEËN [8], are shown.

Displacement values are given by,

$$u^*_{\alpha\beta}(P,S) = \frac{1}{2\pi D}\left\{\frac{2}{1-\nu}\left[B\ \delta_{\alpha\beta} - A\ r,_{\alpha}\ r,_{\beta}\right] + \right.$$

$$\left. - \frac{1}{2}\ \delta_{\alpha\beta}(\ell nz - 1/2) - 1/2\ r,_{\alpha}\ r,_{\beta}\right\}$$

$$u^*_{\alpha 3}(P,S) = -u^*_{3\alpha}(P,S) = \frac{1}{8\pi D(1-\nu)}\ (2\ell nz - 1)rr,_{\alpha} \tag{A-1}$$

$$u^*_{33}(P,S) = \frac{1}{8\pi D(1-\nu)\lambda^2}\left[(1-\nu)z^2(\ell nz-1) - 8\ell nz\right]$$

where r is the distence between the load and the field points, z is equal to λr and A e B are given by,

$$A = K_0(z) + \frac{2}{z}\left[K_1(z) - 1/z\right]$$

$$B = K_0(z) + \frac{1}{z}\left[K_1(z) - 1/z\right] \qquad \text{(A-2)}$$

where $K_0(z)$ and $K_1(z)$ are modified Bessel functions. These functions can be evaluated by polynomials as in ABRAMOVITZ [9]. It can be seen in (A-2) that A is a continuum function while B has a singularity of type $z^{-1}\ell nz$

Traction fundamental solutions can be obtained from (A-1) and are expressed by,

$$p^{*}_{\gamma\alpha}(P,S) = \frac{-1}{4\pi r}\left[\left(4A + 2zK_1(z) + 1-\nu\right)(\delta_{\alpha\gamma}\, r,_n + r,_\alpha\, \eta_\gamma) + \right.$$

$$\left. + (4A+1+\nu)r,_\gamma\eta_\alpha - \left(16A+4zK_1(z)+2+2\nu\right)r,_\alpha\, r,_\gamma\, r,_n\right] \qquad \text{(A-3)}$$

$$p^{*}_{\gamma 3}(P,S) = \frac{\lambda^2}{2\pi}\left[B\eta_\gamma - Ar,_\gamma\, r,_n\right]$$

$$p^{*}_{3\gamma}(P,S) = -\frac{(1-\nu)}{8\pi}\left\{\left[\frac{2(1+\nu)}{(1-\nu)}\ell nz - 1\right]\eta_\alpha + 2r,_\alpha\, r,_n\right\}$$

$$p^{*}_{33}(P,S) = -\frac{1}{2\pi r}\, r,_n$$

The Kernels appeared in the integral representations for bending moments and shear forces, equations (12), are given by,

$$u^{*}_{\alpha\beta\gamma}(p,S) = \frac{1}{4\pi r}\left[\left(4A+2zK_1(z)+1-\nu\right)(\delta_{\beta\gamma}\, r,_\alpha + \delta_{\alpha\gamma}\, r,_\beta) + \right.$$

$$\left. - 2\left(8A+2zK_1(z)+1-\nu\right)r,_\alpha r,_\beta r,_\gamma + (4A+1-\nu)\delta_{\alpha\beta}r,_\gamma\right] \qquad \text{(A-4)}$$

$$u^{*}_{\alpha\beta 3}(p,S) = -\frac{-(1+\nu)}{8\pi}\left\{\left[\frac{2(1+\nu)}{(1-\nu)}\ell nz - 1\right]\delta_{\alpha\beta} + 2r,_\alpha r,_\beta\right\}$$

$$u^{*}_{3\alpha\beta}(p,S) = \frac{\lambda^2}{2\pi}\left[B\ell nz - Ar,_\gamma r,_\beta\right]$$

$$u^*_{3\beta 3}(p,S) = \frac{1}{2\pi r} r,_\beta$$

and

$$p^*_{\alpha\beta\gamma}(p,S) = \frac{D(1-\nu)}{4\pi r^2} \left\{ \left(4A+2zK_1(z)+1-\nu \right) (\delta_{\gamma\alpha}\, \eta_\beta + \delta_{\gamma\beta}\eta_\alpha) + \right.$$

$$+(4A+1+3\nu)\delta_{\alpha\beta}\, \eta_\gamma \quad \left(16A+6zK_1(z)+z^2K_0(z)+2-2\nu \right) \left[(\eta_\alpha r,_\beta + \right.$$

$$\left. +\eta_\beta r,_\alpha) r,_\gamma (\delta_{\gamma\alpha} r,_\beta + \delta_{\gamma\beta} r,_\alpha) r,_n \right] -2 \left(8A+2zK_1(z)+1+\nu \right) (\delta_{\alpha\beta} r,_\gamma r,_n +$$

$$\left. + \eta_\gamma r,_\alpha r,_\beta) + 4 \left(24A+8zK_1(z)+z^2K_0(z)+2-2\nu \right) r,_\alpha r,_\beta r,_\gamma r,_n \right\}$$

(A-5)

$$p^*_{\alpha\beta 3}(p,S) = \frac{D(1-\nu)\lambda^2}{4\pi r} \left[\left(2A+zK_1(z)(r,_\beta\eta_\alpha + r,_\alpha\eta_\beta \right) -2 \left(4A + \right. \right.$$

$$\left. \left. + zK_1(z) \right) r,_\alpha r,_\beta r,_n + 2A\delta_{\alpha\beta} r,_n \right]$$

$$p^*_{3\beta\gamma}(p,S) = \frac{-D(1-\nu)\lambda^2}{4\pi r} \left[\left(2A+zK_1(z) \right) (\delta_{\gamma\beta} r,_n + r,_\gamma \eta_\beta) + 2A\eta,_\gamma r,_\beta + \right.$$

$$\left. - 2 \left(4A+zK_1(z) \right) r,_\gamma r,_\beta r,_n \right]$$

$$p^*_{3\beta 3}(p,S) = \frac{D(1-\nu)\lambda^2}{4\pi r^2} \left[(z^2B+1)\eta_\beta - (z^2A+2) r,_\beta r,_n \right]$$

The Kernels of type with $w^*_{ij}(p,S)$ obtained by transforming the domain integral due to the transversal load into a boundary integral are given by.

$$w^*_{\alpha\beta}(p,S) = \frac{-r}{64\pi} \left\{ (4\ell nz - 3) \left[(1-\nu)(r,_\beta\eta_\alpha + r,_\alpha\eta_\beta) + \right. \right.$$

$$\left. \left. +(1+3\nu)\delta_{\alpha\beta} r,_n \right] + 4 \left[(1-\nu) r,_\alpha r,_\beta + \nu\delta_{\alpha\beta} \right] r,_n \right\} +$$

$$- \frac{\nu}{(1-\nu)\lambda}\, u^*_{\alpha\beta\gamma}(p,S)\eta_\gamma$$

(A-6)

$$w^*_{3\beta}(p,S) = \frac{1}{8\pi} \left[(2\ell nz - 1)\eta_\beta + 2r,_\beta r,_n \right] - \frac{\nu}{(1-\nu)\lambda^2}\, u^*_{3\beta\gamma}(p,S)\eta_\gamma$$

The Kernels in equations (11) and (12) related with the initial moment field were here property derived. For bending moment and shear force representations one has to differentiate equation (11) and apply their definition given in (3).

Taking into account an initial moment field, as in equation (11), the following Kernels can be obtained for the corresponding domain integrals.

$$u^{*}_{\gamma\alpha,\beta}(p,s) = -\frac{1}{4\pi D(1-\nu)r}\left[\left(4A+4zK_1(z)+1-\nu\right)r_{,\beta}\delta_{\alpha\gamma} + \right.$$

$$\left. - \left(16A+4zK_1(z)+2-2\nu\right)r_{,\alpha}r_{,\beta}r_{,\gamma} + (4A+1-\nu)(\delta_{\alpha\gamma}r_{,\beta}+\delta_{\beta\gamma}r_{,\alpha})\right] \quad \text{(A-7)}$$

$$u^{*}_{3\alpha,\beta}(p,s) = \frac{1}{8\pi D}\left[\delta_{\alpha\beta}(2\ell nz-1)+ 2r_{,\alpha}r_{,\beta}\right]$$

After applying bending moment and shear force definition the initial moment integral appear with the following Kernels,

$$e^{*}_{\alpha\beta\gamma\theta}(p,s)= \frac{-1}{8\pi r^2}\left\{\left[16A+12zK_1(z)+2(1-\nu)+4z^2K_0(z)\right](r_{,\theta}r_{,\beta}\delta_{\alpha\gamma}+\right.$$

$$+ r_{,\theta}r_{,\alpha}\delta_{\beta\gamma})+\left[16A+4zK_1(z)+2(1-\nu)\right]\left(2r_{,\beta}r_{,\gamma}\delta_{\alpha\theta} + 2r_{,\alpha}r_{,\beta}\delta_{\gamma\theta}+\right.$$

$$\left.2r_{,\gamma}r_{,\theta}\delta_{\alpha\beta} + 2r_{,\alpha}r_{,\gamma}\delta_{\beta\theta} + r_{,\alpha}r_{,\theta}\delta_{\beta\gamma} + r_{,\beta}r_{,\theta}\delta_{\alpha\gamma}\right) - (4A + 1-\nu)$$

$$(2\delta_{\gamma\theta}\delta_{\alpha\beta} + \delta_{\alpha\theta}\delta_{\gamma\beta} + \delta_{\beta\theta}\delta_{\gamma\alpha}) - \left(4A+4zK_1(z)+1-\nu\right)(\delta_{\alpha\theta}\delta_{\beta\gamma} + \delta_{\alpha\gamma}\delta_{\beta\theta})$$

$$-\left(96A+32zK_1(z)+4z^2K_0(z)+8(1-\nu)\right)(2r_{,\alpha}r_{,\beta}r_{,\gamma}r_{,\theta}) +$$

$$\left. + 4\nu\,\delta_{\alpha\beta}(2r_{,\gamma}r_{,\theta} - \delta_{\gamma\theta})\right\}$$

$$e^{*}_{3\beta\gamma\theta}(p,s) = \frac{-\lambda^2}{2\pi r}\left[\left(A + zK_1(z)\right) r_{,\theta}\,\delta_{\beta\gamma} - \left(4A+zK_1(z)\right)r_{,\beta}r_{,\gamma}r_{,\theta}+\right.$$

$$\left. + A(\delta_{\beta\theta}\, r_{,\gamma} + \delta_{\gamma\theta}\, r_{,\alpha})\right] \quad \text{(A-8)}$$

Due to the differentiation of singular integrals an independent term was obtained as shown in equations (12)

$$g_{\alpha\beta\gamma\theta}(p) = \frac{1}{8}\left[(3-\nu)(\delta_{\alpha\gamma}\,\delta_{\theta\beta} + \delta_{\beta\gamma}\,\delta_{\theta\alpha})-(1-3\nu)\delta_{\alpha\beta}\,\delta_{\gamma\theta}\right] \quad \text{(A-9)}$$

Contact Analysis using BEASY: Theory and Application

S.M. Niku, R.A. Adey, J. Baynham
Computational Mechanics, Ashurst Lodge, Ashurst, Southampton, Hants, SO4 2AA

1 Introduction

Major numerical tools like Finite Elements (FEM) and Boundary Elements (BEM) have advanced to the extent that most engineering analysis problems can be solved. However contact type problems, where two or more objects come together to form a contact zone, have presented considerable difficulties. The main reason for this is that the contact area is not known in advance and in fact the position of the contact edge is part of the required solution.

There are a number of types of contact problems and some special schemes have been developmed to solve them. There are also algorithms which can be used to solve axisymmetric and two dimensional problems but would not be feasable for three dimensional problems. The specification for the BEASY solver required the development of a scheme which could solve two dimensional axisymmetric and three dimensional contact problems. It had to be efficient in computer resources, accurate, reliable and be able to solve a comprehensive range of contact problems including thermal contact. This paper describes the options and decisions made when developing the contact analysis capabilities in the BEASY Boundary Element Analysis System [1].

2 Background

A number of authors have published work on contact analysis using Finite Elements and Boundary Elements. The contact analysis is normally performed in three stages:

- An analysis which computes the displacement and stresses etc.
- A contact analyser which determines the contact conditions from the analysis.

- The implementation of the contact conditions.

In all methods the three stages are repeatedly applied until the contact solution is obtained.

Pascoe [2] has published a survey of techniques used by Finite Elements researchers and described some of the problems and limitations. In implementing a contact algorithm in a large Boundary Element system like BEASY many of the lessons learned by FEM developers can be used as the contact problem is similar. In the contact area both methods have nodes and elements upon which the behaviour of the solution is described. The main differences are that the boundary element model has the surface traction as unknowns compared to the forces in the finite element method, the other factor is that the BEM has obviously elements only on the surface.

Many of the problems which cause problems with FEM are therefore avoided directly, e.g conversion of forces to surface tractions, working with stresses at the gauss point to avoid the –ve stress problems at the corner nodes of three dimensional elements, overlapping displaced shapes depending upon which nodes are used on the element.

In spite of the above many of the procedures required for boundary element contact analysis are similar to those used in finite elements but there are some distinct advantages to the boundary element approach which will be highlighted.

3 Boundary Element Contact Algorithms

A number of authors have published work on BEM contact analysis [3-5] of which these are a small selection. The intention in this paper is not to present a detailed description of the theoretical background of BEM contact analysis but to discuss some of the main issues and describe the results obtained using the BEM system.

3.1 Step 1 The initial analysis and load application

Assuming a problem has been defined which involves the contact of two or more bodies (or different parts of the same body) the first step is to perform a linear analysis. From the linear analysis the possible contact areas can be examined to check if any change in the configuration has occured (e.g areas come into contact or separated). If no change is necessary then the problem is effectively a linear contact problem and the problem is solved.

There are three main approaches to starting the non linear contact algorithm. These are:

1. Applying the full load and then iterating until all the equilibrium and compatibility equation are satisfied. This approach is not satisfactory for most problems as it fails to follow the loading path and causes errors in the contact configuration.

2. The second approach applies the loads incrementally where the size of the load step is determined by the size of load increment necessary to cause a change in the contact condition at any node. While this method accurately follows the development of the contact surface and eliminates the iterations it can cause excessive CPU times for problems with a number of nodes in the contact areas. This is caused by the very small load increments required. For problems with a small number of nodes in the contact area this approach is very successful.

3. The third approach which was chosen for use within BEASY requires the load to be split into a number of load increments. The increments are predefined and adjusted automatically depending upon the rate of convergence of the solution. For each load step the solution is computed and iteration are performed until the compatibility and equilibrium conditions are satisfied.

3.2 Step 2 The Contact Analyser

Once a load increment has been defined and the analysis performed the displacement, traction and stress solutions are obtained at the nodes. The function of the contact analyser is to determine if the surface is under the following conditions:

- separated i.e not in contact
- sticking contact
- frictional sliding contact
- frictionless sliding contact

If for example the current configuration is that the surfaces are in contact but the normal traction on the surface shows a tension stress then the function of the contact analyser is to generate a constraint which applies the separated condition. Alternatively if the tangential stress is greater than the frictional resistance then the frictional sliding contact condition is applied. Similarly the tractions and displacements can be used to determine the correct contact condition for any current values of the loads.

The BEM contact analyser benefits from the directly computed tractions and displacements on the contact surfaces, thus allowing accurate and precise determination of the contact zone and the contact forces necessary for friction calculations. No problems have been observed with interpolation problems with BEM unlike FEM. The accurate tractions and displacements are particularly beneficial on curved surfaces.

3.3 Step 3 Implementation of the Contact Condition

Once a new contact configuration has been identified it has to be tested to determine if the compatibility and equillibrium conditions are satisfied. This requires a solution of the analysis problem under the new contact boundary conditions (constraints). There are a number of approaches which can be used:

3.3.1 1. Penalty function

The penalty function method was investigated within the BEM system and although the method works satisfactorally, two main problems were observed. Firstly the choice of the stiffness co-efficient can cause innacuracies and problems in the solution. To small a value results in overlapping contact surfaces and poor accuracy. Too high a value can cause numerical problems. The second major problem is that major changes occur in the system matrix which require the whole matrix to be resolved for each iteration. While this is partly acceptable for two dimensional problems it is not feasible for three dimensional problems.

3.3.2 2. Constraints

The direct application of the constraints to the system matrix has a number of advantages. The extra equation can be added to the system matrix in such a way that only a partial reduction of the matrix is necessary for each iteration. This provides a major reduction in the computer time necessary and enables three dimensional problems to be solved. The constraint method is also not restricted to node on node contact, and large amounts of sliding can take place. In the current version of BEASY the sliding is restricted to node on node but this can be extended in the future to represent large displacements by introducing more sophisticated algarithms to keep track of the contact surfaces and relative positions etc.

The constraints applied directly to the system matrix are generated by the contact analyzer and they simply express the equilibrium and compatibility conditions on the contact surface.

For example a two dimensional free surface has the following constraints applied:

$$Tn^1 = 0 \qquad Ts^1 = 0$$

$$Tn^2 = 0 \qquad Ts^2 = 0$$

i.e a stress free condition. Whereas a surface in a sticking condition has the following constraints applied:

$$Un^1 = Un^2 \qquad Us^1 = Us^2$$

$$Tn^1 - Tn^2 = 0 \qquad Ts^1 - Ts^2 = 0$$

i.e. displacement continuity traction equilibrium, where:

U = displacements
T = Traction
n = normal
s = tangential
1,2 = refer to the contact surface.

4 Contact Implementation

The above algorithms have been implemented in the BEASY boundary element system and used to perform large scale contact analysis which is described later. It is believed the approach adopted uses the most appropriate techniques developed for both FEM and BEM. A direct constraint approach appears to be the most general as it enables the code to solve most types of contact problems and can be extended in the future. During the development techniques were evaluated not only for their accuracy and ease of implementation but also for their suitability for large scale three dimensional analysis.

4.1 Benchmark Examples

In order to evaluate the performance of the contact analysis a series of test problems we developed to validate the technology. An example of some of these tests are presented in this section.

4.1.1 Example 1. Thermal contact of a cantilever beam

In this example the thermal contact of an aluminium alloy cantilever beam ($E = 10.5 \times 10^6 psi, \alpha = 12.5 \times 10$ in/in^{-0}F) 3in long is investigated (Fig.1).

The beam is initially at a temperature of 70^0F and rests separated by a 0.002in gap from a rigid support. When the beam is heated to 170^0F it is expected to close the gap and contact with the rigid surface. The problem is modelled as shown in Fig 2 in three dimension and the results obtained accurately agree with that obtained by Harris (Ref 6) which suggests that an axial stress of 6250 psi is generated in the beam.

4.1.2 Example 2: Bolt-Lug Connection problem

In this test the analysis is carried out to study the linkage-assemblage problem between a bolt and a lug (Fig 3).

The bolt and the lug are both made of metal but the bolt material is assumed to be much stiffer. The initial clearance between the bolt and the lug surfaces is assumed to be zero. The problem is modelled as a three dimensional problem using linear elements.

The deflected shape (Fig 4) shows the contact configuration and the contact angle between the lug and the bolt.

A similar two dimensional problem has been studied by Anderson (Ref 7) using both Finite Element and Boundary Element methods.

The models used by Anderson (Fig 5) shows the complexity of the Finite Element model compared with the Boundary Element model for this relatively simple problem.

5 Applications

The system has been applied to a number of industrial applications which require contact analysis. Two example applications are described from the aerospace and automotive industries.

5.1 Gas Turbine engine

Fig 6 shows a model of a gas generator turbine disk and seal from an aero-gas turbine engine. The model investigates the condition where there is zero interference between the seal and the disk under the applied rotational loading. The problem is modelled as an axisymmetric geometry. The deflected shape (Fig 7) shows a separation at the engine speed of 30000 rpm.

5.2 Automotive Connecting Rod

This 3D model of one end of a connecting rod makes use of two symmetry planes intersecting the axis of the rod. The bearing is represented simply as a cylindrical pin, with zero initial gap assumed between the pin and the rod.

Loading in the axial direction is applied to the pin and the model is restrained by fixed displacements at the cross section through the rod.

The model is shown in Fig 8 with the exaggerated deformed shape shown in Fig 9.

6 Conclusions

The contact problems are by nature a boundary problem and hence they are best handled by the Boundary Element method.

BEASY employs this technique to solve contact problems. The modelling task is fairly simple, in fact no extra effort is required compared with modelling a multi zone problems.

BEASY can solve contact problems involving thermal, rotational and gravity body forces. Additional efforts also has been made to maximise the performance and minimise the resources required to solve such problems.

7 References

1. BEASY "Boundary Element Analysis System". Computational Mechanics 1991.

2. Pascoe S.K.etal. "A Comparison of Finite Element Techniques for contact Stress Analysis". Ed Mottershead J.E. Proceedings of Modern Practice in Stress and Vibration Analysis, Pergomon 1989.

3. Adey R.A., Niku S.M. "Boundary Element Stress Analysis involving contact using BEASY.". 11th International Conference on Boundary Element Methods. Boston 1989. C.M. Publications.

4. Anderson T. "The Boundary Element Method applied to two dimensional contact problems with Friction". Proceedings 3rd International Seminar on Boundary Element Methods. Irvine, California 1981. CM. Publication.

5. Alcantud M. "Analysis of the Contact Problem between Elastoplastic 2D Bodies by means of BEM". Proceedings. Boundary Elements in Mechanical and Electrical Engineering". Nice, France 1990.

6. Harris C.O. "Introduction to Stress Analysis". The Macmillan Co, New York 1959.

7. Anderson O.T. "Boundary Elements in two dimensional contact and friction". Linkoping Studies in Science and Technology Dissertations No 85, Kinkoping University, Sweden.

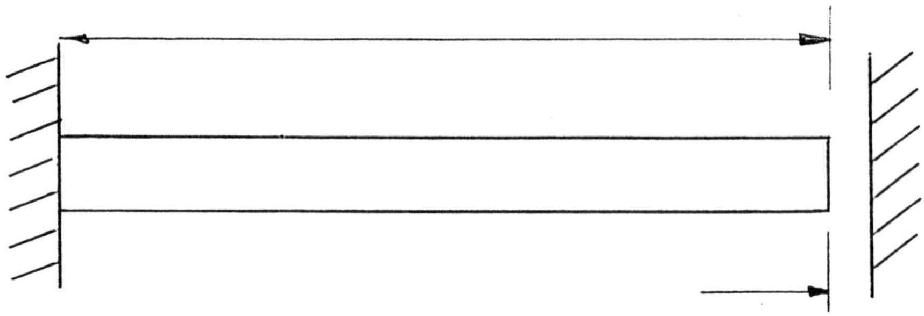

Figure 1: Thermal contact of cantilever beam

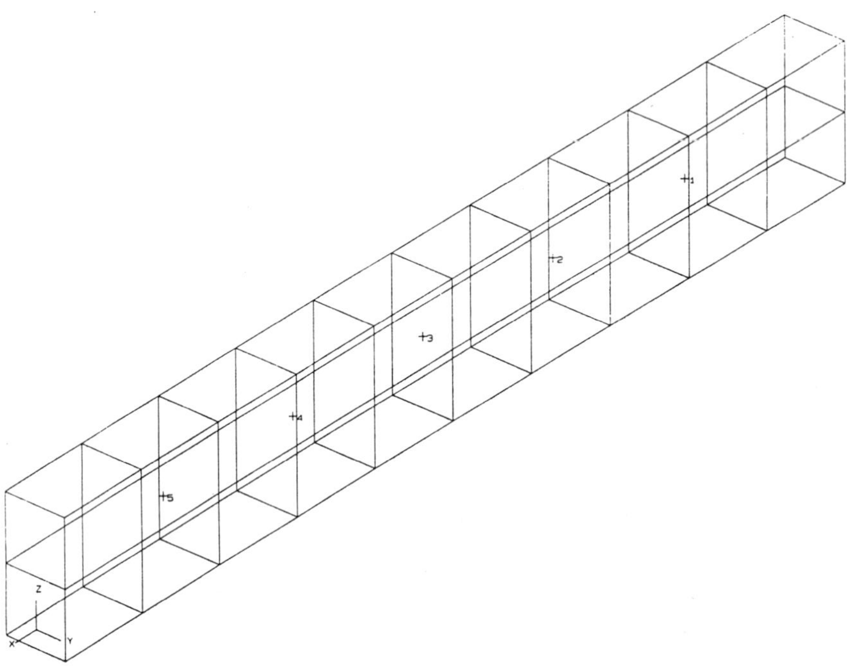

Figure 2: Three dimensional BEM model of beam. The model comprises 64 boundary elements.

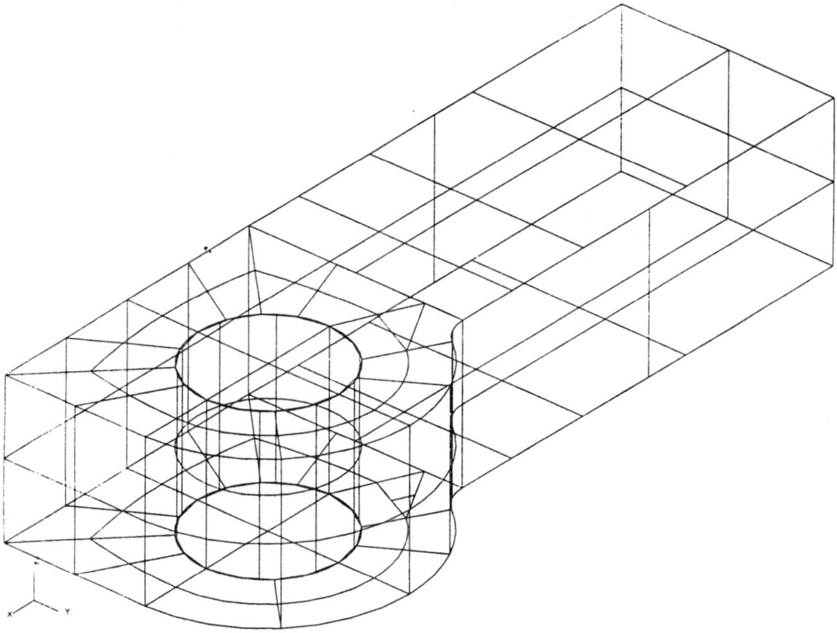

Figure 3: Three dimensional model of bolt lug connection. The main body of the lug is restrained and the bolt is loaded in the x-direction.

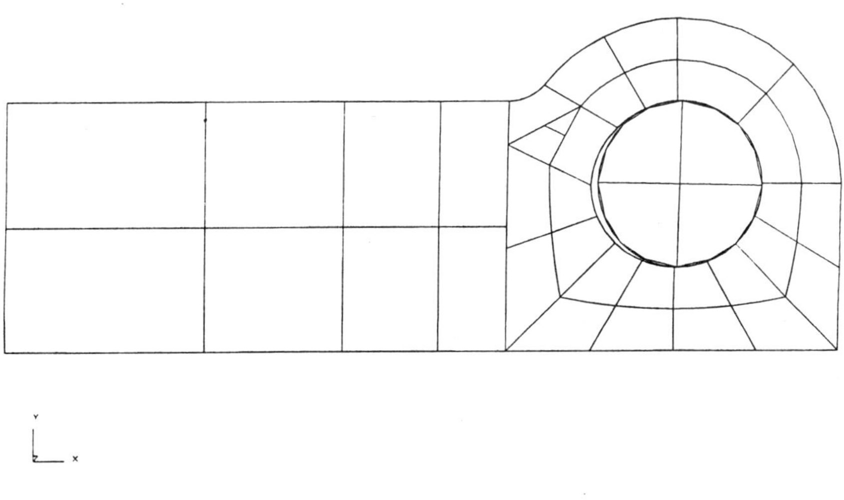

Figure 4: View of lug problem showing the contact area between the bolt and the lug. Note the clear separation between the bolt and the lug.

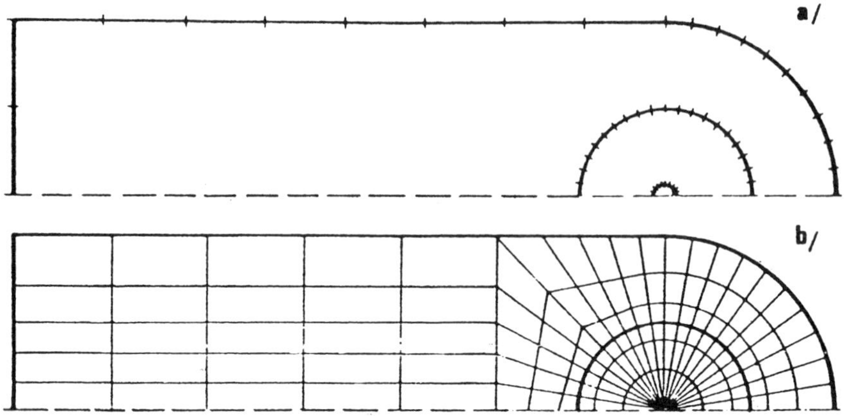

Figure 5: BEM a) and FEM model b) of a linkage assemblage. The lug thickness is equal to the radius of the bolt. (7).

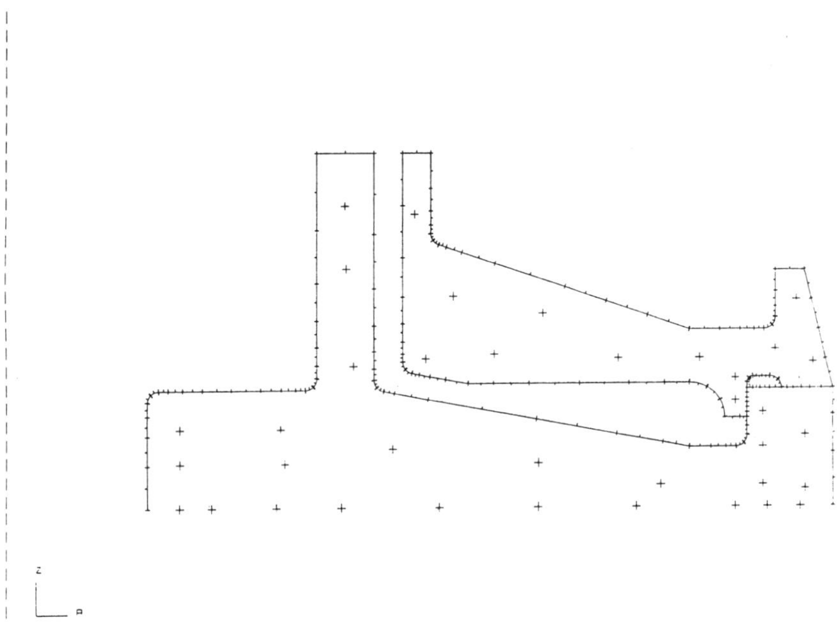

Figure 6: Axisymmetric model of real and turbine disk. Symmetry is assumed about 2 = 0.

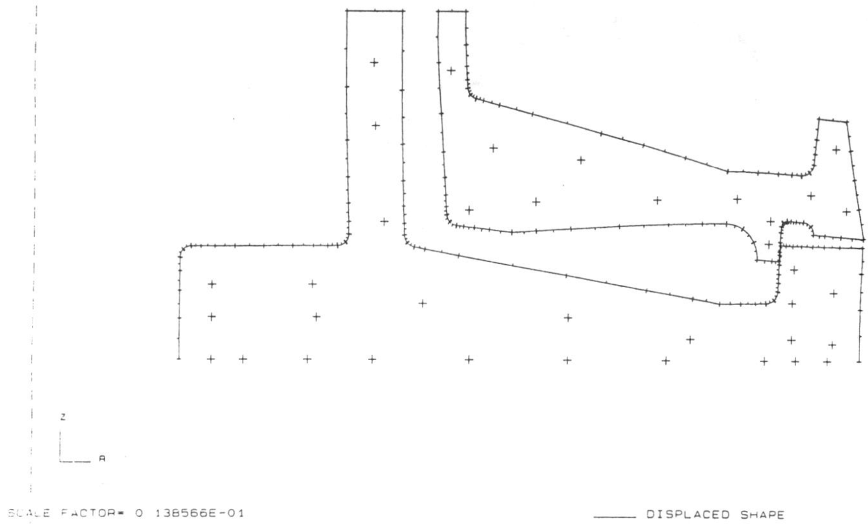

Figure 7: Deformed shape of turbine disk showing separaton and contact areas at a speed of 30000 rpm.

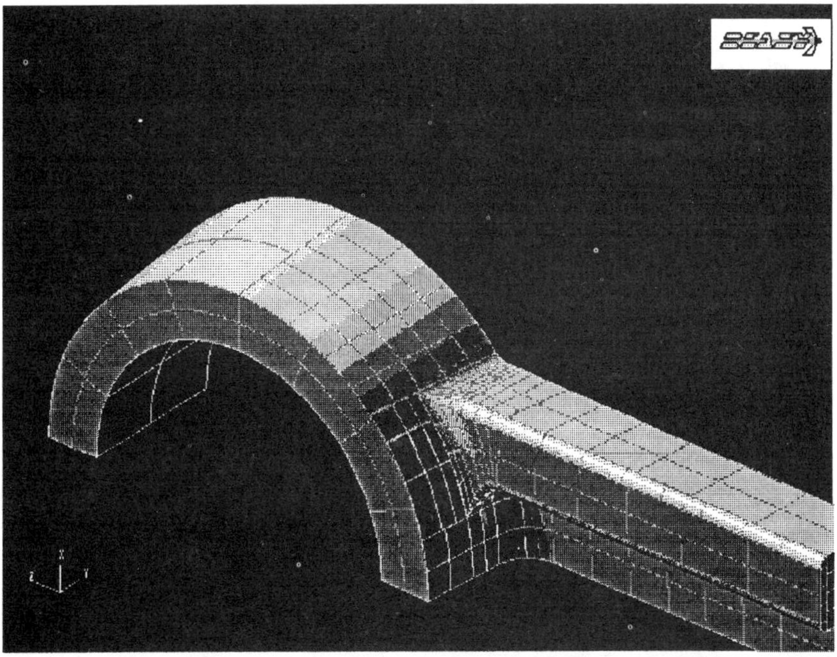

Figure 8: BEM model of connecting rod with pin removed. Note symmetry planes.

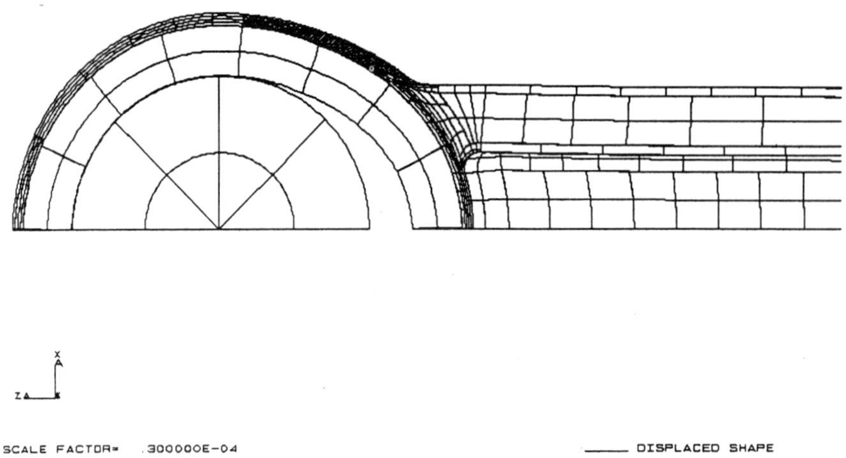

Figure 9: Cross sectional view of the connecting rod showing the contact areas.

Transient Dynamic Elastoplastic Analysis by the Boundary Element Method

J.A.M. Carrer, J.C.F. Telles

COPPE - Universidade Federal do Rio de Janeiro, Civil Engineering Department, Caixa Postal 68506, CEP: 21945, Fax (5521) 290-6626, Rio de Janeiro-RJ., Brazil

ABSTRACT

The analysis of transient dynamic elastoplastic problems is performed with the use of the elastostatic fundamental solution. This implies the maintenance of the inertial domain integral in the formulation. Therefore, as opposed to static analysis, not only the internal plastic region but the entire domain needs discretization.

The initial stress problem is solved, for each time-step, by using efficient implicit procedures recently developed.

This alternative implementation is justified by the simplicity and universality of its formulation. In addition, it also produces accurate results, as demonstrated by the example included here.

INTRODUCTION

The search for adequate Boundary Element formulations with the aim of solving time-dependent problems, has received a great deal of attention during the last few years. This is demonstrated by the number of works recently published using different alternative formulations. In order to reform dynamic elastic analysis, three general formulations have been succesfully employed. The first formulation uses time-dependent fundamental solutions and is particularly useful to solve infinite domain problems. Its efficiency and accuracy was demonstrated in the solution of several problems [1].

The second formulation uses static fundamental solutions but the approximate transformation of the inertial domain integral into a boundary integral, allows for only boundary discretization [2].

The third formulation also uses static fundamental solutions but does not eliminate the explicit inertial domain integral. Thus, the dynamic elastic analysis is carried out with the entire domain discretization.

This alternative is called by Kontoni and Beskos [3] Domain Boundary Element Method.

This formulation is developed here to solve dynamic elastic and elastoplastic problems. Therefore, as opposed to static analysis, not only the internal plastic region but the entire domain is discretized by means of internal cells. The time-dependent problem is solved by the direct integration method (Houbolt scheme). The initial stress problem is solved, for each time-step, with the use of the implicit procedures previously presented by the authors in references [4,5,6].

In the end of this work, an example solved by this procedure is presented.

BOUNDARY ELEMENT FORMULATION

The starting boundary integral equation (Somigliana's identity) associated to the initial stress formulation is defined as follows ($\xi \in \Gamma$):

$$\begin{aligned} c_{ij}(\xi)u_j(\xi;t) = & \int_\Gamma u^*_{ij}(\xi;X)p_j(X;t)d\Gamma(X) - \\ & - \int_\Gamma p^*_{ij}(\xi;X)u_j(X;t)d\Gamma(X) - \\ & - \rho \int_\Omega u^*_{ij}(\xi;X)\ddot{u}_j(X;t)d\Omega(X) + \\ & + \int_\Omega \varepsilon^*_{jki}(\xi;X)\sigma^p_{jk}(X;t)d\Omega(X) \end{aligned} \qquad (1)$$

In the equation above, t is time, Ω represents the domain of the body, Γ its boundary and c_{ij} is the usual free coefficient of elastic analysis. The subscripts vary between 1-3 and 1-2 for 3-D and 2-D problems, respectively. The following notation is used:

u^*_{ij}, P^*_{ij} and ε^*_{jki}: displacement, traction and strain components at point X (field point) due to a unit concentrated load applied in "i" direction at point ξ (source point);

u_j, P_j and σ^P_{jk} : displacement, traction and initial (residual) stress components;

$\ddot{u}_j = \dfrac{\partial^2 u_j}{\partial t^2}$: acceleration components and

ρ is the mass density.

The stress integral equation is obtained by combining the correct derivatives of equation (1) with respect to the coordinates of $\xi \in \Omega$ ($c_{ij}(\xi) = 1$), according to the procedure presented in [7]. The final expression is the following:

$$\begin{aligned}\sigma_{ij}(\xi;t) = & \int_\Gamma u^*_{ijk}(\xi;X)P_k(X;t)d\Gamma(X) - \\ & - \int_\Gamma P^*_{ijk}(\xi;X)u_k(X;t)d\Gamma(X) - \\ & - \rho \int_\Omega u^*_{ijk}(\xi;X)\ddot{u}_k(X;t)d\Omega(X) + \\ & + \int_\Omega \varepsilon^*_{ijk\ell}(\xi;X)\sigma^p_{k\ell}(X;t)d\Omega(X) + g_{ij}(\sigma^p_{k\ell})\end{aligned} \tag{2}$$

where the initial stress domain integral must be evaluated in the principal value sense and the "free" term $g_{ij}(\sigma^p_{k\ell})$ appears due to the correct derivative of the initial stress domain integral in equation (1).

The boundary stress components are directly calculated as functions of the derivative of the interpolated displacements and tractions over Γ.

NUMERICAL ANALYSIS

In order to perform the numerical analysis, equations (1) and (2) are discretized employing linear elements on the boundary and linear triangular cells over the domain. Due to the direct time integration method, both the boundary unknowns (displacements and tractions) and the displacements at internal points are calculated simultaneously. The application of the discretized version of equation (1) to all boundary nodes and internal points produces the matrix equation represented below:

$$\begin{bmatrix} \underset{\sim}{H}^{bb} & \underset{\sim}{0} \\ \underset{\sim}{H}^{ib} & \underset{\sim}{I} \end{bmatrix} \begin{Bmatrix} \underset{\sim}{u}^{b} \\ \underset{\sim}{u}^{i} \end{Bmatrix} = \begin{bmatrix} \underset{\sim}{G}^{bb} & \underset{\sim}{0} \\ \underset{\sim}{G}^{ib} & \underset{\sim}{0} \end{bmatrix} \begin{Bmatrix} \underset{\sim}{p}^{b} \\ \underset{\sim}{0} \end{Bmatrix} - \begin{bmatrix} \underset{\sim}{M}^{bb} & \underset{\sim}{M}^{bi} \\ \underset{\sim}{H}^{ib} & \underset{\sim}{M}^{ii} \end{bmatrix} \begin{Bmatrix} \underset{\sim}{\ddot{u}}^{b} \\ \underset{\sim}{\ddot{u}}^{i} \end{Bmatrix} + \begin{bmatrix} \underset{\sim}{Q}^{bb} & \underset{\sim}{Q}^{bi} \\ \underset{\sim}{Q}^{ib} & \underset{\sim}{Q}^{ii} \end{bmatrix} \begin{Bmatrix} \underset{\sim}{\sigma}^{pb} \\ \underset{\sim}{\sigma}^{pi} \end{Bmatrix} \quad (3)$$

in which the superscripts b and i denote, respectively, boundary and domain associated coefficients and unknowns. The double superscript is to be interpreted as follows: the first one corresponds to the source point, and the second, to the field point positions. The identity matrix $\underset{\sim}{I}$ represents the coefficients $c_{ij}(\xi)$ related to the internal points.

The matrix equation (3) can be written in a concise manner as

$$\underset{\sim}{H}\underset{\sim}{u}(t) = \underset{\sim}{G}\underset{\sim}{p}(t) - \underset{\sim}{M}\underset{\sim}{\ddot{u}}(t) + \underset{\sim}{Q}\underset{\sim}{\sigma}^{p}(t) \quad (4)$$

The computation of the stress components at selected boundary nodes and internal points is accomplished by the unified matrix equation:

$$\underset{\sim}{\sigma}(t) = \underset{\sim}{G}'\underset{\sim}{p}(t) - \underset{\sim}{H}'\underset{\sim}{u}(t) - \underset{\sim}{M}'\underset{\sim}{\ddot{u}}(t) + (\underset{\sim}{Q}' + \underset{\sim}{E}')\,\underset{\sim}{\sigma}^{p}(t) \quad (5)$$

In equations (4) and (5), matrices $\underset{\sim}{H}$, $\underset{\sim}{G}$, $\underset{\sim}{H}'$ and $\underset{\sim}{G}'$ correspond to the boundary integrals, matrices $\underset{\sim}{M}$ and $\underset{\sim}{M}'$, to the inertial domain integrals and matrices $\underset{\sim}{Q}$ and $\underset{\sim}{Q}'$, to the initial stress domain integrals. The free terms in equation (2) are represented by matrix $\underset{\sim}{E}'$.

The time-dependent problem is solved by the direct integration method with the use of the Houbolt integration scheme. Since this scheme is an implicit one, equations (4) and (5) are written for time $(t + \Delta t)$ as functions of the previous solution values. The following finite difference expansions are therefore employed for the velocities and accelerations:

$$\dot{\underset{\sim}{u}}(t + \Delta t) = \frac{1}{6\Delta t}\Big[11\underset{\sim}{u}(t + \Delta t) - 18\underset{\sim}{u}(t) + 9\underset{\sim}{u}(t - \Delta t) - - 2\underset{\sim}{u}(t - 2\Delta t)\Big] \tag{6}$$

$$\ddot{\underset{\sim}{u}}(t + \Delta t) = \frac{1}{\Delta t^2}\Big[2\underset{\sim}{u}(t + \Delta t) - 5\underset{\sim}{u}(t) + 4\underset{\sim}{u}(t - \Delta t) - - \underset{\sim}{u}(t - 2\Delta t)\Big] \tag{7}$$

The substitution of expression (7) in equation (4) gives

$$\Big[2\underset{\sim}{M} + \Delta t^2 H\Big]\ \underset{\sim}{u}(t + \Delta t) = \Delta t^2 \underset{\sim}{G}\underset{\sim}{p}(t - \Delta t) + + \underset{\sim}{M}\Big[5\underset{\sim}{u}(t) - 4\underset{\sim}{u}(t - \Delta t) + \underset{\sim}{u}(t - 2\Delta t)\Big] + + \Delta t^2 \underset{\sim}{Q}\underset{\sim}{\sigma}^p(t + \Delta t) \tag{8}$$

or, in a compact form:

$$\underset{\sim}{H}^{*}\underset{\sim}{u}(t + \Delta t) = \underset{\sim}{G}^{*}\underset{\sim}{p}(t + \Delta t) + \underset{\sim}{h}_1(t) + \underset{\sim}{h}_2(t - \Delta t) + + \underset{\sim}{h}_3(t - 2\Delta t) + \underset{\sim}{Q}^{*}\underset{\sim}{\sigma}^p(t + \Delta t) \tag{9}$$

After the application of the boundary conditions, equation (9) reads

$$\underset{\sim}{A}^{*}\underset{\sim}{y}(t + \Delta t) = \underset{\sim}{f}^{*}(t + \Delta t) + \underset{\sim}{Q}^{*}\underset{\sim}{\sigma}^{p}(t + \Delta t) \tag{10}$$

with vector $\underset{\sim}{f}^{*}(t + \Delta t)$ given by

$$\underset{\sim}{f}^{*}(t + \Delta t) = \underset{\sim}{f}(t + \Delta t) + \underset{\sim 1}{h}(t) + \underset{\sim 2}{h}(t - \Delta t) + \\ + \underset{\sim 3}{h}(t - 2\Delta t) \tag{11}$$

where the contribution of the prescribed values, at time $(t + \Delta t)$, is included in vector $\underset{\sim}{f}(t + \Delta t)$. The remaining vectors contain the contribution of the displacements previously determined.

Equation (6), can be written as

$$\underset{\sim}{\sigma}(t + \Delta t) = -\underset{\sim}{A}'\underset{\sim}{y}(t + \Delta t) + \underset{\sim}{f}'(t + \Delta t) - \\ - \underset{\sim}{M}'\underset{\sim}{\ddot{u}}(t + \Delta t) + (\underset{\sim}{Q}' + \underset{\sim}{E}')\underset{\sim}{\sigma}^{p} \tag{12}$$

in which the contribution of the prescribed values is included in vector $\underset{\sim}{f}'(t + \Delta t)$. The solution of equation (4) for the boundary unknowns and displacements at internal points, included in vector $\underset{\sim}{y}$, gives

$$\underset{\sim}{y}(t + \Delta t) = \underset{\sim}{R}\underset{\sim}{\sigma}^{p}(t + \Delta t) + \underset{\sim}{m}(t + \Delta t) \tag{13}$$

and, after the substitution of (13) in (12), the stress components are calculated as follows:

$$\underset{\sim}{\sigma}(t + \Delta t) = \underset{\sim}{S}\underset{\sim}{\sigma}^{p}(t + \Delta t) + \underset{\sim}{n}(t + \Delta t) \tag{14}$$

In the above equations, $\underset{\sim}{m}$ represents the dynamic elastic solution including boundary unknowns and internal displacements, and vector $\underset{\sim}{n}$ the elastic dynamic stress components. Furthermore:

$$\begin{aligned} \underset{\sim}{m} &= \underset{\sim}{A}^{*-1}\underset{\sim}{f}^{*} \quad ; \quad \underset{\sim}{n} = \underset{\sim}{f}' - \underset{\sim}{A}'\underset{\sim}{m} \\ \underset{\sim}{R} &= \underset{\sim}{A}^{*-1}\underset{\sim}{Q} \quad ; \quad \underset{\sim}{S}(\underset{\sim}{Q}' + \underset{\sim}{E}') - \underset{\sim}{A}'\underset{\sim}{R} \end{aligned} \tag{15}$$

The time-step length was considered acceptable if defined by the following semi-empirical relation

$$\Delta t = \ell/3c_L \tag{16}$$

where ℓ is the smallest boundary element length and c_L is the longitudinal wave propagation velocity.

CONSTITUTIVE EQUATIONS

The incremental stress-strain relations for inviscid plasticity problems can be written in the form [7]:

$$d\sigma_{ij} = C^{ep}_{ijk\ell}\, d\varepsilon_{k\ell} \tag{17}$$

where $d\varepsilon_{k\ell}$ is the total strain increment and within the context of associated isotropic work hardening theory

$$C^{ep}_{ijk\ell} = C_{ijk\ell} - \frac{1}{\gamma'} C_{ijmn}\, a_{mn}\, a_{op}\, C_{opk\ell} \tag{18}$$

where

$$\begin{aligned} C_{ijk\ell} &= \frac{2G\nu}{1-2\nu}\,\delta_{ij}\,\delta_{k\ell} + G\left(\delta_{ik}\,\delta_{j\ell} + \delta_{i\ell}\,\delta_{jk}\right) \\ a_{ij} &= \partial\sigma_e/\partial\sigma_{ij} \\ \gamma' &= a_{ij}\,C_{ijk\ell}\,a_{k\ell} + H' = a_{ij}\,d_{ij} + H' \\ H' &= d\sigma_0/d\varepsilon^p_e \end{aligned} \tag{19}$$

where σ_e and ε^p_e are the equivalent (or effective) stress and plastic strain, respectively, σ_0 is the uniaxial yield stress and δ_{ij} is the Kronecker delta.

For the implementation of the initial stress formulation, equation (17) can be conveniently simplified by introducing a fictitious elastic stress defined by

$$d\sigma^e_{ij} = C_{ijk\ell}\, d\varepsilon_{k\ell} \tag{20}$$

Equation (17) can then be written in the following form:

$$d\sigma_{ij} = d\sigma^e_{ij} - \frac{1}{\gamma'} C_{ijmn}\, a_{mn}\, a_{k\ell}\, d\sigma^e_{k\ell} \tag{21}$$

which means that true stresses can be computed from the corresponding elastic stresses in incremental form. In addition, the increments of initial stress can be calculated by the relation:

$$d\sigma^p_{ij} = d\sigma^e_{ij} - d\sigma_{ij} = \frac{1}{\gamma'} C_{ijmn}\, a_{mn}\, a_{k\ell}\, d\sigma^e_{k\ell} \qquad (22)$$

Hence, the increments of elastic stress can be computed directly from equation (2) after the substitution of the free coefficient g_{ij} by $\bar{g}_{ij}$, given by

$$\bar{g}_{ij}(\sigma^p_{k\ell}) = g_{ij}(\sigma^p_{k\ell}) + \sigma^p_{ij} \qquad (23)$$

For numerical purposes, the increments of elastic stress can be computed directly from equation (5) if matrix $\underset{\sim}{E}'$ is replaced by

$$\bar{\underset{\sim}{E}} = \underset{\sim}{E}' + \underset{\sim}{I} \qquad (24)$$

where $\underset{\sim}{I}$ is the identify matrix. This gives (in incremental form):

$$\Delta\underset{\sim}{\sigma}^e(t + \Delta t) = \bar{\underset{\sim}{S}}\, \Delta\underset{\sim}{\sigma}^p(t + \Delta t) + \Delta\underset{\sim}{n}(t + \Delta t) \qquad (25)$$

in which

$$\bar{\underset{\sim}{S}} = (\underset{\sim}{Q}' + \bar{\underset{\sim}{E}}) - \underset{\sim}{A}'\underset{\sim}{R} \qquad (26)$$

and

$$\Delta\underset{\sim}{n}(t + \Delta t) = \underset{\sim}{n}(t + \Delta t) - \underset{\sim}{n}(t) \qquad (27)$$

SOLUTION TECHNIQUES

The implicit procedures presented in what follows were succesfully used to solve static elastoplastic problems [5,6]. The application of equation (22) to the boundary nodes and internal points gives

$$\Delta\underset{\sim}{\sigma}^p = \underset{\sim}{D}^p \Delta\underset{\sim}{\sigma}^e \qquad (28)$$

in which $\underset{\sim}{D}^p$ is a quasi-diagonal matrix.

At a fixed time $(t + \Delta t)$, the substitution of equation (28) in the accumulated form of equation (25) allows for the application of an iterative elastoplastic algorithm whose step "k" can be defined by:

$$^{(k-1)}\underset{\sim}{\sigma}^{e} + {}^{(k)}\Delta\underset{\sim}{\sigma}^{e} = \overline{\underset{\sim}{S}} \left[{}^{(k-1)}\underset{\sim}{\sigma}^{p} + {}^{(k)}\underset{\sim}{D}^{p} \; {}^{(k)}\Delta\underset{\sim}{\sigma}^{p} \right] + $$

$$+ \underset{\sim}{n}(t) + \Delta\underset{\sim}{n}(t + \Delta t) \tag{29}$$

A residual vector is defined as follows

$$^{(k)}\underset{\sim}{\psi} = \overline{\underset{\sim}{S}}\,{}^{(k-1)}\underset{\sim}{\sigma}^{p} + \underset{\sim}{n}(t) - {}^{(k-1)}\underset{\sim}{\sigma}^{e} \tag{30}$$

Substituting it in equation (29) and rearranging, yields

$$^{(k)}\underset{\sim}{S}^{p} \; {}^{(k)}\Delta\underset{\sim}{\sigma}^{e} = \Delta\underset{\sim}{n}(t + \Delta t) + {}^{(k)}\underset{\sim}{\psi} \tag{31}$$

where the nonlinear matrix $^{(k)}\underset{\sim}{S}^{p}$ is given by

$$^{(k)}\underset{\sim}{S}^{p} = \underset{\sim}{I} - \overline{\underset{\sim}{S}}\,{}^{(k)}\underset{\sim}{D}^{p} \tag{32}$$

($\underset{\sim}{I}$ is the identity matrix).

Equation (31) can now be solved for each time step requiring, in principle, no iterations. In practice, however, if the time step length is not sufficiently small, a few iterations would have to be performed.

In this work, the solution schemes tested are the following:

(a) Pure Incremental - Equation (31) is applied only once for each load increment without iterations. Matrix $\underset{\sim}{S}^{p}$ is always updated as a function of the previous solution.

(b) Newton-Raphson - Matrix $\underset{\sim}{S}^{p}$ is updated after each iteration (by definition). For the first iteration of a given time step, equation (31) is written as

$$\underset{\sim}{S}^{p} \; \Delta\underset{\sim}{\sigma}^{e} = \Delta\underset{\sim}{n}(t + \Delta t) + \underset{\sim}{\psi}(t) \tag{33}$$

and for the subsequent iterations, until $\underset{\sim}{\psi}$ becomes vanishingly small (i.e. $\underset{\sim}{\psi} \longrightarrow \underset{\sim}{0}$ within some pre-determined tolerance), written as

$$\tilde{S}^p \, \Delta\tilde{\sigma}^e = \tilde{\psi}(t + \Delta t) \qquad (34)$$

In practice, however, only alternative versions of the procedure described above have been utilized. The most common version and, perhaps, the most efficient, is that known by Modified Newton-Raphson, in which matrix $\tilde{S}^p$ is updated only once at each time step (it is re-assembled as a function of the converged previous solution) and kept constant for all the iterations.

(c) Standard Initial Stress - This case is the explicit one extensively tested in reference [7], being treated here like a particular case of (b) in which $\tilde{S}^p$ is never updated (i.e. $\tilde{S}^p = \tilde{I}$).

EXAMPLE

Deep beam

In this problem, a simply supported deep beam is submitted to a suddenly applied uniform load. The perfectly plastic material is assumed to obey the Von Mises yield criterion, with the following parameters:

$$E = 100 \quad ; \quad \sigma_0 = 0.16 \quad ; \quad \nu = 1/3 \quad ; \quad \rho = 150$$

The discretization utilized is shown in Figure 1. In order to verify the results obtained by the B.E.M., the problem was analysed by the Finite Element Method. The mesh is shown in Figure 2.

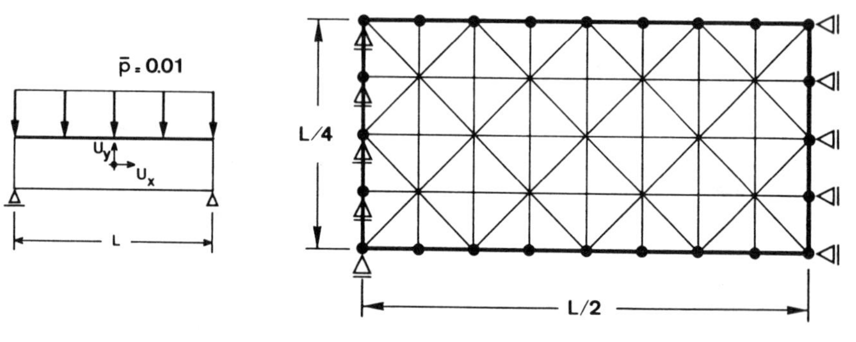

Figure 1 - Boundary element and internal cell discretization.

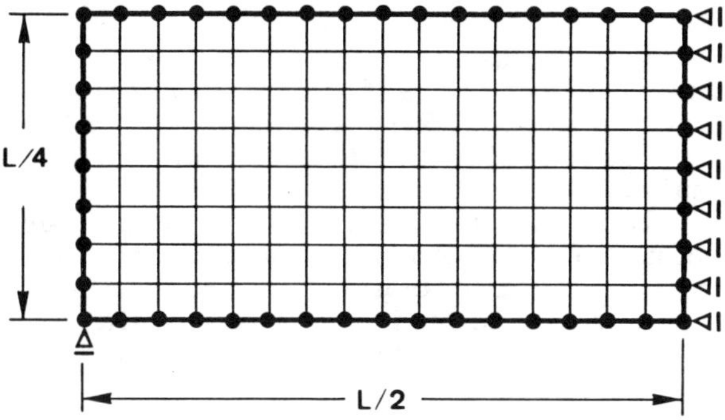

Figure 2 - Finite Element discretization.

The boundary element analysis was performed by employing the techniques described above. The results obtained by the SIS technique for the vertical displacement at the centroid of the centre section are represented in Figure 3. These results are in agreement with those produced by the Modified Newton-Raphson technique, therefore only one BE result is presented.

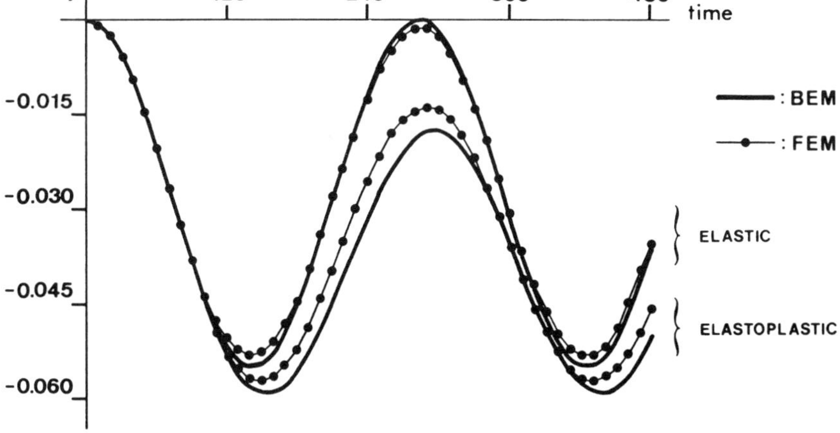

Figure 3 - Vertical displacement for elastic and elastoplastic analyses.

CONCLUSIONS

In order to perform static elastoplastic analysis by the Boundary Element Method, only the internal plastic region needs to be discretized. This advantage of the method is not taken into account in the formulation developed here, since it is necessary to discretize the entire domain to perform the analysis. This disadvantage is, however, counterbalanced by the simplicity of the inertial domain integral. Furthermore, alternative procedures, such as dual-reciprocity, can be attempted to remove the domain integral of the acceleration term.

With respect to the Houbolt integration scheme, it seems (based on the authors' previous experience) to be an adequate technique, perhaps ideal for dynamic BE solutions. However, it is important not only to use the techniques taken directly from the Finite Element Method but also to search for alternative ones, which could be more appropriate for the BEM.

The implicit procedures have already had their efficiency demonstrated before [5,6], but for dynamic analysis they need careful definition of loading/unloading situations at each stress point in order to predict the appropriate tangent matrix $\tilde{S}^p$. This is currently under study and will be reported in a future paper.

ACKNOWLEDGEMENTS

Special thanks are due to TOPTECH Consultoria e Projetos Especiais Ltda. for the Finite Element analysis.

REFERENCES

1. Mansur, W.J. A Time-stepping Technique to Solve Wave Propagation Problems Using the Boundary Element Method, Ph.D. Thesis, University of Southampton, 1983.

2. Nardini, D. and Brebbia, C.A. A New Approach to Free Vibration Analysis Using Boundary Elements, Boundary Element Methods in Engineering, Springer-Verlag, Berlin and New York, 1982.

3. Kontoni, D.P.N. and Beskos, D.E. B.E.M. Dynamic Analysis of Materially Nonlinear Problems,(Ed. Brebbia, C.A.), pp. 3-119 to 3-132, Proceedings of the 10th International Conference on the Boundary Element Method, Southampton, 1984.

4. Telles, J.C.F. On Inelastic Analysis Algorithms for Boundary Elements, in Advanced Topics in Boundary Element Analysis, (Ed. Cruse, T.A., Pifko, A.B. and Armen, H.), pp. 35 to 44, AMD - Vol. 72, 1985.

5. Telles, J.C.F. and Carrer, J.A.M. Implicit Solution Techniques for Inelastic Boundary Element Analysis, in 10th Int. Conf. on Boundary Element Methods in Engineering (Ed. Brebbia, C.A.) pp. 3-3 to 3-15, Springer-Verlag, 1988.

6. Carrer, J.A.M. and Telles, J.C.F. The Boundary Element Method Applied to the Solution of Elastoplastic Problems with the use of Implicit Procedures, (Ed. Carlomagno, G.M. and Brebbia, C.A.), pp. 35 to 46, Proceedings of The Fourth International Conference on Computational Methods and Experimental Measurements, Capri, Italy, 1989.

7. Telles, J.C.F. The Boundary Element Method Applied to Inelastic Problems, Lecture Notes in Engineering, Vol. 1, Springer-Verlag, Berlin and Heidelberg, 1983.

A General Adjoint Approach to Shape Design Sensitivity Analysis

J. Unzueta, E. Schaeidt, A. Longo, J.J. Anza
Analysis & Design Department, LABEIN, Cuesta de Olabeaga, 16, 48013 BILBAO, Spain

ABSTRACT

A Shape Design Sensitivity Analysis formulation is presented for 2D+ Axisymetric potential problems, using a continuous approach with boundary representation and the material derivative method to account for shape variation. The functions to be derived are boundary functionals involving the gradient of the state variable. The evaluation of these derivatives requires the response of an associated problem to the original one, following an adjoint technique. BEM is chosen as the analysis method for solving both problems. A numerical example is presented, illustrating the capabilities of the proposed solution procedure.

INTRODUCTION

In optimum design methodologies, the sensitivity analysis is defined as the effect of a change in the current design on a response functional. This calculus is a central step in formulating the mathematical model of a optimization problem; it can be used for approaching the cost function and constraints in terms of the design variables, providing the essential gradient information to compute a search direction in the optimization process.

The gradients are also important in their own right as they represent trends for the problem performance. Thus, the sensitivity analysis has recently emerged as a powerful design tool, not only from the standpoint of the role of derivatives in optimization, but also from the point of view of using sensitivity information in a computer-aided engineering environment for interactive design.

Many papers have come up over the past few years in the field of shape design sensitivity analysis (see Haftka [1] and references quoted therein). There have been two commonly approaches contributing to this topic, namely the discrete approach and the continuous approach.

The first approach, a natural adaptation of sizing design sensitivity, derives the implicit discretized equilibrium equation. Since there is not an explicit relation of shape variables for the terms appearing in that equation, generally speaking, the calculus of derivatives can be more expensive than the original analysis. Moreover, the derivatives obtained have a spurious component, which reflects the changing accuracy solution when the mesh is distorted.

Using the BEM as analysis tool (Defourny [2], Kane[3], Saigal [4]), this technique requires the numerical integration of a new class of sensitivity Kernels, which exhibit a singular behaviour similar to that of the analysis fundamental solution. Nevertheless, the implementation of this approach requires a significant programming effort to develop the numerical integration.

The second approach uses a continuous model of the problem and the material derivative method of continuum mechanics to obtain computable expressions for the effect of shape variation on the functionals arising in the shape design problem.

Two equivalent formulations, either on the boundary or in the domain, can be obtained for the continuous approach depending upon the way of calculating the derivatives of the functionals.The expressions reached through the boundary approach are functions of the partial shape derivatives of state variables. The domain approach, on the other hand, includes the total shape derivatives.

To evaluate the sensitivity equation derived by the boundary approach, only the design velocity field along the varied boundary is required. This represents a considerable computational saving compared to the domain approach, in which the design velocity field over the entire domain needs to be specified.

As it is done in this paper, the boundary approach is usually employed for shape sensistivity analysis in conjunction with the boundary element method for analysis. Thus, inherent numerical difficulties in MEF regarding lack of accuracy for the response over the boundaries can be all avoided.

The continuous shape sensitivity is obtained in terms of state variables, their shape derivatives and geometric changes (velocity field). The evaluation of those derivatives requires the response of an associated problem, for each shape variable by direct differentiating the governing equations - direct method -, or and adjoint problem for each functional which removes the shape derivatives from the formulation - adjoint method.

Since it has been concluded that the adjoint method is more efficient than the direct method if the number of functionals is less than the number of design variables, and active constraints selection algorithms are included in most of optimization codes, developments in shape sensitivities analysis were oriented towards the adjoint method.

The adjoint method, which has its origin in optimal control theory, is basically a Lagrangian variational technique which leads to build an adjoint problem for each functional (expressed in an integral form) and, using the state variables, getting the desired sensitivity.

Haug, Choi et al. (Choi [5], Haug [6]) developed a theory of structrual design sensitivity analysis for lineal elastic structures, differentiating the variational state equation. They use domain approach and, therefore, domain functionals.

Meric (Meric [7], Meric [8], Meric [9]) employs the adjoint variable technique deriving an augmented functional constructed by adjoining the system state equations with the original functional. Here, the adjoint variable exactly represents the lagrange multipliers associated with the state equations and the functional.

The most general result was obtained by Dems and Mróz (Dems [10], Mróz [11]), which forms the adjoint problem removing from the formulation the unknown-shape derivatives in each boundary. Their formulation is equally applicable to any numerical or analytical solution technique.

Kwak and Choi (Kwak [12], Choi [13]) obtain the same results with a less general conception of the problem. BEM is not only used as a tool for the solution of the state equations, but as a mean for sensitivity developing. Despite the fact BEM formulation is not in a symmetric form, the adjoint variable method is successfully applied by taking advantage of relations between direct and indirect BEM.

Although the formal adaptation of the adjoint technique is conceptually straightforward, major computational difficulties arise when evaluating sensitivities of functionals at discrete boundary points, or sensitivities of boundary functionals involving spatial gradients of state variables. The problem stems from the fact that the adjoint solution can not be expressed in terms of the boundary element formulation, because they give rise to infinite integrals (i.e., in elasticity the adjoint solution would correspond to a concentrated force and moment for displacement and stress sensitivities, respectively).

To work out this singular adjoint solution, some empirical guidelines are found in reference Mota Soares [14] in which the adjoint solution is represented by a statically equivalent triangular load and in Zhao [15], using the fundamental solution of semi-plane spaces. In this paper, another very efficient alternative is proposed.

MATERIAL DERIVATIVE FOR SHAPE SENSITIVITY ANALYSIS

Since the domain Ω is to be varied, it is treated as a continuum moving with a "time like" parameter t in the material derivative formulation of shape design sensitivity analysis. The variations of a point $\underline{x}$ in the domain are expressed in terms of a velocity field $\underline{V}$, which defines the direction of movement of $\underline{x}$ on the nominal domain Ω to points $\underline{x}_t$ on a deformed domain Ω_t, given by the transformation,

$$\underline{x}_t = \underline{x} + t\underline{V}(\underline{x})$$

Then, the variation of a general differentiable function $\emptyset$ with respect to a shape variation is expressed as the material derivative or total derivative of $\emptyset$ at $t = 0$,

$$\dot{\phi}=\frac{d\phi_t(\underline{x}_t)}{dt}\Big|_{t=0}=\lim_{t\to 0}\frac{\phi_t(\underline{x}_t)-\phi(\underline{x})}{t}=\phi'(\underline{x})+\underline{\nabla}\phi\underline{V}(\underline{x})$$

where,

$\phi_t(\underline{x}_t)$ is the solution of the boundary value problem on Ω_t, evaluated at a point $\underline{x}_t$ that moves with t, and

$$\phi'(\underline{x})=\lim_{t\to 0}\frac{\phi_t(\underline{x})-\phi(\underline{x})}{t}$$

namely local derivative, is the variation of ϕ at point $\underline{x}$.

<u>Material derivatives of integral functionals</u>

Let ψ_t be a functional defined over an open boundary Γ_{At} piece-wise C^{k+1} regular on the varied domain Ω_t in R^2.

$$\psi_t=\int_{\Gamma_{At}}g_t(\underline{x}_t)d\Gamma_t \qquad (1)$$

Defining χ as a characteristic function,

$$\chi(\underline{x})\big|_\Gamma=\begin{cases}1,\underline{x}\in\Gamma_A\\ 0,\underline{x}\notin\Gamma_A\end{cases}$$

the expression (1) has the form

$$\psi_t=\int_{\Gamma_t}\chi_t(\underline{x}_t)g_t(\underline{x}_t)d\Gamma \qquad (2)$$

where Γ_t is $\partial\Omega_t$.

The material derivative of ψ at Γ is,

$$\psi'=\int_{\Gamma_A}[g'+(\underline{\nabla}g\underline{n}+gH)V_n]d\Gamma+\sum^{corners\Gamma_A}[\![gV_s]\!]+gV_s\big|_A^B \qquad (3)$$

being H the curvature of Γ, $[\![\,]\!]$ a jump operator and A, B, the initial and final points of Γ_A.

In many cases, it will be interesting to formulate functionals weighted with the measure

m_A of Γ_A,

$$\psi_t = \frac{1}{m_{At}} \int_{\Gamma_{At}} g_t(\underline{x}_t) d\Gamma_t \qquad (4)$$

also defined over an open boundary Γ_{At} piece-wise C^{k+1} regular in R^2. The material derivative of ψ at Γ is,

$$\psi' = \frac{1}{m_A} [\int_{\Gamma_A} [g' + (\nabla g \underline{n} + (g - \psi)H) V_n] d\Gamma +$$

$$+ \sum^{corners\Gamma_A} [(g - \psi)V_s] + (g - \psi)V_s|_A^B] \qquad (5)$$

SHAPE SENSITIVITY IN ELECTROSTATIC PROBLEMS

Let Ω be an heterogeneous domain in R^2 over which an electrostatic phenomenon is defined, with Dirichlet, Neumann and interface boundary conditions. The governing equations are:

$$\begin{aligned} \nabla^2_k \phi &= 0 && \text{in } \Omega_i \\ \phi &= \phi_o && \text{on } \Gamma^i_\phi \qquad (6) \\ \underline{\nabla}_k \phi \underline{n} &= q_o && \text{on } \Gamma^i_q \end{aligned}$$

$$\left.\begin{aligned} \phi^i &= \phi^j \\ q^i + q^j &= 0 \end{aligned}\right\} \quad \text{on } \Gamma^{i/j}_I \qquad i, j = 1 \ldots\ldots n$$

being,

n = number of subregions of Ω

$\Gamma^i = \partial\Omega^i$

$\Gamma^i_\phi \cup \Gamma^i_q \cup \Gamma^i_I = \Gamma^i$

$\underline{\nabla}_k = (K_x\, \partial/\partial x,\ K_y\, \partial/\partial y)$

K_x, K_y = orthotropic permitivities in Ω^i.

The functionals determinating a design in electrostatics (and in most of physical phenomenons) are functions of the state variable gradient on the boundaries, that is,

$$\psi = \frac{1}{m_A} \int_{\Gamma_A} g(\underline{\nabla}\phi) d\Gamma \qquad (7)$$

which has the form of equation (4). Therefore, its material derivative is like equation (5), where

$$g' = g_{\underline{\nabla}\phi} \underline{\nabla}\phi'$$

Equation (7) is also used to express functionals at isolated points, that is,

$$\psi = g(\underline{\nabla}\phi) \sim \frac{1}{m_A} \int_{\Gamma_A} g(\underline{\nabla}\phi) d\Gamma \qquad , \quad m_A \to 0$$

Thus, the smaller m_A value is taken, the better approximation will be obtained for ψ and therefore, for ψ'.

For calculating (5) an adjoint problem is created. This problem obtains the integral part, denoted by P, containing the local derivatives of the state variable,

$$P = \int_\Gamma \chi g_{\underline{\nabla}\phi} \underline{\nabla}\phi' d\Gamma \qquad (8)$$

It is necessary to express $\underline{\nabla}\phi'$ in terms of ϕ' and/or $\underline{\nabla}_k\phi'\mathrm{n}$,

$$\underline{\nabla}\phi' = (\underline{\nabla}_k \phi' \underline{n}) \underline{u} + \frac{\partial \phi'}{\partial s} \underline{v} \qquad (9)$$

being,

$\underline{u} = \underline{n}/(\underline{n}_k\underline{n})$

$\underline{v} = \underline{s}_{1/k} / \underline{n}_k\underline{n}$

$\underline{s}_{1/k} = (K_y s_x,\ K_x s_y)$

Substituting (9) into (8) and taken into account that,

$$f(s_o) = \int_\Gamma \delta(s, s_o) f(s) d\Gamma$$

the expression (8) reaches the form after manipulation,

$$P=\sum_i [\int_{\Gamma^i}\chi_A[g_{\underline{\nabla}\phi}\underline{u}(\underline{\nabla}_k\phi'\underline{n})-\phi'\frac{\partial}{\partial s}(g_{\underline{\nabla}\phi}\underline{v})]d\Gamma+$$

$$+\sum^{corners\Gamma_A}\int_{\Gamma^i}\delta(s,s_o)\phi'[g_{\underline{\nabla}\phi}\underline{v}]d\Gamma+\int_{\Gamma^i}\delta(s,s_o)\phi' g_{\underline{\nabla}\phi}\underline{v}|_A^B d\Gamma]$$

where s_o is the parameter of Γ at the corners of Γ_A or at the initial and final points A, B of Γ_A, and superscript i, the subregion index of Ω.

The adjoint variable is introduced adding to P the next null quantity,

$$\sum_i [\int_{\Omega^i}(\underline{\nabla}\phi^a\underline{\nabla}_k\phi'-\underline{\nabla}_k\phi^a\underline{\nabla}\phi')d\Omega]=0$$

Applying Green formula in each subdomain Ω^i, taking into consideration that

$\nabla_k^2\,\phi' = 0$ in Ω^i

due to the equilibrium equation of (6) and imposing that

$\nabla^2_k\,\phi^a = 0$ in Ω^i (10)

as equilibrium equation of the adjoint problem, and calling

$$\chi_A g_{\underline{\nabla}\phi}\underline{u}=a_\gamma \qquad ,on\ \Gamma_\gamma^i$$

$$-\chi_A\frac{\partial}{\partial s}(g_{\underline{\nabla}\phi}\underline{v})+\sum^{corners\Gamma_A\cap\Gamma_\gamma^i}\delta(s,s_o)[g_{\underline{\nabla}\phi}\underline{v}]+\delta(s,s_o)g_{\underline{\nabla}\phi}\underline{v}|_A^B=b\gamma \qquad ,on\ \Gamma_\gamma^i$$

where, now A, B are the initial and final points of $\Gamma_A \cap \Gamma_\gamma^i$, and doing that,

$\phi_a = -a_\phi$ on Γ_ϕ^i (11)

$\underline{\nabla}_k\phi^a\underline{n} = bq$ on Γ_q^i (12)

$a_I^i + \phi^{ia} = a_I^j + \phi^{ja}$

on $\Gamma_I^{i/j}$ (13)

$q^{ia} + q^{ja} = b_I^i + b_I^j$

as boundary conditions of the adjoint problem, P can be rewritten in the form,

$$P=\int_{\Gamma_A} g_{\underline{\nabla}\phi}\underline{\nabla}\phi' d\Gamma=\sum_i [\int_{\Gamma_\phi^i}(\chi_A \frac{\partial}{\partial s}(g_{\underline{\nabla}\phi}\underline{v})+\underline{\nabla}_k\phi^a\underline{n})\underline{\nabla}\phi\underline{V}d\Gamma-$$

$$-\sum^{corners\,\Gamma_A\cap\Gamma_\phi^i}[g_{\nabla\phi}\underline{v}](\underline{\nabla}\phi\underline{V})-g_{\nabla\phi}\underline{v}(\underline{\nabla}\phi\underline{V})|_A^B+$$

$$+\int_{\Gamma_q^i}(\chi_A g_{\underline{\nabla}\phi}\underline{u}+\phi^a)(\frac{\partial V_n}{\partial s}\underline{\nabla}_k\phi\underline{s}-\underline{n}_k D^2\phi\underline{V}-V_s\underline{\nabla}_k\phi\frac{\partial\underline{n}}{\partial s})d\Gamma+$$

$$+\int_{\Gamma_I^i}[(\chi_A\frac{\partial}{\partial s}(g_{\underline{\nabla}\phi}\underline{v})+q^a)\underline{\nabla}\phi\underline{V}-(\chi_A g_{\underline{\nabla}\phi}\underline{u}+\phi^a)(\underline{V}D^2\phi\underline{n}_k+$$

$$+\underline{\nabla}_k\phi(V_s\frac{\partial\underline{n}}{\partial s}-\frac{\partial V_n}{\partial s}\underline{s}))]d\Gamma-\sum^{corners\,\Gamma_A\cap\Gamma_I^i}[g_{\underline{\nabla}\phi}\underline{v}](\underline{\nabla}\phi\underline{V})-g_{\underline{\nabla}\phi}\underline{v}(\underline{\nabla}\phi\underline{V})|_C^D$$

which allows the evaluation of (5), after having solved the adjoint problem defined by eq. (10-13).

Comparing the adjoint system with eq. (6), it can be seen that only a second member has to be computed. If a Gauss triangularization has been employed, a lot of time will be spared if the upper triangular has been kept on disk.

For axisymetric problems, the null term to add is:

$$\sum_i [\int_{\gamma^i}\phi^a\underline{\nabla}_k\phi'\underline{n}r d\gamma-\int_{\gamma^i}\underline{\nabla}_k\phi^a\underline{n}\phi' r d\gamma]=0$$

where γ^i is the curve Γ^i (r, z, θ_o), θ_o = const. Following the same steps than before, the obtaining of P and adjoint boundary conditions is straightforward.

Numerical treatment of the singular adjoint boundary conditions

Eq. (11-13) show the singularity introduced in the adjoint problem by the b_γ boundary conditions. They have a finite part which does not cause any complication and a singular part to be tackled. The procedure proposed consists of removing from the boundary conditions the singular part of coeficients b_γ and couping with it into the B.E.M. formulation. Thus, it is possible to calculate the exact effect of singular boundary conditions for all nodes except the one on which b_γ is on

$$\int_\Gamma W(s,s_l)\delta(s,s_o)d\Gamma = W(s_o,s_l) \qquad ,s_l \neq s_o \qquad on\Gamma$$

where W is the fundamental solution, s_l the source point and s_o the functional point.

When $s_l = s_o$, an approximation of Dirac delta is used,

$$\delta(s,s_o) \sim \frac{1}{\Pi}\frac{\varepsilon}{[r-r_o]^2+\varepsilon^2}+h \qquad , \quad \varepsilon,h \to 0$$

where $r_o = r(s_o)$ and h a remaining value which makes the integral

$$\int_\Gamma \delta(s,s_o)d\Gamma = 1$$

be satisfied. The election of ε has influence on the accuracy of results. Sensitivity values presented in Figure 2 correspond to

$$\varepsilon = \frac{2\log_e(m+1)}{\Pi p} \qquad , \quad p=10$$

being m the measure of the boundary element on which b_γ is, and p the ratio of the δ height to m.

EXAMPLE

Figures 1 to 3 show a real life design application example. Figure 1 shows the design models, both the initial geometry (a) and final geometry (b) (regarding tidiness the boundary elements analysis models have not been presented). It is an axisymetric insulator, which only half of it has been modelled, surrounded by air. The electrode is conected to 75 Kv. Dimensions selected as design variables are displayed. The optimization problem consists on determining V1, V2, V3, V4 and V5 in order to minimize,

$$\frac{1}{m_c}\int_{\Gamma_c}(|\nabla\phi|)^2 d\Gamma$$

where Γ_c is the curved interface boundary close to ground. There are two behaviour constraints at interface points A and B,

$$|\underline{\nabla}\phi| - \frac{1}{mp}\int_{\Gamma_p} |\underline{\nabla}\phi| d\Gamma \begin{array}{l} \leq 400 v/mm \;\; at\ A \\ \leq 350 v/mm \;\; at\ B \end{array}$$

where boundary Γ_p is the anterior and posterior boundary element to points A or B.

Figure 2 shows sensitivities at initial geometry obtained with the adjoint method (a) compared to finite differences (b).

Figure 3 shows objective function (a) and maximum constraint violation (b) history. Notice that even though one constraint (A) is violated at initial shape, there is an objective function decrease during the optimization process. Constraint A is active at the optimum.

CONCLUSIONS

In this paper, shape sensitivity analysis of boundary functionals defined as, either integrals involving the gradient of the state variable or gradient functionals applied on isolated points of the boundaries, has been presented. The main claim of the developments reached is, besides the treatment of the functionals mentioned above, its generality. The domain can show C^k regularity (piecewise C^{k+1} regular boundaries) because the tangential velocity is considered, there are no restrictions on the movements of the boundaries (Dirichlet, Neumann and interfaces), functionals can be applied over whichever varied boundary and a closed or open piece of it containing non-singular corners can be used to define the integrals. Punctual functionals can be also placed at non-singular points on smooth boundaries. Finally, the formulation takes into account the possibility of dealing with orthotropic materials.

REFERENCES

1. Haftka R.T. and Adelman H.M. Recent Developments in Structural Sensitivity Analysis. Third International Conference on CAD/CAM, Robotics and Factories of the Future, Southfield, Michigan, August 14-17, 1988.
2. Defourny M. Optimization Techniques and Boundary Element Method. Proc. Boundary Elements X. Vol 1: Mathematical and Computational Aspects. Springer-Verlag, 1988.
3. Kane J.H., Saigal S. Design Sensitivity Analysis of Solids Using BEM. Journal of Engineering Mechanics. Vol 114, pp 1703-1722, 1988.

4. Saigal S., Borggaard J.T., Kane J.H. Boundary Element Implicit Differentiation Equations for Design Sensitivity of Axisymetric Structures. International Journal for Solids and Structures. Vol 25, No 5, pp. 527-538, 1989.
5. Choi K.K., Seong H.G. A Domain Method for Shape Deisgn Sensitivity of Built-up Structures. Computer Methods in Applied Mechanics and Engineering, 57, pp. 1-15, 1986.
6. Haug E.J., Choi K.K. Komkov V. Design Sensitivity Analysis of Structural Systems. Academic Press. New York, 1986.
7. Meric R.A. Boundary Elements in Shape Design Sensitivity Analysis of Thermoelastic Solids. Computer Aided Optimal Design: Structural and Mechanical Systems. Ed: C.A. Mota Soares. Springer-Verlag, Berlin, Heidelberg, 1987.
8. Meric R.A. Shape Design Sensitivity Analysis for Non-Linear Anisotropic Heat Conducting Solids and Shape Optimization by the BEM. International Journal for Numerical Methods in Engineering. Vol 26, pp 109-120, 1988.
9. Meric R.A. Boundary Elements for Exact and Optimal Shape Inverse Problems. Boundary Element Techniques: Applications in Engineering. Ed: C.A. Brebbia, N.G. Zamani, CMP, 1989.
10. Dems K. Sensitivity Analysis in Thermoelasticity Problems. Computer Aided Optimal Design: Structural and Mechanical Systems. Ed: C.A. Mota Soares, Springer-Verlag, Heidelberg, 1987.
11. Mróz Z. Sensitivity Analysis and Optimal Design with Account for Varying Shape and Support Conditions. Computer Aided Optimal Design: Structural and Mechanical Systems. Ed: C.A. Mota Soares, Springer-Verlag, 1987.
12. Kwak B.H., Choi J.H. Design Sensitivity Analysis Based on Boundary Integral Equation Method Considering General Shape Variations. Part I: For Self-Adjoint Elliptic Operator Problems. KSME Journal. Vol 1, No 1, pp. 70-73, 1987.
13. Choi J.H., Kwak B.H. Shape Design Sensitivity Analysis of Elliptic Problems in Boundary Integral Equation Formulation. Mech. Struct. & Mach. 16 (2), 147-165, 1988.
14. Mota Soares C.A., Choi K.K. Boundary Elements in Shape Optimal Design of Structures. The Optimum Shape: Automated Structural Design, Ed: J.A. Bennet, M.E. Botkin, Plenum, New York, pp. 199-231, 1986.
15. Zhao Z, Adey R.A. The Accuracy of the Variational Approach to Shape Design Sensitivity Analysis. Boundary Elements in Mechanical and Electrical Engineering. Ed: C.A. Brebbia, Springer-Verlag, 1990.

Acknowledgement: This research was carried out under project OCIDE 132135.

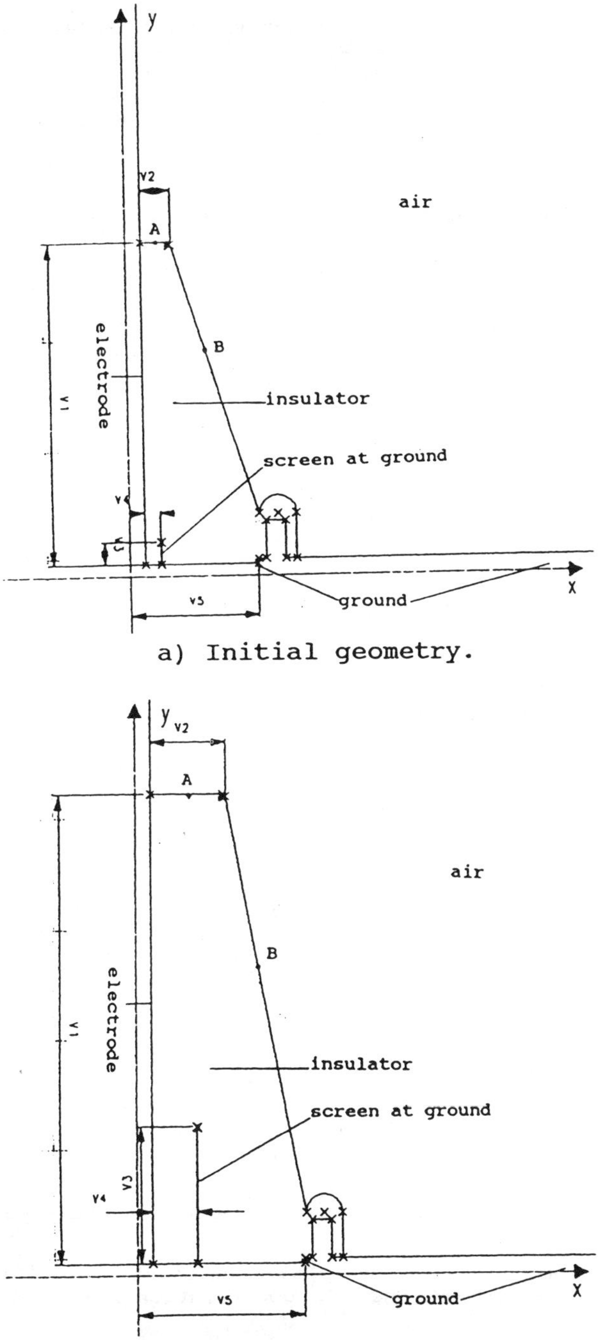

a) Initial geometry.

b) Final geometry.

Figure 1. Insulator: Design model.

	OBJECT. FUNCT.	CONSTRAINT 1	CONSTRAINT 2
VARIABLE 1	213.595520	-0.001832	-0.004619
VARIABLE 2	608.465140	-0.020491	-0.009060
VARIABLE 3	-1342.447805	0.000994	0.002245
VARIABLE 4	-891.677456	0.000549	0.001226
VARIABLE 5	-1820.679334	0.002596	0.002306

a) Adjoint Method.

	OBJECT. FUNCT.	CONSTRAINT 1	CONSTRAINT 2
VARIABLE 1	213.917890	-0.003756	-0.004525
VARIABLE 2	627.283961	-0.020767	-0.009000
VARIABLE 3	-1209.870590	0.000993	0.002247
VARIABLE 4	-779.077121	0.000547	0.001220
VARIABLE 5	-1973.503866	0.002191	0.001726

b) Finite differences.

Figure 2. Insulator: Sensitivity values.

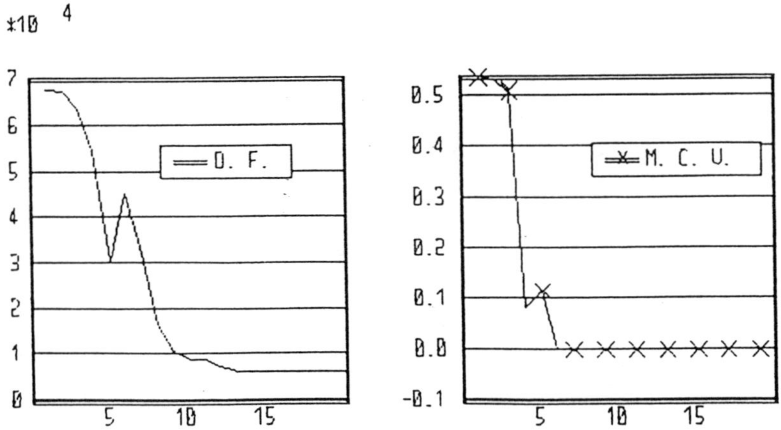

a) Objective function V. optimization iterations.

b) Maximum constraint violation V. optimization iterations.

Figure 3. Insulator: Optimization history.

SECTION 5: NUMERICAL AND MATHEMATICAL TECHNIQUES

A new B.E. Formulation and its Applications in Engineering

C.A. Brebbia, T. DeFigueiredo
Wessex Institute of Technology, U.K.

INTRODUCTION

This paper describes the derivation of a numerical model for the solution of boundary-value problems in linear elasticity. The formulation is based on a generalized variational principle rather than the integral formula or weighted residual techniques used in classical BEM. The variational principle involves three independent variables and can be applied for potential problems as well as in elasticity. For potential problems the variables are the potential in the domain and the potential and its normal derivative on the boundary. In the case of elasticity the variables are the displacement in the domain and the displacement and traction on the boundary. For this reason and by analogy with the assumed displacement hybrid finite element formulation, initially proposed by Tong [1] in 1970, the present model is called a hybrid displacement model.

The final system of equations to be solved is similar to that found in a stiffness-type formulation. The equivalent stiffness matrix for this model is symmetric and can be evaluated by performing only integrations along the boundary. The final unknowns are boundary potentials in potential problems or boundary displacements in elasticity problems and all other variables can be obtained from them.

The variational approach presented here is original but other variational formulations have been proposed in the past. In particular, the Galerkin weighted residual approach has been used as a BEM discretization technique. McDonald et al. [2] and Wexler [3,4] have also proposed numerical approaches with a variational basis for the solution of the boundary integral equations. It has been shown that the resulting discretized equations in [2] and [4] are identical to those resulting from the Galerkin weighted residual method.

Modified Galerkin boundary formulations have also appeared in the literature. Hsiao, Kopp and Wendland [5] proposed the so-called Galerkin collocation method in potential problems. Their approach demonstrated the existence of a solution of algebraic system of equations, as well as the asymptotic convergence of the approximate series.

In elasticity the Galerkin method was applied by Polizzotto [6] to derive a symmetric boundary element formulation. In reference [7] by Onishi, the Galerkin method is applied to the solution of the boundary integral equations in heat conduction, and the convergence and stability of the approach are proved.

The utilization of variational principles to generate boundary integral equations is relatively recent. In the late eighties, some authors have developed boundary element formulations based on generalized principles, such as the Hellinger-Reissner or the Wu-Washizu principles.

Dumont [8,9] has proposed a hybrid stress boundary element formulation based on the Hellinger-Reissner principle. The formulation considers the stress in the domain and the displacement on the boundary as independent functions. The resulting coefficient matrix is symmetric. The classical point force fundamental solution is used in this formulation to approximate the domain variable.

DeFigueiredo and Brebbia [10] presented the hybrid boundary formulation for the solution of potential problems. It is based on the generalized variational principle, involving three independent variables. The approach uses the same type of approximation for the domain variables applied in [8] and [9]. This hybrid boundary formulation has been more recently extended to elastostatic problems [11].

Recently, Felippa [12] proposed a formulation for boundary elements in elastostatics, which is based on his parametric displacement generalized hybrid variational principle [13]. In this approach distinct boundary element models can then be obtained by setting different values to a parameter. Therefore it can be considered a systematic way of deriving boundary element formulations.

In this paper, the hybrid boundary element formulation of DeFigueiredo and Brebbia will be applied for the solution of linear elastostatics problems. The independent field variables are the displacement and traction on the boundary and the displacement inside the domain.

After introducing three approximate field variables into the expression for the boundary hybrid formulation and making its first variations equal to zero, a set of three matrix equations is generated. The final system of algebraic equations is afterwards written in terms of a symmetric stiffness matrix and nodal boundary values of displacements as unknowns.

MODIFIED VARIATIONAL PRINCIPLE

The generalized variational principle applied in this paper is

$$\Pi(u_i, \tilde{u}_i, \tilde{p}_i) = \int_\Omega (\frac{1}{2}\sigma_{ij}\varepsilon_{ij} - b_i u_i) d\Omega + \int_\Gamma \tilde{p}_i(\tilde{u}_i - u_i) d\Gamma - \int_{\Gamma_2} \bar{p}_i\, \tilde{u}_i\, d\Gamma \quad (1)$$

plus the boundary condition

$$\tilde{u}_i = \bar{u}_i \qquad \text{on } \Gamma_1 \quad (2)$$

where the $\sigma_{ij}, \varepsilon_{ij}$ are the components of the stress and strain tensors, u_i and b_i represents the displacement and body force components and the surface tractions are given by p_i. Γ_1 and Γ_2 are the parts of the boundary where the displacements or tractions are prescribed. Notice that the total boundary is $\Gamma = \Gamma_1 + \Gamma_2$.

The bars on the variables indicate a known value and the tilde represents a field which is only defined on the boundary.

The stationary conditions for the functional Π can be obtained by setting to zero its first variation, which results by taking variations with respect to the three independent variables $u_i, \tilde{u}_i$ and $\tilde{p}_i$ in equation (1) as follows,

$$\begin{aligned}\delta\Pi(u_i, \tilde{u}_i, \tilde{p}_i) &= \int_\Omega (\sigma_{ii}\delta\varepsilon_{ij} - b_i\, \delta u_i) d\Omega - \int_\Gamma \tilde{p}_i\, \delta u_i\, d\Gamma \\ &+ \int (\tilde{u}_i - u_i)\delta\tilde{p}_i\, d\Gamma + \int_{\Gamma_2} (\tilde{p}_i - p_i)\delta\tilde{u}_i\, d\Gamma \quad (3)\end{aligned}$$

After integrating by parts and regrouping, expression (3) can be written as:

$$\begin{aligned}\delta\Pi(u_i, \tilde{u}_i, \tilde{p}_i) &= -\int_\Omega (\sigma_{ij,j} + b_i)\delta u_i\, d\Omega - \int (\tilde{u}_i - u_i)\delta\tilde{p}_i\, d\Gamma \\ &+ \int (p_i - \tilde{p}_i)\delta u_i\, d\Gamma + \int_{\Gamma_2} (\tilde{p}_i - \bar{p}_i)\delta\tilde{u}_i\, d\Gamma \quad (4)\end{aligned}$$

The Euler equations for the functional can then be obtained from formula (4) as follows,

$$\sigma_{ij,j} + b_i = 0 \qquad \text{in } \Omega \quad (5)$$

$$\tilde{u}_i - u_i = 0 \qquad \text{in } \Gamma \quad (6)$$

$$p_i - \tilde{p}_i = 0 \qquad \text{in } \Gamma \quad (7)$$

$$\tilde{p}_i - \bar{p}_i = 0 \qquad \text{on } \Gamma \quad (8)$$

The functional can now be written as

$$\begin{aligned}\Pi(u_i, \tilde{u}_i, \tilde{p}_i) &= \int_\Gamma \frac{1}{2} p_i u_i d\Gamma + \int_\Gamma \tilde{p}_i(\tilde{u}_i - u_i)\, d\Gamma \\ &- \int_{\Gamma_2} \bar{p}_i\tilde{u}_i\, d\Gamma - \int_\Omega b_i u_i\, d\Omega - \int_\Omega \frac{1}{2}\sigma_{ij,j}\, u_i\, d\Omega \quad (9)\end{aligned}$$

DERIVATION OF THE BEM MODEL

One can now proceed to generate the algebraic system of equations for the problem. This is achieved by representing the three independent field variables through approximate functions and substituting them into the generalised variational principle.

The approximations for the boundary variables is made by using interpolation functions of the type normally adopted in finite or boundary elements. The displacement field in the domain, which is an independent variable is approximated in a different way; the so-called fundamental solution is applied here to generate the approximate internal displacement field. The 'l' component of the displacement vector at a point x inside the domain Ω - i.e. $u_l(x)$ - will be approximated as a series of fundamental solutions and unknown parameters as follows,

$$u_l(x) = \sum_{i=1}^{i=N} u_{kl}^{*^i}(x)\gamma_s \tag{10}$$

where s is given by $s = 2i - 2 + k$; the unit load is considered to be at node i in the k direction; k and l have a range of 2 or 3, in two or three dimensions respectively. The total number of nodes is denoted by N. The asterisk * indicates the fundamental solution function.

The parameter γ_s can be interpreted as fictitious concentrated loads in an infinite domain at points which correspond to boundary points in the actual body.

Equation (10) can be represented in matrix notation as,

$$\mathrm{u} = \mathbf{U}^{*,T}\gamma \tag{11}$$

where **u** is the displacement's vector at a point in the domain, γ is a vector of fictitious γ loads and the fundamental solution matrix $\mathbf{U}^*$ is (3 x N) x 3 or (2 x N) x 2 in three and two dimensions respectively.

Similarly, one can write a traction matrix $\mathbf{P}^*$ associated with the matrix of fundamental solution $\mathbf{U}^*$ and of the same dimensions.

The boundary displacement and traction vectors denoted by $\tilde{u}$ and $\tilde{p}$ are represented in terms of the values of those variables at a series of boundary nodes. They can be written as products of known interpolation functions and unknown parameters, i.e.

$$\left.\begin{aligned} \tilde{\mathrm{u}}(x) &= \mathbf{\Phi}^T \mathrm{u} \\ \mathrm{p}(x) &= \psi^T \mathrm{p} \end{aligned}\right\} \quad \text{on } \Gamma \tag{12}$$

u and **p** are vectors whose components are nodal values for boundary displacements and tractions respectively. Observe that they should have had a tilde but this has been dropped for simplicity.

FINAL SYSTEM OF EQUATIONS

The final system of algebraic equations is obtained as the stationary condition of the functional (9) after introducing the approximate field variables $u, \tilde{u}$ and $\tilde{p}$ as defined by equations (11) and (12). This gives

$$\begin{aligned} \Pi &= \int_\Gamma \frac{1}{2}\gamma^T \mathbf{U}^* \mathbf{P}^{*,T}\gamma \, d\Gamma + \int_\Gamma \mathrm{p}^T \psi(\mathbf{\Phi}^T \mathrm{u} - \mathbf{U}^{*,T}\gamma) d\Gamma \\ &\quad - \int_{\Gamma_2} \mathrm{u}^T \, \mathbf{\Phi}\bar{\mathrm{p}} \, d\Gamma - \int_\Omega \gamma^T \mathbf{U}^* \, \mathrm{b} \, d\Omega \end{aligned} \tag{13}$$

Defining

$$\begin{aligned} \mathbf{F} &= \textstyle\int_\Gamma \mathbf{U}^* \, \mathbf{P}^{*,T} \, d\Gamma \\ \mathbf{G} &= \textstyle\int_\Gamma \mathbf{U}^* \, \psi^T \, d\Gamma \\ \mathbf{L} &= \textstyle\int_\Gamma \psi \, \mathbf{\Phi}^T \, d\Gamma \\ \bar{\mathbf{P}} &= \textstyle\int_\Gamma \mathbf{\Phi} \, \bar{\mathrm{p}} \, d\Gamma \\ \mathbf{B} &= \textstyle\int_\Gamma \mathbf{U}^* \, \mathrm{b} \, d\Gamma \end{aligned} \tag{14}$$

one can write,

$$\Pi = \frac{1}{2}\gamma^T \, \mathbf{F} \, \gamma - \mathrm{p}^T \mathbf{G}^T \mathrm{u} + \mathrm{p}^T \mathbf{L} \, \mathrm{u} - \mathrm{u}^T \, \bar{\mathbf{P}} - \gamma^T \mathbf{B} \tag{15}$$

The first variation of Π can now be obtained as follows,

$$\begin{aligned} \delta\Pi &= (\delta\gamma)^T[\mathbf{F}\gamma - \mathbf{G} \, \mathrm{p} - \mathbf{B}] + \delta \mathrm{p}^T[-\mathbf{G}^T\gamma + \mathbf{L} \, \mathrm{u}] \\ &\quad + (\delta \mathrm{u})^T[\mathbf{L}^T \mathrm{p} - \bar{\mathbf{P}}] \end{aligned} \tag{16}$$

The stationary conditions for the Π functional are then,

$$\mathbf{F}\gamma - \mathbf{G}\mathrm{p} - \mathbf{B} = 0 \tag{17}$$

$$-\mathbf{G}^T\gamma + \mathbf{L}\mathrm{u} = 0 \tag{18}$$

$$\mathbf{L}^T \mathrm{p} - \bar{\mathrm{p}} = 0 \tag{19}$$

The matrix $\mathbf{G}^T$ is non singular, hence equation (18) produces

$$\gamma = (\mathbf{G}^T)^{-1}\mathbf{L} \, \mathrm{u} \tag{20}$$

Substituting this result into (17) gives,

$$\mathrm{p} = \mathbf{G}^{-1} \, \mathbf{F}(\mathbf{G}^T)^{-1} \, \mathbf{L} \, \mathrm{u} - \mathbf{G}^{-1} \, \mathbf{B} \tag{21}$$

Finally, one can replace γ from (20) and **p** from (21) into equation (19) to obtain the following system:

$$\mathbf{K} \, \mathrm{u} - \mathbf{Q} = 0 \tag{22}$$

where

$$\begin{aligned} \mathbf{K} &= \mathbf{R}^T \mathbf{F}\, \mathbf{R} \\ \mathbf{R} &= (\mathbf{G}^T)^{-1} \mathbf{L} \\ \mathbf{Q} &= \bar{\mathbf{P}} + \mathbf{R}^T \mathbf{B} \end{aligned} \tag{23}$$

It is possible to conclude from equation (22) that the hybrid displacement boundary formulation leads to an equivalent stiffness approach; the matrix **K** can be interpreted as a symmetric boundary stiffness matrix and the vector **Q** as a consistent load vector. Solution of equation (22) after introducing the kinematic boundary conditions produces all the boundary displacements.

Boundary tractions can afterwards be obtained from equation (21) and displacements at internal points can be found after solving for γ and applying equation (20), i.e.

$$u_l^{\text{int}}(x) = u_{kl}^*(x)\gamma_k \tag{24}$$

Stresses at internal points are also computed in a similar way using the expression given by the fundamental solution, i.e.

$$\sigma_{ij}(x) = \sigma_{ij}^{*k}(x)\gamma_k \tag{25}$$

NUMERICAL ASPECTS

The main difficulty of the above procedure is in the computation of the singular integrals in matrices **F** and **G**. These computations can be carried out by applying analytical procedures or by using special numerical integration schemes. The matrix **F** does not vary with the type of approximation used for the boundary variables. It depends only on the fundamental solution adopted and on the geometry of the element. The matrix **G** in this approach is similar to the one that appears in the classical direct boundary element method. Therefore, its evaluation including the treatment of the singular integrals, follows the same lines as in conventional boundary elements. The elements of the matrix **L** and vector $\bar{\mathbf{P}}$ do not include any kind of singularity and hence are simpler to integrate.

When body forces are present in the problem, the domain integrals corresponding to the coefficients of the vector **B** are given by the same domain integrals that have to be computed in the classical direct boundary element method. The computations can be attempted using particular integrals as shown in the Dual Reciprocity Method [14].

Notice that in this approach, the evaluation of internal displacements and stresses does not require working with any domain integrals.

EXAMPLES

The applications of the new formulation are gereby demonstrated by a series of representative examples and results are compared with conventional boundary elements and analytical solutions. These examples are

i) Stress concentration in a square plate with a circular hole
ii) Cylinder under compressive forces
iii) Cantilever beam

Example i) Stress concentration in a square plate with a circular hole

The first example considered here is the classical stress concentration problem of a square plate of width $2l$, with a small central circular hole of radius a (Figure 1). The plate is subjected to a uniform tensile stress σ in the x_2 direction. The ratio of the diameter of the hole to the width of the plate is taken to be $a/l = 0.1$ and the Poisson's ration for the material is $\nu = 0.25$.

Figure 2 shows the variation of normal stresses in the x_2 direction along the x_1 axis with distance form the hole. Both the conventional BEM and the hybrid BEM solutions were obtained by using the same 29 boundary constant elements mesh, shown in Figure 1. K_t is the stress concentration factor defined as $K_t = \sigma_{22}/\sigma$.

The analytical solution curve corresponds to the case of an infinite plate and is taken from Timoshenko [15] which gives a stress concentration factor of 3.000. This value increases slightly to 3.024 for the finite width plate [16].

The results for the hybrid and conventional BEM are in good agreement and are both accurate when compared with the exact solution. Although the constant elements used here are not recommended for modelling curved boundaries and problems where the solution varies rapidly, the numerical results follow the analytical curve and simple extrapolation of the boundary element values gives an accurate estimate of the stress concentration factor.

Example ii) Cylinder under compressive forces

The problem of a cylinder under compression will now be considered. The cylinder is supposed to be long enough for a cross section (at a reasonable distance from the extremities) to be considered in a plane strain.

The exact solution for the stress field is presented in reference [15]. The convergence for stresses at the internal points shown in Figure 4 have been studied. Four meshes have been used with 4, 8, 16 and 32 equal elements, respectively.

Figure 5 demonstrates that the numerical results converge towards the analytical solution. Table 1 shows the results for the normal stresses in the x_2 direction at the internal points for the 16 element mesh, and they present a good degree of accuracy. It is possible to see that the results for the internal points, when they are close to the singularities (boundary nodes), are less accurate than the results for points away from the singularities.

In this example, the concentrated loads have been introduced as boundary conditions. This is an interesting feature of this hybrid BEM formulation: it allows simple representation of concentrated loads.

Example iii) Cantilever Beam

As the last example, the solution of the cantilever beam, which is shown in Figure 6, will be presented. It can be considered a plane stress problem in which the elastic constants are assumed to be the same as those in example 9, i.e. G = 80000 and $\nu = 0.25$.

The hybrid BEM results for the vertical displacement at the free end (point A) is presented in table 2, where it is compared against the conventional BEM results. The elastic beam theory gives for this problem a free end deflection of

$$u_2^A = -\frac{128}{2EI}qa^4 = 0.0102 \tag{26}$$

Since the ration between the depth of the beam and its length is equal to 5, there is a small contribution in the total deflection due to the shear which is not considered in the beam theory.

When solving bending problems with quadratic elements, the hybrid-displacement BEM requires a more refined mesh than the conventional BEM, to give the same accuracy. The results show that the quadratic element is a bit stiff. This behaviour was unexpected because it has been noticed that the opposite occured for the constant element where the hybrid BEM had a very good performance when compared to the conventional BEM [17].

In the hybrid finite element theory [1] it is reported that the elements can be made more flexible if polynomials of lower order than the ones used for the displacement were used to represent the tractions. Based on this statement, in the case of this hybrid BEM formulation, one is lead to believe that the results for the quadratic element could be improved by using a lower order polynomial to expand the tractions. Another possibility for improving the results is to use discontinuous elements, as is frequently done in the conventional BEM. In that method, it is known that for bending problems discontinuous elements are more accurate than the continuous ones.

CONCLUSIONS

In this paper a novel boundary element formulation called the hybrid-displacement BEM is presented based on a multifield variational principle.

The functional that is used in the generation of the formulation has three independent variables, i.e. displacements and tractions on the boundary and displacements in the domain. The formulation becomes a boundary only approach by integrating by parts and defining the independent domain variable in terms of the elasticity fundamental solution. The boundary is then divided into elements as in the classical BEM approach.

The stationary conditions of the functional produce a system of equations which can be reduced to a symmetric stiffness type matrix of the type used in FEM. Thus the unknowns are only boundary displacements and the equations can easily be solved.

The numerical performance of the new approach is excellent as demonstrated by a series of examples. The major drawback of the technique is that it is more time consuming than the classical BEM. The high demand of computer time is due to the need to evaluate a series of singular integrals and invert a large matrix.

REFERENCES

1. Tong, P. "New Displacement Hybrid Finite Models for Solid Continua" Int. J. Num. Meth. Eng. Vol.2, 1970, pp.73-88.

2. McDonald, B.H., Friedman, M. and Wexler, A. "Variational Solution of Integral Equations" IEEE Trans. Microwave Theory Tech., Vol.1 MTT 22, 1974, pp.237-248.

3. Jeng, G. and Wexler, A. "Isoparametric, Finite Element, Variational Solution of Integral Equations for Three-Dimensional Fields" Int. J. Num. Meth. Eng., Vol.11, 1977, pp.1455-1471.

4. Jeng, G. and Wexler, A. "Self-Adjoint Variational Formulation of Problems having Non-Self-Adjoint Operators" IEEE Trans. Microwave Theory Tech., Vol. MTT-26, 1978, pp.91-94.

5. Hsiao, G.C., Kopp, P. and Wendland, W.L. "The Synthesis of the Collocation and the Galerkin Method applied to some Integral Equations of the First Kind" New Developments in Boundary Element Methods (C.A. Brebbia, Ed.), CML Publications, Southampton, UK, 1980, pp.122-136.

6. Polizzotto, C. "A BEM Approach to the Bending Techniques" Boundary Elements VII (C.A. Brebbia, Ed.) Springer Verlag, Berlin, 1985, pp.13/103-114.

7. Onishi, K. "Heat Conduction Analysis using Single Layer Heat Potential" Boundary Element Methods in Applied Mechanics (M. Tanaka and T.A. Cruse, Eds.) Pergamon Press, 1988.

8. Dumont, N.A. "The Variational Formulation of the Boundary Element Method" Boundary Element Techniques: Applications in Fluid Flow and Computational Aspects (C.A. Brebbia and W.S. Venturini, Eds.) CML Publications, Southampton, 1987.

9. Dumont, N.A. "The Hybrid Boundary Element Method" Boundary Elements IX (C.A. Brebbia, W.L. Wendland and G. Kuhn, Eds.) Vol.1, Springer-Verlag, Berlin, 1987, pp.117-130.

10. DeFigueiredo, T.G.B. and Brebbia, C.A. "A Hybrid Displacement Variational Formulation of BEM" Boundary Elements X (C.A. Brebbia, Ed.) Vol.1, CML Publications, Southampton, Springer-Verlag, Berlin, 1988, pp.33-42.

11. DeFigueiredo, T.G.B. and Brebbia, C.A. "A New Hybrid Displacement Variational Formulation of BEM for Elastostatics" Advances in Boundary Elements (C.A. Brebbia and J.J. Connor, Eds.) Vol.1, CML Publications, Southampton, Springer-Verlag, Berlin, 1988, pp.47-57.

12. Felippa, C.A. "A Parametric Representation of Symmetric Boundary Finite Elements" Topics in Engg., Vol.7 (C.A. Brebbia and M. Tanaka, Eds.) CML Publications, 1990.

13. Felippa, C.A. "Parameterized Multifield Variational Principles in Elasticity: II Hybrid Functionals and the Free Formulation" Communications in Applied Numerical Methods, Vol.5, 1989, pp.89-98.

14. Partridge, P., Brebbia, C.A. and Wrobel, L. "The Dual Reciprocity Boundary Element Method" Computational Mechanics Publications, Southampton, 1991.

15. Timoshenko, S.P. and Goodier, J.N. "Theory of Elasticity" McGraw Hill, NY, 1982.

16. Roark, R.J. and Young, W.C. "Formulas of Stress and Strain" McGraw Hill, London, 1975.

17. DeFigueiredo, T. "A New Boundary Element Formulation and its Applications in Engineering" PhD Thesis, Wessex Institute of Technology, Southampton, 1991. To be published in Lecture Notes in Engineering, Springer-Verlag, Berlin, 1991.

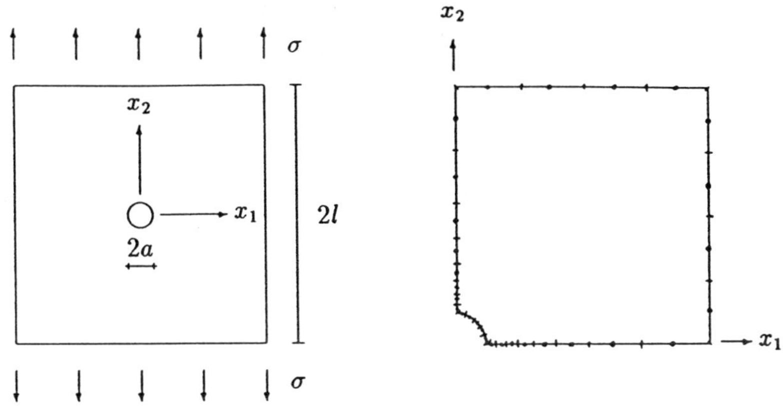

Figure 1: Example i) definition of the problem and discretization

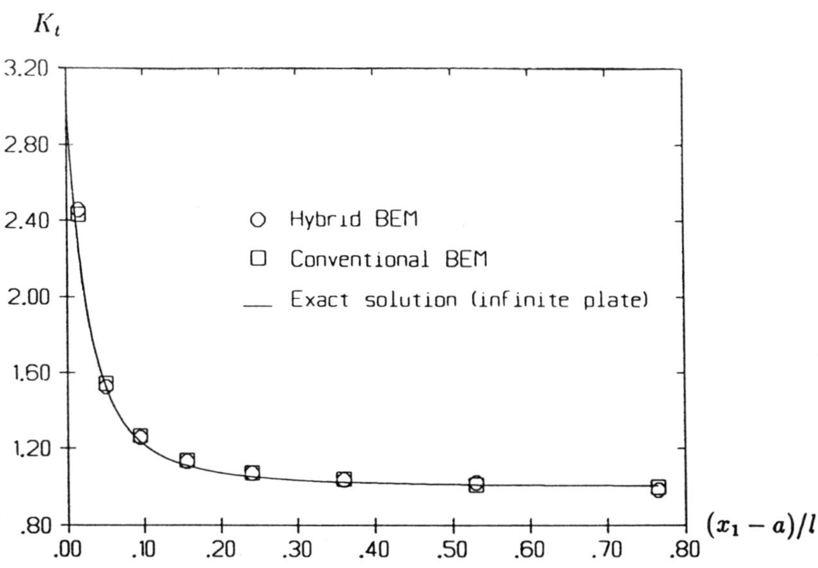

Figure 2: Variation of stresses with distance from the hole

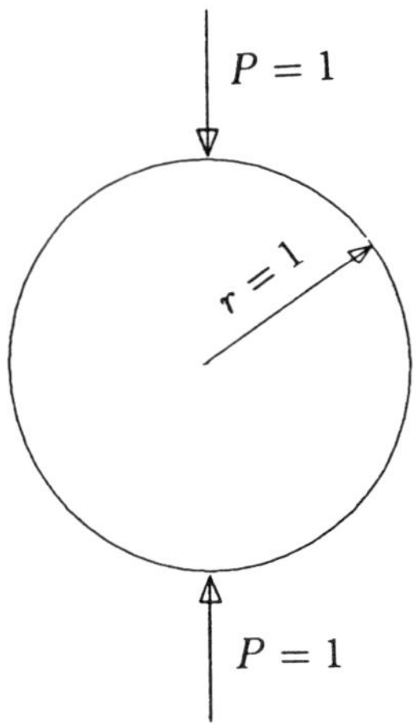

Figure 3: Cylinder under compression: definition of the problem

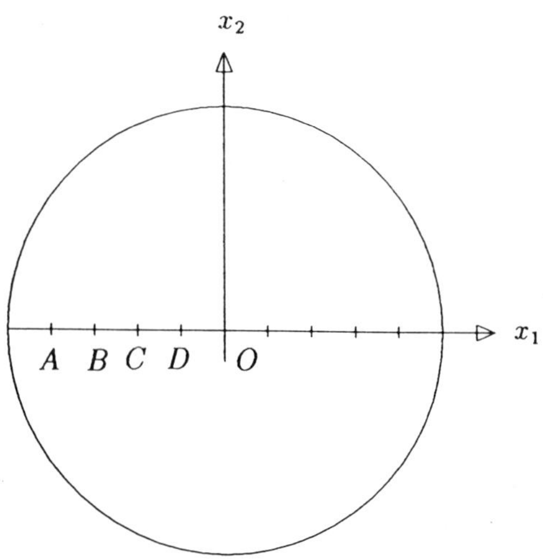

Figure 4: Cylinder under compression: internal points

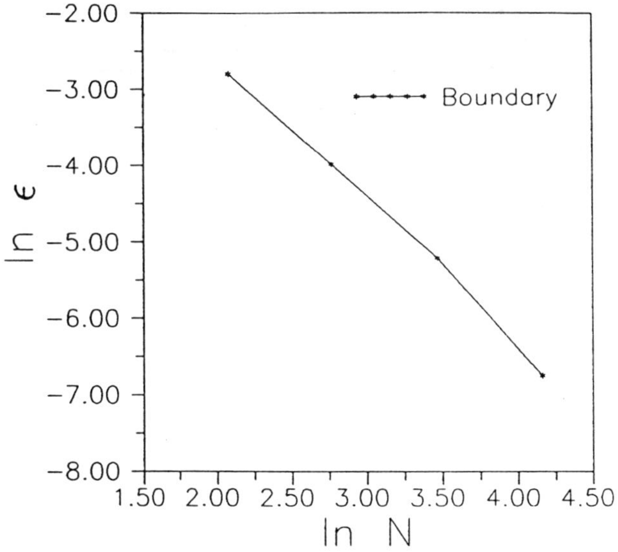

Figure 5: Example ii) convergence of the stresses σ_{22} at internal points

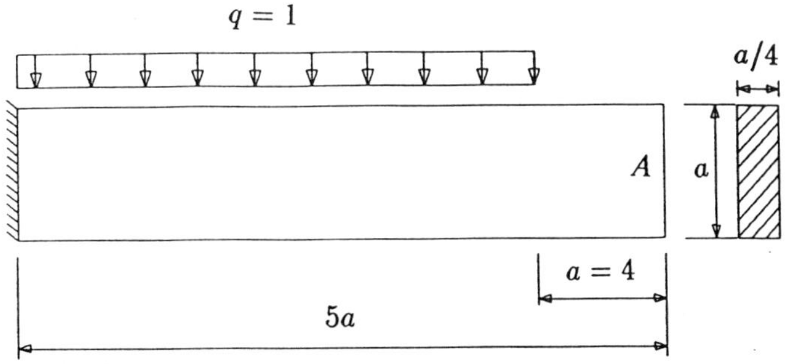

Figure 6: Example iii) cantilever beam

Point	Exact	Hyb. BEM	Error %
A	-0.15551	-0.1579	+1.8
B	-0.3701	-0.3733	+0.9
C	-0.6279	-0.6281	+0.0
D	-0.8588	-0.8549	+0.5
O	-0.9549	-0.9490	+0.6

Table 1: Example iii) stresses σ_{22} at internal points

Displacement component	12 elements		24 elements	
	Hyb. BEM	Conv. BEM	Hyb. BEM	Conv. BEM
u_2^A	0.0881	0.0105	0.0103	0.0106

Table 2: Cantilever beam: deflection at the free end

Boundary Elements for Square-Root Singularities on Curves

J.C. Mason

Royal Military College of Science, Shrivenham, Swindon, SN6 8LA

ABSTRACT

Square-root singularities occur frequently in fracture mechanics and in potential and fluid flow problems. Boundary element technology for singularities is well developed in the case of straight cracks and boundaries, but in the case of curved cracks or boundaries, practical implementations have to our knowledge been limited to the use of a short quarter-point or similar element at the crack tip, based on the assumption that a curve resembles a straight boundary on a small element. However, there do exist cubic boundary elements which, by suitable placement of internal nodes, match singularities exactly on curves. In particular we prove that resulting cubic elements are unique on convex boundary segments, and may always be determined by applying a bisection method to solve a single nonlinear algebraic equation. We also give a necessary and sufficient condition to test for a forbidden element. Quartic elements are also of interest, since, by introducing an extra degree of freedom, we permit in principle the imposition of a user-specified constraint. However, quartic elements are not necessarily unique, and we show interesting examples where forbidden elements or coalescent nodes occur. The determination of nodal positions involves the solution of one or two nonlinear algebraic equations, and to solve these we have successfully adopted either a bisection method or a Newton iteration. Many numerical examples are given to illustrate the generation of eligible nodes, and algorithms are generally successful for sensibly chosen element spacings.

INTRODUCTION

It is well known that singularities in the solution of boundary element formulations of continuum mechanics and other problems may be accurately modelled by placing internal boundary nodes at special positions and adopting isoparametric elements. This approach, which has become extremely popular amongst engineers, is rooted in the "quarter-point" quadratic element developed by Henshell and Shaw [1] and discussed by Mitchell and Wait [2]. The purpose of the present paper is to extend this idea from straight boundaries to curved boundaries. We first discuss "nearly straight" boundaries on which it is possible to use quadratic elements, and then consider general curved boundaries on which it is necessary to adopt cubic or quartic elements.

Although our treatment is based on a boundary element formulation, it is in fact applicable in finite element methods also.

STRAIGHT AND NEARLY STRAIGHT QUADRATIC BOUNDARY ELEMENTS

Suppose that an $r^{\frac{1}{2}}$-type singularity occurs at a point A and that an isoparametric quadratic boundary element is placed along a straight boundary, with end point nodes A ($x = \xi = x_1 = 0$) and C($x = \xi = x_3 = 0$) and internal node B ($x = x_2$). Then the x coordinate is related to the parameter ξ by

$$x = c_1 \xi + c_2 \xi^2.$$

If the node $x_2 = \frac{1}{4}$ is chosen to correspond to the parameter $\xi = \frac{1}{2}$, then the coefficient c_1 vanishes and x is related to ξ by

$$x = \xi^2 \text{ and } \xi = r^{\frac{1}{2}}, \tag{1}$$

where $r = x$ is the distance along the boundary from A.

The quadratic element approximation to the required problem solution u therefore takes the form

$$u = d_0 + d_1 \xi + d_2 \xi^2 = d_0 + d_1 r^{\frac{1}{2}} + d_2 r \tag{2}$$

and the appropriate $r^{\frac{1}{2}}$-type singularity is included. If a solution is sought with an $r^{-\frac{1}{2}}$ singularity, as for example in the case of a traction solution t in elasticity, then it is simply a matter of introducing a multiplicative factor of ξ^{-1}. This leads to the so-called "traction-singular" quadratic element (Cruse and Wilson [3]), given by

$$t = \xi^{-1}(e_0 + e_1 \xi + e_2 \xi^2) = r^{-\frac{1}{2}}(e_0 + e_1 r^{\frac{1}{2}} + e_2 r). \tag{3}$$

Clearly these ideas extend easily to higher order elements. For example, in the cubic case, two internal nodes may be placed at

$x_2 = 1/9$ and $x_3 = 4/9$ (with end point nodes at $x_1 = 0$ and $x_4 = 1$) so that the points x_1, x_2, x_3, x_4 correspond to $\xi = 0, \frac{1}{3}, \frac{2}{3}, 1$. Then (1) holds as before, and a cubic element approximation takes the appropriately singular form

$$u = d_0 + d_1 \xi + d_2 \xi^2 + d_3 \xi^3 = d_0 + d_1 r^{1/2} + d_2 r + d_3 r^{3/2} \quad (4)$$

However, other choices of x_2 and x_3 are also possible. Indeed it is sufficient that (at each node) x should take the form

$$x = c \xi^2 + (1-c) \xi^3 \quad (5)$$

where c is an arbitrary non-zero constant. Then, since r = x, ξ has an infinite expansion in powers of $r^{1/2}$. The family of nodal positions corresponding to (5) which leads to (1) for c = 1, is given by

$$x_2 = (2c+1)/27, \quad x_3 = 4(c+2)/27. \quad (6)$$

In the case of a curved quadratic element, there is no position for the internal node which exactly generalises the quarter-point for a straight boundary. However, for "nearly straight" curved boundaries we may place the internal node exactly or approximately a quarter of the distance along the element, and this approach has been rather successfully used by Mason and Smith [4] in the context of curved crack problems. Indeed, by adopting a relatively short length for the crack-tip element, so that this element is essentially a straight one, accurate solutions may be obtained for a wide variety of problems.

CURVED CUBIC BOUNDARY ELEMENTS

The technique that we adopt in constructing elements for general curves y = y(x) is, as before, to set to zero the leading coefficients in ξ- representations of x and y, so that both x and y behave as ξ^2.

Nodal Positions

In the cubic case, x and y have representations

$$x = a_1 \xi + a_2 \xi^2 + a_3 \xi^3, \quad y = b_1 \xi + b_2 \xi^2 + b_3 \xi^3, \quad (7)$$

where the end point nodes may be taken to be A: $(x_1, y_1) = (0,0)$ and D: $(x_4, y_4) = (1,0)$, corresponding to $\xi = 0$ and $\xi = 1$, and where the internal nodes B: (x_2, y_2) and C: (x_3, y_3) correspond to $\xi = \frac{1}{3}$ and $\frac{2}{3}$. (See Figure 1).

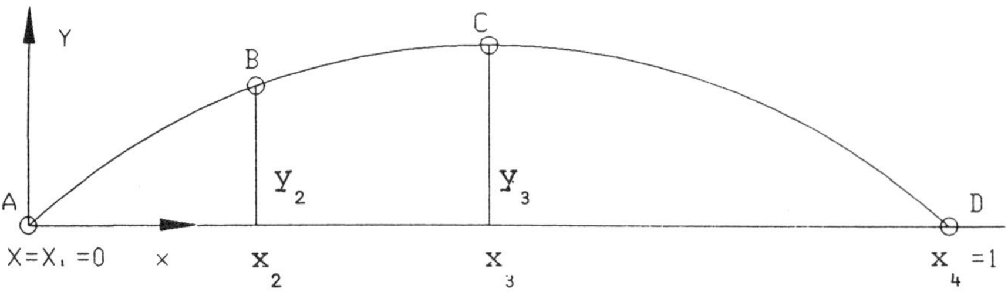

Figure 1 Isoparametric Cubic Element

If a_1 and b_1 are zero in (7), then x and y, and hence the radial distance r, have series expansions beginning with terms in ξ^2. Thus ξ has an expansion in integral powers of $r^{\frac{1}{2}}$, and the appropriate $r^{\frac{1}{2}}$-type behaviour is obtained in the element approximation to the required solution

$$u = c_0 + c_1 \xi + c_2 \xi^2 + c_3 \xi^3.$$

Setting $\xi = \frac{1}{3}, \frac{2}{3}, 1$ at $x = x_2, x_3, 1$ in (7), leads to the equations

$$\begin{aligned} 27x_2 &= 9a_1 + 3a_2 + a_3 \\ 27x_3 &= 18a_1 + 12a_2 + 8a_3 \\ 1 &= a_1 + a_2 + a_3 \end{aligned} \tag{8}$$

By Cramer's rule, (denoting the system determinant by Δ), we deduce that

$$\Delta . a_1 = \begin{vmatrix} 27x_2 & 3 & 1 \\ 27x_3 & 12 & 8 \\ 1 & 1 & 1 \end{vmatrix} = 0,$$

and hence that

$$x_3 = 2x_2 + 2/9 \tag{9}$$

Similarly

$$\Delta . b_1 = \begin{vmatrix} 27y_2 & 3 & 1 \\ 27y_3 & 12 & 8 \\ 0 & 1 & 1 \end{vmatrix} = 0$$

and hence

$$y_3 = 2y_2 \tag{10}$$

Straight Boundary

If y(x) happens to be zero at either x_2 or x_3, then, by (10), both y_2 and y_3 must be zero, and so (9) is the only active equation. This equation leads to the general solution (6) above for x on the element, and we have infinitely many solutions for the possible positions of the internal nodes x_2, x_3. In this case the boundary is modelled as a straight line.

However, if y(x) is non-zero at both x_2 and x_3, so that the boundary is curved, then we expect there to be only one value of c in (6) for which both (9) and (10) are satisfied. We now consider this general case.

Curved Boundary

Assuming that $y_2 \neq 0$ and $y_3 \neq 0$ on a given curve $y=y(x)$, then (9) and (10) combine to give a single nonlinear equation for x_2:

$$f(x) \equiv y(2x_2 + 2/9) - 2y(x_2) = 0 \tag{11}$$

Once x_2 has been determined from (11), then x_3 is given by (9).

Suppose now that y(x) is positive, continuous and convex for $0 \leq x \leq 1$, as in Figure 1. (This is a realistic assumption in practice). Then

$$f(0) = y(2/9) > 0$$

and

$$f(7/18) = y(1) - 2y(7/18) = -2y(7/18) < 0$$

Hence equation (11) has <u>at least one</u> solution x_2 between 0 and 7/18. Also, from (11) and (9), the gradient of f at x_2 is given by

$$f'(x_2) = 2(y'(x_3) - y'(x_2)) \tag{12}$$

and so f is strictly decreasing (since y′ is decreasing on a positive convex curve). Hence equation (11) has a <u>unique</u> solution x_2 in the range [0,7/18]. Moreover since f is continuous and changes sign on [0,7/18], a bisection method based on [0,7/18] is guaranteed to converge to the solution x_2 of (11).

One-to-One Mappings

Once x_2 and x_3 have been determined from (11) and (9), it is essential to verify that the resulting mapping (7) of x to ξ is one-to-one. Otherwise we obtain a "forbidden element". Now the values of a_2 and a_3 may be determined explicitly by Cramer's rule from (8). If these values are then written in terms of x_2 (from (9)) and substituted into (7), it follows that

$$x = \tfrac{1}{2}(27x_2 - 1)\,\xi^2 + \tfrac{3}{2}(1 - 9x_2)\,\xi^3 \tag{13}$$

It is hence easy to verify, by requiring that $x'(\xi)$ is non-negative on [0,1], that a necessary and sufficient condition for a one-to-one mapping is

$$1/27 \le x_2 \le 7/27 \tag{14}$$

Numerical Examples

Numerical solutions x_2 and x_3 of the equations (11) and (9) have been obtained for a variety of functions $y(x)$, and the results are given in Table 1. Note that (14) is satisfied in every case, and so each $x - \xi$ mapping is one-to-one. This is even true for the last function $\sqrt{(x-x^2)}$, which corresponds to a semi-circular element with an infinite gradient at $x=0$.

Table 1 Cubic Nodal Positions on Curves $y = y(x)$

$y(x)$	x_2:	x_3:
$x-x^2$	.1463	.5148
$x-x^3$	.1987	.6096
$\sin \pi x$	.1643	.5509
$\log (1+x-x^2)$	.1367	.4957
$-\log (1+x^2-x)$	.1581	.5385
$-1+\exp (x-x^2)$	.1566	.5354
$1-\exp (x^2-x)$	.1356	.4934
$\sqrt{(8x+1)}-(2x+1)$	.0838	.3899
$\sqrt{(x-x^2)}$	.0598	.3419

However, as a counter-example, we note that, for the parabola $y = x^{1/2} - x$, the solution of (11) and (9), namely $x_2 = 0.0214$, $x_3 = 0.2650$, does not satisfy (14).

CURVED QUARTIC BOUNDARY ELEMENTS

In the case of quartic elements, the isoparametric mapping takes the form

$$x = a_1\xi + a_2\xi^2 + a_3\xi^3 + a_4\xi^4, \quad y = b_1\xi + b_2\xi^2 + b_3\xi^3 + b_4\xi^4, \tag{15}$$

where the end point nodes are $A:(x_1,y_1) = (0,0)$ and $E:(x_5,y_5) = (1,0)$ and the internal nodes are $B:(x_2,y_2)$, $C:(x_3,y_3)$ and $D:(x_4,y_4)$ (see Figure 2). It is again realistic to assume that the boundary curve $y = y(x)$ from A to E is positive, continuous and convex. In this case the nodes A, B, C, D, E must be chosen to correspond, respectively, to $\xi = 0, \frac{1}{4}, \frac{1}{2}, \frac{3}{4}, 1$ and a_1 and b_1 must be set to zero.

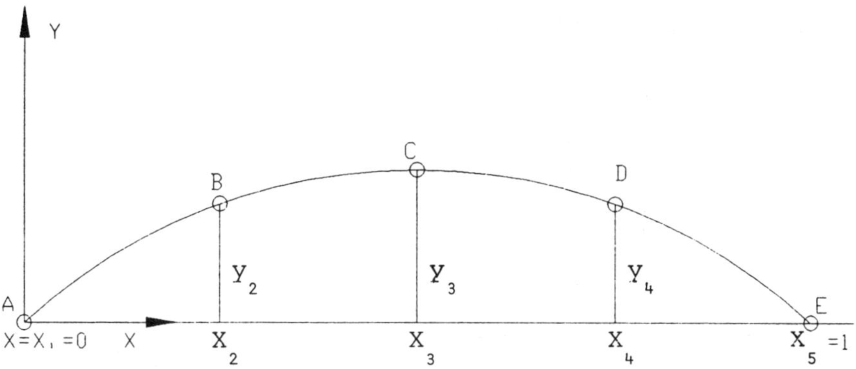

Figure 2 Isoparametric Curved Quartic Element

This ensures that the radial distance r behaves like ξ^2, so that ξ has an (infinite) expansion in powers of $r^{\frac{1}{2}}$, thus leading to an $r^{\frac{1}{2}}$-type behaviour in the element approximation

$$u = c_0 + c_1 \xi + c_2 \xi^2 + c_3 \xi^3 + c_4 \xi^4.$$

Setting $\xi = \frac{1}{4}, \frac{1}{2}, \frac{3}{4}, 1$ at $x = x_2, x_3, x_4, x_5$ in (15) leads to the equations

$$\begin{aligned} 256x_2 &= 64a_1 + 16a_2 + 4a_3 + a_4 \\ 16x_3 &= 8a_1 + 4a_2 + 2a_3 + a_4 \\ 256x_4 &= 192a_1 + 144a_2 + 108a_3 + 81a_4 \\ 1 &= a_1 + a_2 + a_3 + a_4 \end{aligned} \tag{16}$$

By Cramer's rule (denoting the system determinant by Δ), we deduce that

$$\Delta . a_1 = \begin{vmatrix} 256x_2 & 16 & 4 & 1 \\ 16x_3 & 4 & 2 & 1 \\ 256x_4 & 144 & 108 & 81 \\ 1 & 1 & 1 & 1 \end{vmatrix} = 0$$

and hence that

$$12x_2 - 9x_3 + 4x_4 = 3/4 \tag{17}$$

By a similar procedure we deduce that

$$12y_2 - 9y_3 + 4y_4 = 0 \tag{18}$$

However, only 2 equations, (17) and (18), have been obtained for determining x_2, x_3 and x_4.

In order to determine the nodes uniquely, it is necessary either to fix one of the nodes or to impose a further constraint on the element. We consider both possibilities below, but first we consider the requirement to avoid forbidden elements.

One-to-One Mapping

It is essential to check that any nodes obtained from (17) and (18) give a one-to-one mapping of x to ξ. By applying Cramer's rule to determine a_2, a_3, a_4 from (16), substituting in (15), and expressing x_4 in terms of x_2, x_3 from (17), we deduce that

$$x = \tfrac{1}{3}(C\,\xi^2 + 6B\,\xi^3 + 8A\,\xi^4) \tag{19}$$

where $A = 4(8x_2 - 3x_3) + 1$, $B = 4(-16x_2 + 5x_3) - 1$, $C = 8(16x_2 - 3x_3) + 1$. Clearly we must require that $C > 0$, and that $dx/d\xi > 0$ on (0,1) of ξ. By considering the sign of the discriminant

$$D = 81\,B^2 - 64\,AC \tag{20}$$

of the quadratic factor in

$$\frac{dx}{d\xi} = \tfrac{1}{3}\,\xi(32\,A\,\xi^2 + 18\,B\,\xi + 2\,C),$$

we obtain the necessary and sufficient conditions for a one-to-one mapping, namely

$$C > 0$$

and either

$$D < 0 \tag{21}$$

or

$$D \geq 0,\ 9B + 2C \geq 0 \text{ and } 16A + 9B + C \geq 0$$

Fixing a Node

Suppose that one internal node is to be fixed. Then the most obvious node to choose is the central one x_3, and we might choose a value equal to or close to .5. Eliminating x_4 between (4.3) and (4.4),

$$f(x_2) \equiv 12\,y(x_2) - 9\,y(x_3) + 4\,y(2.25\,x_3 - 3x_2 + 0.1875) = 0 \tag{22}$$

For x_3 fixed, (22) is a nonlinear algebraic equation for the determination of x_2. In order to guarantee the convergence of a secant method for (22) it is necessary to have a universal way of choosing an interval in which a solution must lie. However, for a general function f, we cannot yet give such an explicit interval.

In Table 2 are shown the solutions x_2, x_3, x_4 for fixed values of x_3 corresponding to the same range of functions as Table 1 above. The last function in Table 1, for which we were unable to find a useful solution, has been replaced by the almost semicircular curve (ix). All results given adhere to conditions (21) above and therefore correspond to 1-1 mappings. However, for some functions, x_3 has needed to be reduced below .5 to ensure a useful solution. For example, for x_3 = .5 the nodes on curve (viii) in Table 1 become

$$x_2 = .1246, \qquad x_4 = .9388;$$

then conditions (21) are not satisfied, and a forbidden element is obtained.

Table 2 Quartic Nodal Positions - x_3 Fixed

	y(x):	x_2:	x_3:	x_4:
(i)	$x - x^2$	.1650	.5	.8175
(ii)	$x - x^3$	.1819	.5	.7667
(iii)	$\sin \pi x$	.1791	.5	.7751
(iv)	$\log(1 + x-x^2)$	.1581	.5	.8382
(v)	$-\log(1 + x^2-x)$	.1742	.5	.7900
(vi)	$-1 + \exp(x-x^2)$	.1729	.5	.7938
(vii)	$1 - \exp(x^2-x)$	.1573	.5	.8407
(viii)	$\sqrt{(8x+1)} - (2x+1)$	.1021	.4	.7812
(ix)	$-.05+\sqrt{(.0025+x-x^2)}$	.0498	.3	.7131

Adding a Further Constraint

One further equation may (for example) be obtained by exactly reproducing the value of $y'(0)$, namely the gradient of the boundary curve at the singularity A. In the case of a fracture mechanics problem, this imposes the correct crack-tip direction on the element. Let us assume that this gradient has the specified value

$$y'(0) = K^{-1}, \quad (0 \le K < \infty) \tag{23}$$

so that a zero gradient is excluded but an infinite gradient is possible (for $K = 0$). Since a_1 and b_1 are zero, it follows from (15) that

$$a_2 = K\, b_2.$$

By applying Cramer's rule to determine a_2 from (16), and by adopting a similar procedure to determine b_2, we deduce that

$$\begin{vmatrix} 64 & 256x_2 & 4 & 1 \\ 8 & 16x_3 & 2 & 1 \\ 192 & 256x_4 & 108 & 81 \\ 1 & 1 & 1 & 1 \end{vmatrix} = K \begin{vmatrix} 64 & 256x_2 & 4 & 1 \\ 8 & 16y_3 & 2 & 1 \\ 192 & 256y_4 & 108 & 81 \\ 1 & 0 & 1 & 1 \end{vmatrix}$$

and hence that

$$(52x_2 - 57x_3 + 28x_4 - 11/2) = K\,(52y_2 - 57y_3 + 28y_4) \tag{24}$$

On any given curve $y = y(x)$, (17), (18) and (24) combine to give a pair of simultaneous nonlinear equations for x_2 and x_3, namely

$$\begin{aligned} f_1(x_2,x_3) &\equiv 12y(x_2) - 9y(x_3) + 4y(x_4) = 0 \\ f_2(x_2,x_3) &\equiv -32x_2+6x_3-1/4-K\{52y(x_2)-57y(x_3)+28y(x_4)\} = 0 \end{aligned} \tag{25}$$

where

$$x_4(x_2,x_3) = 9x_3/4 - 3x_2 + 3/16. \tag{26}$$

If values of $y'(x)$ can be determined everywhere on $[0,1]$, then Newton's method may be adopted to solve these equations.

Initial values $x_2^{(0)}$ and $x_3^{(0)}$ need to be specified, and, unless otherwise stated, it is proposed to adopt the nodal positions for the mapping $x = \xi^2$ on a straight boundary, namely

$$x_2^{(0)} = 1/16, \; x_3^{(0)} = 1/4. \tag{27}$$

If $y'(x)$ cannot readily be calculated, then an appropriate variant of Newton's method might be adopted. Alternatively, a slow but reliable search procedure might be used initially.

In Table 3 are given values of x_2, x_3, x_4 obtained from equations (25) for the same selection of curves as in Table 2, but with the last example modified. In each case, Newton's method converged rapidly from the starting values (27). Note that the nodes obtained for the parabola $y = x - x^2$ correspond, as expected, to the exact parametrisation

$$x = \xi^2, \; y = \xi^2 - \xi^4.$$

In every example in Table 3, conditions (21) were satisfied, and so the mapping of x to ξ was one-to-one.

However, curves formed by the family of circular arcs

$$y = y(x) = -a + \surd(a^2 + x-x^2), \tag{28}$$

which pass through (0,0) and (1,0) and have centre at (-a,0), present a greater challenge, and indeed forbidden elements have been obtained for $a \leq .33$, as well as more than one solution. For example, for $a = .2$ we obtained the pair of forbidden solutions

$$x_2, x_3, x_4 = .00938, .1156, .4194$$
$$x_2, x_3, x_4 = .1472, .5207, .9175.$$

Moreover, for $a = 0$, a solution of (25) exists with internal nodes at $x_2 = 0$, $x_3 = 1/24$, $x_4 = 9/32$ and hence, since $x_1 = 0$ there are a pair of coincident nodes at the left end point.

For $a > .34$, we were able to obtain valid elements and one such (for $a = .4$) is included in Table 3.

Table 3 Quartic Nodal Positions - Fixed Crack-Tip Direction

$y(x)$:	x_2:	x_3:	x_4:
$x-x^2$	.0625	.2500	.5625
$x-x^3$	.1172	.3690	.6662
$\sin \pi x$	.0385	.2059	.5352
$\log(1 + x-x^2)$	.0435	.2095	.5284
$-\log(1+x^2-x)$	.0922	.3059	.5991
$-1+\exp(x-x^2)$	.0293	.1873	.5211
$1-\exp(x^2-x)$	.0428	.2075	.5259
$\surd(8x-+1) - (2x+1)$	.0178	.1146	.3920
$-.4 + \surd(.16 + x - x^2)$	.0265	.1692	.4885

BOUNDARY SHAPE

The elements that we have obtained here lead to cubic and quartic boundary approximations with unequally spaced nodes. Traditional elements (which do not include singularities) would also give inexact boundaries, but nodes would be spaced for convenience. We have made many numerical comparisons of the shapes obtained from our singular elements with those obtained from non-singular elements for the numerical examples considered in Table 1, and have not found that a significant sacrifice in shape occurs in order to achieve a singularity. Somewhat more surprisingly, in the case of quartics, we have not found there to be a significant improvement in shape by using a fixed central node (equation (22)) rather than a node determined by fixing the crack-tip direction (equations (25)), even though the spacing of nodes seems at first sight to be more reasonable in Table 3 than in Table 2.

APPLICATIONS TO FINITE ELEMENTS

Although our discussion has so far been limited to boundary elements, we note that the special nodal positions that we have developed may in principle be adopted also on the curved sides of triangular or quadrilateral finite elements for applications to partial differential equations.

ACKNOWLEDGEMENTS

We thank the School of Defence Management at RMCS (Cranfield), the Mathematics Department at Australian Defence Force Academy and the Centre for Mathematical Analysis at Australian National University for providing academic and computing support.

REFERENCES

1. Henshall, R.D. and Shaw, K.G., Crack Tip Finite Elements are Unnecessary. Int. J. Num. Meth. Enging, Vol 9, pp 495-507, 1975.

2. Mitchell, A.R. and Wait, R., The Finite Element Method in Partial Differential Equations. Wiley, Chichester, 1977.

3. Cruse, T.A. and Wilson, R.B., Boundary Integral Equation Method for Elastic Fracture Mechanics. AFSOR-TR-78-0355, pp 10-11, 1977.

4. Mason, J.C. and Smith, R.N.L., Boundary Integral Equation Methods for a Variety of Curved Crack Problems, In: Treatment of Integral Equations by Numerical Methods, (Ed. Baker, C.T.H. and Miller, G.F.), pp 239-252, Academic Press, London, 1982.

Improving the Accuracy of BEM in the use of Non-equidistant Elements

M. Karamanoglu, G.E. Beswick
School of Mechanical & Manufacturing Engineering, Middlesex Polytechnic, London N11 2NQ, U.K.

ABSTRACT

This paper investigates the accuracy of surface displacements of a solid sphere under uniform pressure. Three dimensional quadratic elements were used to discretize the surface and accurate results were obtained. This was done by using equally spaced elements.

The same problem was solved with non-equidistant elements where a small and a large element were adjacent to each other without any special treatment of elements. This caused distortion of an unacceptable magnitude in the results. The error was greater in the presence of singularity. The use of two separate integration schemes were made to deal with singular and non-singular elements. An ordinary two dimensional Gauss [1] quadrature rule was used for non-singular, and Gauss and Kutt [2] integration rules were used for singular elements by applying a composite transformation.

This paper explains the method used for correcting the inaccuracy caused by having a mixture of coarse and fine mesh close together by giving examples with quadratic elements. The method proved to be very effective and it was implemented without significant changes to the existing code.

INTRODUCTION

In the Boundary Element Method, it is known that determining the system matrices requires considerable CPU time. This is due to the large amount of numerical integrations being carried out during the determination of these coefficients [3-5]. It is also true to say that the singular integrals that are present in

3-D applications, pose a separate problem themselves. As the field point gets closer to the source point, that is the distance between these two points gets very small, the integral becomes singular and causes computer programs to fail.

To avoid the singularity problem, several methods have been proposed by a number of researchers [3-8]. These methods are very similar in principle. The element under consideration is divided into several small areas where integration is carried out separately and the contributions from each area are added together.

This is an iterative process where the singular point is isolated and several iterations or sub-divisions are carried out until the total contribution converges within a pre-determined tolerance. In this method the element may be divided into rectangular or triangular sub-elements. With rectangular elements, some researchers have taken divisions half the size of the elements and others have taken more severe divisions to accelerate the convergence.

One other method used by Telles [5], is to use quadrature rules that include the singularity. In this case, the element is divided into triangles and each sub-element is integrated only once. The total contribution is again the sum of the individual results from each sub-element.

There is also a problem of having different size elements adjacent to each other. In this paper, a solid sphere under uniform external pressure of 1 MPa is taken to illustrate the above points.

BEM MODEL OF SOLID SPHERE

In general, for elastic problems [9] in BEM equations take the form of:

$$c^i u_l^i + \int_S u_k p_{lk}^* \, dS = \int_S \mathrm{P}_k w_{lk}^* \, dS + \int_\Omega b_k w_{lk}^* \, d\Omega \tag{1}$$

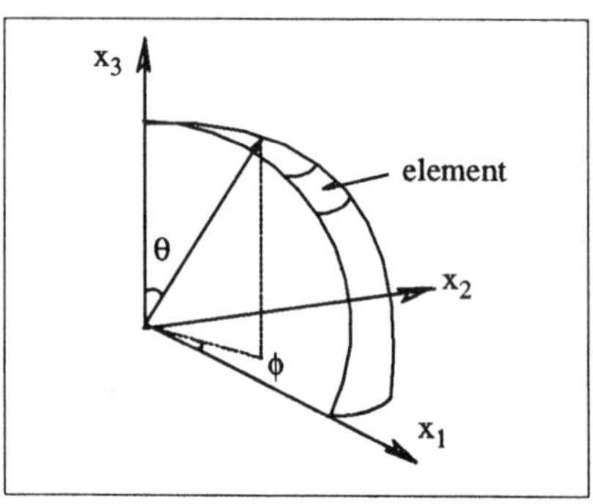

Figure 1. Coordinate system used for the model.

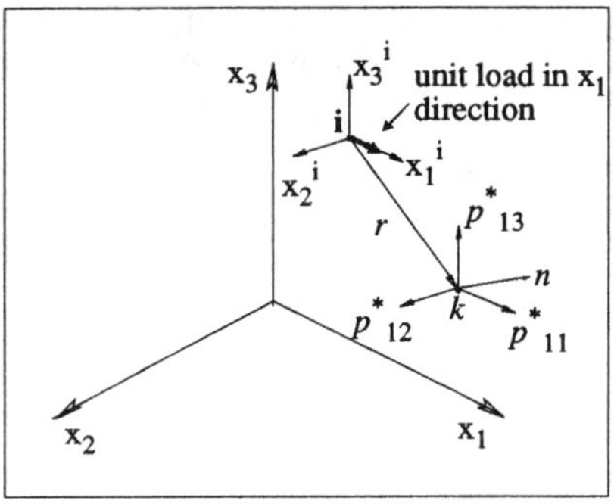

Figure 2. Coordinate system for the fundamental solution.

For a smooth boundary $c^i =$ 0.5 and without any body forces equation (1) becomes:

$$\frac{1}{2} u_l^i + \int_S u_k p^*_{lk}\, dS = \int_S \mathrm{P_k}\, w^*_{lk}\, dS \tag{2}$$

where p^*_{lk} and w^*_{lk} represent tractions and displacements in the k direction due to a unit force acting in the l direction. For 3-D isotropic body:

$$w^*_{lk} = \frac{1}{16\,\pi\, G\,(1-\nu)\, r}\left[(3-4\nu)\,\delta_{lk} + 3\,\frac{\partial r}{\partial x_l}\frac{\partial r}{\partial x_k}\right] \tag{3}$$

$$p^*_{lk} = \frac{1}{8\,\pi\,(1-\nu)\, r^2}\left[\frac{\partial r}{\partial n}\left\{(1-2\nu)\,\delta_{lk} + 3\frac{\partial r}{\partial x_l}\frac{\partial r}{\partial x_k}\right\} - (1-2\nu)\left\{\frac{\partial r}{\partial x_l} n_k - \frac{\partial r}{\partial x_l} n_l\right\}\right] \tag{4}$$

where n is the outward normal to the surface of the body, δ_{lk} is the Kroneker delta, r is the distance from the source point to the field point under consideration and n_k and n_l are the direction cosines (see Figure 2)

$$\begin{aligned} u_1 &= V_1 \cos\phi \\ u_2 &= V_1 \cos\phi \\ u_3 &= V_3 \end{aligned} \tag{5}$$

In this application, equation (2) is better suited to be expressed in a spherical coordinate system. As the pressure is uniform the u_1 and u_2 surface displacements can be expressed in terms of the resultants (see Figure 3).

Because of the symmetry, only the top half of the sphere is analyzed. By placing source points on $\phi = 0°$ and at $0° \leq \theta \leq 90°$ this drastically reduces the number of equations. When the coefficients of the system matrices are formed, the whole sphere is considered.

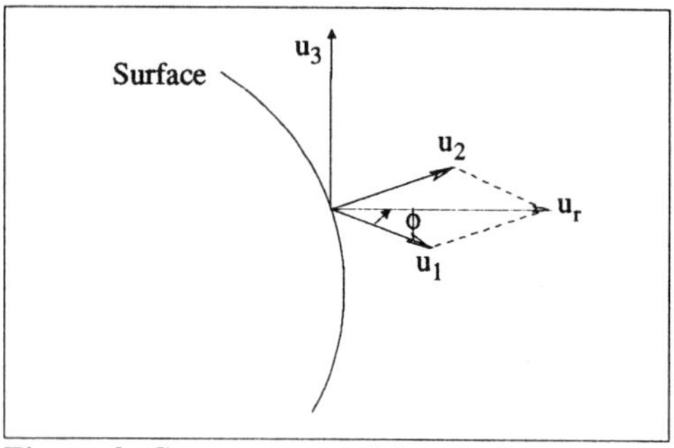

Figure 3. Components of the displacements.

Rewriting equation (2) in spherical coordinates:

$$\frac{1}{2}V_l^i + R^2\int_S \left(V_1 \cos\phi \, p_{l1}^* + V_1 \sin\phi \, p_{l2}^* + V_3 p_{l3}^* \right) \sin\theta \, d\phi \, d\theta = \qquad (6)$$

$$R^2\int_S \left(w_{l1}^* \mathrm{P}_1 + w_{l2}^* \mathrm{P}_2 + w_{l3}^* \mathrm{P}_3 \right) \sin\theta \, d\phi \, d\theta$$

Applying the following Lagrangian interpolation functions shown below:

$$V_1 = L_1 V_1^i + L_2 V_1^{i+1} + L_3 V_1^{i+2} \qquad (7)$$

$$V_3 = L_1 V_3^i + L_2 V_3^{i+1} + L_3 V_3^{i+2}$$

where (8)

$$L_1 = \frac{(\theta - \theta_{j+1})\,(\theta - \theta_{j+2})}{(\theta_j - \theta_{j+1})\,(\theta_j - \theta_{j+2})}$$

$$L_2 = \frac{(\theta - \theta_j)\,(\theta - \theta_{j+2})}{(\theta_{j+1} - \theta_j)\,(\theta_{j+1} - \theta_{j+2})}$$

$$L_3 = \frac{(\theta - \theta_j)\,(\theta - \theta_{j+1})}{(\theta_{j+2} - \theta_j)\,(\theta_{j+2} - \theta_{j+1})}$$

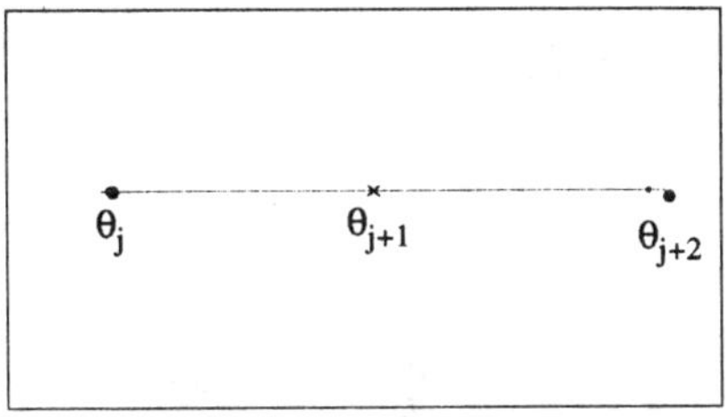

Figure 4. Angles used for interpolation functions.

For any field point on the surface S, the components of the applied pressure load can be expressed in terms of the resultant pressure, P, by making use of the spherical coordinate system (see Figure 5).

$$\begin{aligned} P_1 &= P \sin\theta \cos\phi \\ P_2 &= P \sin\theta \cos\phi \\ P_3 &= P \cos\theta \end{aligned} \tag{9}$$

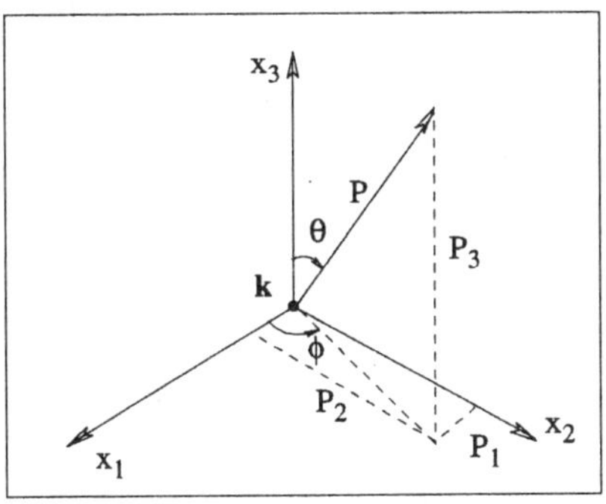

Figure 5. Components of the pressure load used.

Equation (10) shows the BEM formulation that has been used in the analysis. Having all the unknowns on the left-hand side and all the knowns on the right-hand side, produces system matrices in the form of **Ax=b.**

$$\tfrac{1}{2}V_1^j + R^2\int_S \left(V_1^j A_l + V_1^{j+1} B_l + V_1^{j+2} C_L + V_3^j D_l + V_3^{j+1} E_l + v_3^{j+2} F_l\right)\sin\theta d\phi\, d\theta =$$

$$P\, R^2 \int_S \left(w^*_{l1} \sin\theta \cos\phi + w^*_{l2} \sin\theta \sin\phi + w^*_{l3} \cos\phi\right)\sin\theta\, d\phi\, d\theta \tag{10}$$

where (11)

$$
\begin{aligned}
A_l &= L_1 K_l \\
B_l &= L_2 K_l \\
C_l &= L_3 K_l \\
D_l &= L_1 p^*_{l3} \\
E_l &= L_2 p^*_{l3} \\
F_l &= L_3 p^*_{l3} \\
K_l &= p^*_{l1} \cos\phi + p^*_{l2} \sin\phi
\end{aligned}
$$

DETERMINING A COMPOSITE MAPPING

The intention of employing a composite mapping, Figure 6, was to avoid sub-divisioning of singular elements for integration purposes. One-dimensional integration points and weights for Gaussian type finite part quadrature had been developed by Kutt [4]. The formula in general terms is:

$$\int_0^b \frac{f(y)}{y}\, dy \approx \sum_{i=1}^{n} f(b\,\xi_i)\,\omega_i + f(0)\ln(b) \tag{12}$$

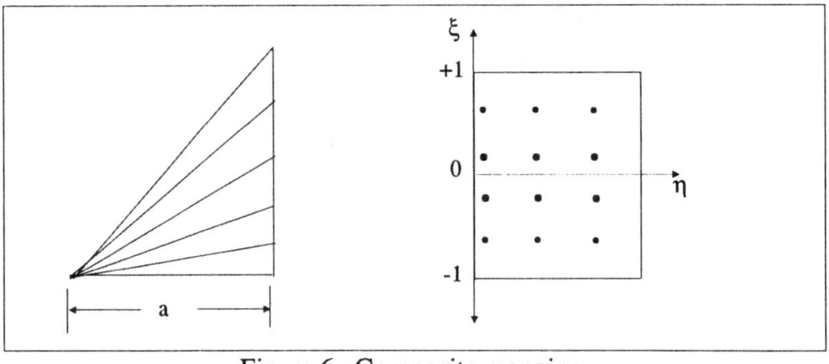

Figure 6. Composite mapping.

To apply Kutt, the rectangular elements were divided into triangles as shown in Figure 7.

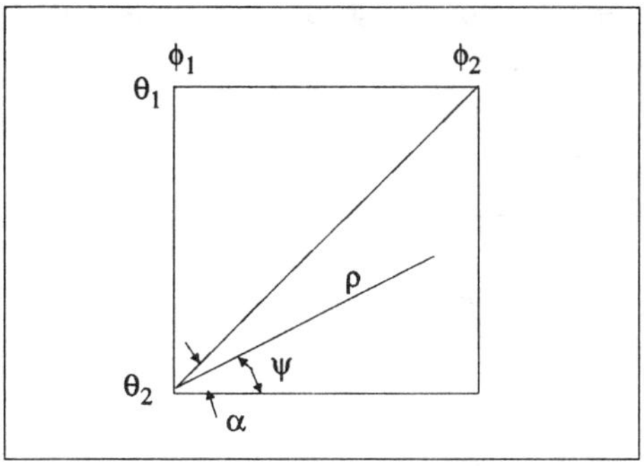

Figure 7. Composite mapping parameters.

$$\psi = \frac{\alpha}{2}(\xi + 1) \tag{13}$$

$$\rho = \frac{a}{\cos\psi}\eta \tag{14}$$

$$\theta = \rho \sin\psi \tag{15}$$

$$\phi = \rho \cos\psi \tag{16}$$

In order to use the Kutt type formulation, the integral equations needed to be modified. Integrals in BEM formulation were in the form of:

$$\int_0^\alpha\int_0^\rho f(\rho,\psi)\ \rho\, d\rho\, d\psi \tag{17}$$

To adopt inner integration to Kutt, it must be in the form :

$$\int_0^\rho \frac{f(\rho\,,\psi)}{\rho}\, d\rho \tag{18}$$

Therefore the top and bottom of the BEM equations were multiplied by ρ. This gave equation (19).

$$\int_0^\alpha\int_0^\rho \frac{f(\rho\,,\psi)}{\rho}\,\rho^2 d\rho d\psi \tag{19}$$

The inner integral was evaluated using Kutt quadrature, mapped to (0,1) space and the outer integral was done by using Gauss quadrature in (-1,1) space. This is shown in Figure 6.

The reason for using the Kutt formulation was that, in BEM equations, there is a $1/r^2$ factor which causes singularity as r values get very small. The value of r is approximated to ρ, therefore they cancel each other out and the effect of singularity disappears. This is clearly shown in equation (20) if the $1/r^2$ is taken outside of $f(\rho,\psi)$ and written as;

$$\int_0^{\alpha}\int_0^{\rho} \frac{f'(\rho,\psi)}{\rho r^2}\,\rho^2\, d\rho d\psi \qquad (20)$$

where

$$f(\rho,\psi) = \frac{f'(\rho,\psi)}{r^2} \qquad (21)$$

As ρ^2 will cancel r^2, this will lead to a formulation without a singularity built in as the remaining $1/\rho$ term is taken care of by Kutt transformation. Results obtained by this method are given in Table 1.

The results shown in Table 1, only show the displacements on the top half of the sphere as the symmetry of the structure was used. However, the integrals were evaluated all over the surface. Two different orders of quadrature were used for the singular elements, Kutt order of 10 and Gauss order of 12, and for all the other elements 12x12 Gauss quadrature was used.

Table 1: External displacements using quadratic elements. (Theoretical result is 0.2×10^{-5})

Node Number	Angle θ	Resultant Displacements $/10^{-5}$
1	0.00	0.200488
2	5.62	0.200127
3	11.25	0.199773
4	16.87	0.199732
5	22.50	0.199384
6	28.12	0.199506
7	33.75	0.199108
8	39.37	0.199350
9	45.00	0.198907
10	50.62	0.199239
11	56.25	0.198763
12	61.87	0.199162
13	67.50	0.198665
14	73.12	0.199113
15	78.75	0.198608
16	84.37	0.199089
17	90.00	0.198589

It was also found that the contribution of the log term of equation (12) was insignificantly small in this application, so that it was considered during the formulation. The results obtained also confirmed this, but this is not always the case.

Further work was carried out to determine the internal displacements and stresses using equations (22) and (23). These equations do not take into account the body forces since in this application these do not apply.

$$u_l^i = \int_S \mathrm{P}_k w_{lk}^* - \int_S u_k p_{lk}^* \, dS \tag{22}$$

$$\sigma_{ij} = \int_S D_{kij} \mathrm{P}_k \, dS - \int_S S_{kij} u_k \, dS \tag{23}$$

where the third order tensor components D_{kij} and S_{kij} for 3-D case are [9] :

$$D_{kij} = \frac{1}{r^2\, 8\pi(1-\nu)} \left\{ (1-2\nu)\,(\delta_{ki}\, r_{,j} + \delta_{kj}\, r_{,i} - \delta_{ij}\, r_{,k}) + 3 r_{,i}\, r_{,j}\, r_{,k} \right\} \tag{24}$$

$$S_{kij} = \frac{G}{r^3 4\pi(1-\nu)} \left\{ 5\frac{\partial r}{\partial n}\left[(1-2\nu)\,\delta_{ij}\, r_{,k} + \nu(\delta_{ik}\, r_{,j} + \delta_{jk}\, r_{,i}) - 5\, r_{,i}\, r_{,j}\, r_{,k} \right] + \right. \tag{25}$$

$$\left. 5\nu(n_i\, r_{,j}\, r_{,k} + n_j\, r_{,i}\, r_{,k}) + (1-2\nu)(5 n_k\, r_{,i}\, r_{,j} + n_j \delta_{ik} + n_i \delta_{jk}) - (1-4\nu) n_k \delta_{ij} \right\}$$

Table 2 and Table 3 show the results for internal displacements and stresses. The results show the well known limitation of BEM as the field points get very close to the boundary.

Table 2 : Internal displacements using equally spaced elements.

Node Number	Radius m	Horizontal Component	Vertical Component	Resultant	Theoretical / 10^{-5} m
2	0.10	0.1410E-06	0.1415E-06	0.0199E-05	0.02
3	0.20	0.2820E-06	0.2830E-06	0.0399E-05	0.04
4	0.30	0.4230E-06	0.4244E-06	0.0599E-05	0.06
5	0.40	0.5641E-06	0.5658E-06	0.0799E-05	0.08
6	0.50	0.7052E-06	0.7071E-06	0.0999E-05	0.10
7	0.60	0.8464E-06	0.8482E-06	0.1198E-05	0.12
8	0.70	0.9877E-06	0.9893E-06	0.1398E-05	0.14
9	0.80	0.1129E-05	0.1130E-05	0.1598E-05	0.16
10	0.90	0.1270E-05	0.1270E-05	0.1797E-05	0.18
11	0.95	0.1341E-05	0.1341E-05	0.1897E-05	0.19

Table 3 : Internal stresses using equally spaced quadratic elements.

Node No	Radius m	Internal Stresses /Nm^{-2}					
		σ_{11}	σ_{12}	σ_{13}	σ_{22}	σ_{23}	σ_{33}
1	0.00	0.998E+06	-0.655E-11	-0.856E-11	0.998E+06	-0.142E-10	0.999E+06
2	0.10	0.998E+06	-0.662E-11	0.524E-12	0.998E+06	-0.131E-10	0.999E+06
3	0.20	0.998E+06	0.948E-12	-0.579E-11	0.998E+06	-0.815E-11	0.999E+06
4	0.30	0.998E+06	-0.640E-11	-0.775E-11	0.998E+06	-0.141E-10	0.999E+06
5	0.40	0.998E+06	0.216E-12	-0.173E-10	0.998E+06	-0.414E-11	0.999E+06
6	0.50	0.998E+06	-0.703E-11	-0.978E-11	0.998E+06	0.375E-11	0.999E+06
7	0.60	0.999E+06	-0.639E-11	-0.245E-10	0.999E+06	0.243E-10	1.000E+06
8	0.70	0.999E+06	0.108E-10	-0.225E-10	0.999E+06	0.359E-10	1.000E+06
9	0.80	0.999E+06	0.236E-10	-0.295E-10	0.999E+06	-0.297E-10	1.000E+06
10	0.90	1.000E+06	-0.377E-10	-0.135E-09	1.000E+06	-0.255E-10	1.003E+06
11	0.95	1.000E+06	0.377E-10	-0.192E-09	1.000E+06	0.282E-10	1.010E+06

INVESTIGATION OF USING SMALL AND LARGE ELEMENTS ADJACENT TO EACH OTHER:

One of the BEM's limitations has been the sudden change in the spacing of elements. If two adjacent elements were different in size, this resulted in fluctuations of the results. Several cases had been looked at and the following results were obtained for external displacements.

Quadratic elements with equal spacing were used to determine the surface displacements. Since the elements were of the same size, this did not require special attention. Figure 8 shows the integration scheme used. The results were in good agreement with the theoretical values (see Figure 10).

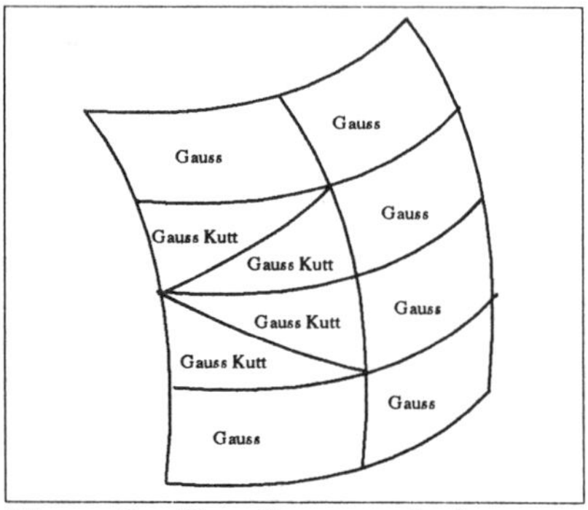

Figure 8 . Equally spaced elements. No distortion in the results.

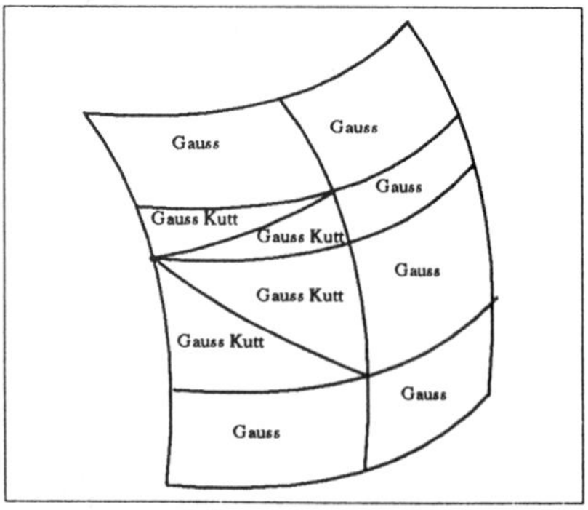

Figure 9 . Unequally spaced elements. Significant distortion in the results.

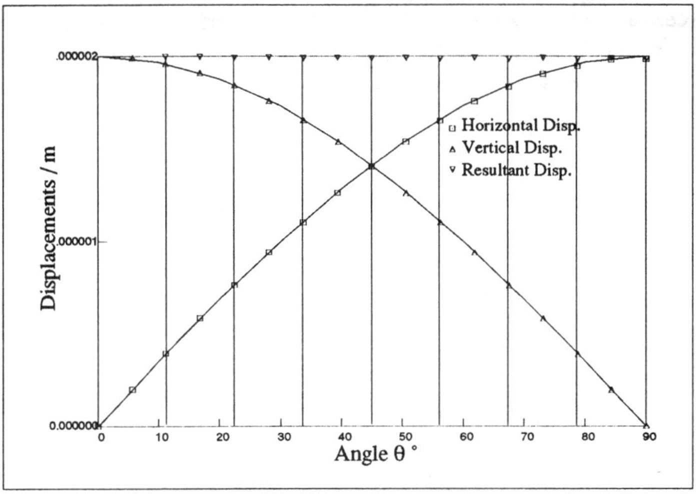

Figure 10 . Displacements of equally spaced quadratic elements.

Figure 9 shows the integration scheme used with different sizes of elements without having any special treatment. Changing the size of the elements, so that long and short elements are adjacent to each other, caused unacceptable results as shown in Figure 11. This problem was solved by sub-dividing the elements when the integration was performed. Before an integration was performed for a singular element, the two adjacent elements were compared in size and if necessary they were made equal by dividing the longer element into two parts.

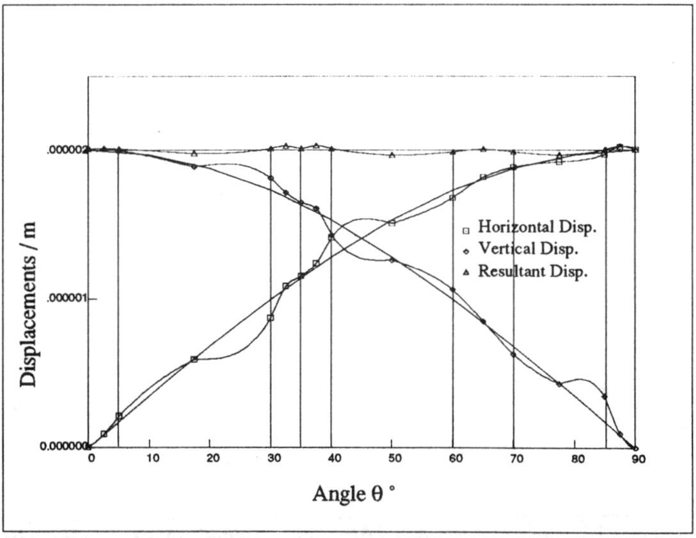

Figure 11 . Displacements of unevenly distributed quadratic elements without element subdivisioning.

The excess area was integrated using Gauss quadrature and the singular area using the Composite Mapping method (see Figure 12). The results were improved considerably as shown in Figure 13.

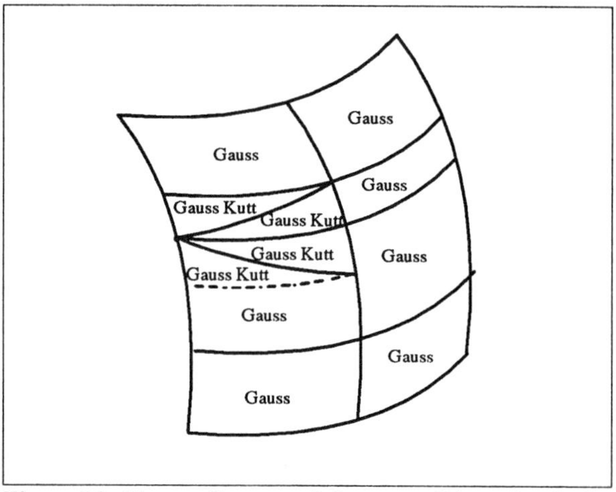

Figure 12 . Unequally spaced elements. Distortion in the results are removed by the use of sub-divisioning

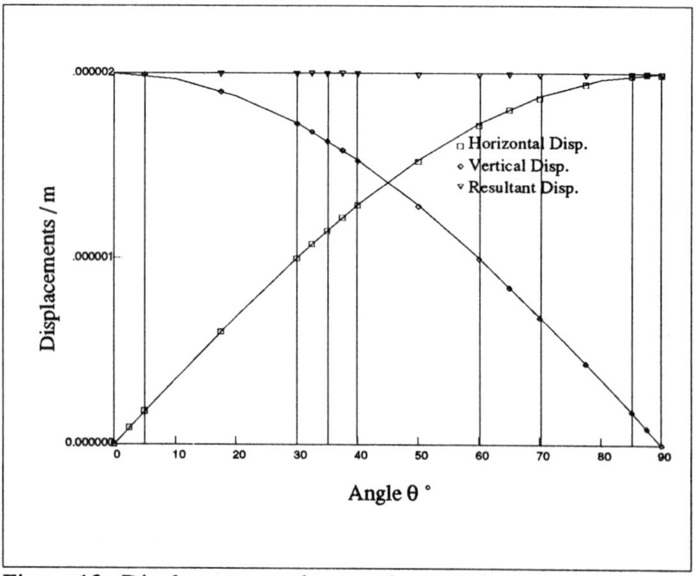

Figure 13 . Displacements of unevenly spaced quadratic elements with sub-divisioning present.

CONCLUSION:

This paper has clearly shown that BEM is very sensitive to the nature of numerical integration, particularly when integrating over singular elements. It is also shown that using Kutt type quadrature, integration over a singular element can easily be done in one step. However, when dealing with a mixture of different size elements, it is necessary to check the size of the two neighbouring elements at the singular point. If they are of different size, then integration has to be done over equal regions, as shown before.

It is also found that, during the assembly of the system matrices, some of the coefficients did not converge. If a Kutt type formulation was used, the validity of the numerical answers was uncertain. In such cases, the sub-divisioning method would be more advantageous because it would be possible to monitor the convergence of such coefficients as integration is based on using a refinement routine. A disadvantage of this solution is large CPU time.

REFERENCES :

1 - Stroud, A.H. and Secrest, D. Gaussian Quadrature Formulae, Prentice Hall, Engelwood Cliffs, NJ, 1966

2 - Kutt, H.R. Quadrature formulas for finite-part integrals, Report Wisk 178, The National Institute for Mathematical Sciences, Pretoria, 1975.

3 - Hayami, K., Brebbia, C.A., A New Coordinate Transformation Method for Singular and Nearly Singular Integrals over General Curved Boundary Elements. Boundary Element Techniques: Application in Stress Analysis and Heat Transfer, (Ed. Brebbia, C.A., Venturini, W.S.), pp. 375-399, CML Publications, 1987.

4 - Jun, L., Beer, G. and Meek, L., The Application of Double Exponential Formulas in Boundary Element Method, in Boundary Elements Method VII (Ed. Brebbia, C.A., Maier, G.) Vol.2, pp. 13-6 to 13-16, Proceedings of the 7th Int. Conf. on Boundary Elements, Springer Verlag, 1985.

5 - Telles, J.C.F., A Self Adaptive Coordinate Transformation for Efficient Numerical Evaluation of General Boundary Element Integrals, International Journal for Numerical Methods in Engineering, Vol. 24, pp. 959-973, (1987).

6 - Higashimachi, T., Ezawa, Y., Okamoto, N., Aizawa, T., Highly Accurate Boundary Element Method and Its Application to Structural Analysis, JSME Bulletin, Vol. 29, pp. 1096-1103, 1986.

7 - Labeyrie, J., Blanc, M., A New Integration Scheme in the B.E.M. to Fracture Mechanics Problems, in Boundary Elements Method VII (Ed. Brebbia, C.A., Maier, G.) Vol.2, pp. 8-47 to 8-56, Proceedings of the 7th Int. Conf. on Boundary Elements, Springer Verlag, 1985.

8 - Lachat, J.C., Watson, J.O., Effective Numerical Treatment of Boundary Integral Equations: A Formulation for the Three-Dimensional Elastostatics, International Journal for Numerical Methods in Engineering, Vol. 10, pp. 991-1005, (197).

9 - Brebbia, C.A. The Boundary Element Method For Engineers, Pentech Press, London, 1984.

High Precision Numerical Integration Method for 3-D BEM and its Error Analysis using Complex Function Theory

K. Hayami

C&C Information Technology Research Laboratories, NEC Corporation, 4-1-1, Miyazaki, Miyamae, Kawasaki, 213 Japan

The Boundary Element Method (BEM) has long been established as a powerful method for numerical simulation in engineering and science. Numerical integration plays a key role in BEM, since the accurate and efficient calculation of the boundary integrals governs the accuracy and computation time of the whole analysis, and analytical integration is not possible for general integral kernels over curved boundary elements.

The numerical integration becomes particularly challenging for nearly singular integrals which arise when the source point is very near the boundary over which the integration is performed. This situation frequently occurs in engineering when analysing thin structures or gaps, or when using boundary elements with high aspect ratio and when calculating the potential or flux at a point very near the boundary. The author proposed the Projection and Angular & Radial Transformation (PART) method for the accurate and efficient calculation of these nearly singular integrals [1,2].

For simplicity, let us take the three dimensional potential problem as an example. (The method can also be applied to general problems.) The boundary integral equation is given by

$$c(\mathbf{x}_s)\, u(\mathbf{x}_s) = \int_S (q\, u^* - u\, q^*) dS \tag{1}$$

where $\mathbf{x}_s$ is the source point, $u(\mathbf{x})$ is the potential, $q(\mathbf{x}) \equiv \partial u/\partial n$ is the derivative of u along the unit outward normal $\mathbf{n}$ at $\mathbf{x}$ on the boundary S . S is the boundary of the region V of interest and boundary conditions concerning u and q are given on S . $c(\mathbf{x}_s) = 1$ when $\mathbf{x}_s \in V$ and $c(\mathbf{x}_s) = 1/2$ when $\mathbf{x}_s$ is on a smooth boundary. The fundamental solution u^* and q^* are defined by

$$u^*(\mathbf{x}, \mathbf{x}_s) = \frac{1}{4\pi r}, \quad q^*(\mathbf{x}, \mathbf{x}_s) = -\frac{(\mathbf{r}, \mathbf{n})}{4\pi r^3} \tag{2}$$

where $\mathbf{r} = \mathbf{x} - \mathbf{x}_s$ and $r = |\, \mathbf{r}\, |$.

The flux at a point $\mathbf{x}_s \in V$ is given from the potential gradient

$$\frac{\partial u}{\partial \mathbf{x}_s} = \int_S \left(q \frac{\partial u^*}{\partial \mathbf{x}_s} - u \frac{\partial q^*}{\partial \mathbf{x}_s} \right) dS \tag{3}$$

where

$$\frac{\partial u^*}{\partial \mathbf{x}_s} = \frac{\mathbf{r}}{4\pi r^3}, \quad \frac{\partial q^*}{\partial \mathbf{x}_s} = \frac{1}{4\pi} \left\{ \frac{\mathbf{n}}{r^3} - \frac{3\mathbf{r}(\mathbf{r}, \mathbf{n})}{r^5} \right\} \tag{4}$$

Equations (1) and (3) are discretized on the boundary S by boundary elements defined by interpolation functions. The integral kernels of equations (1) and (3) become nearly singular when the distance d between $\mathbf{x}_s$ and S is small compared to the size of the boundary elements which discretize the boundary S.

The PART method is briefly described as follows:
For the nearly singular integral

$$I = \int_S \frac{f}{r^\alpha} dS \quad , \quad (\alpha \in \mathbf{N}) \tag{5}$$

1. Source projection

   Find the nearest point $\bar{\mathbf{x}}_s = \mathbf{x}(\bar{\eta}_1, \bar{\eta}_2)$ on S to $\mathbf{x}_s$ (e.g. using the Newton-Raphson method).

2. Approximate projection of S

   Approximately project the curved element S onto the polygon $\bar{S}$ in the plane tangent to S at $\bar{\mathbf{x}}_s$.

3. Introduce polar coordinates (ρ, θ) in $\bar{S}$, centred at $\bar{\mathbf{x}}_s$ to get

$$I = \int_0^{2\pi} d\,\theta \int_0^{\rho_{max}(\theta)} \frac{f}{r^\alpha} J \rho d\,\rho \tag{6}$$

   where J is the Jacobian of the mapping from Cartesian coordinates on $\bar{S}$ to curvilinear coordinates on S .

4. Apply Radial variable transformation : $R(\rho)$ in order to weaken the near singularity due to $1/r^\alpha$, which is essentially related to the radial variable ρ only .

5. Apply the Angular variable transformation:

$$t(\theta) = \frac{h_j}{2} \log \left\{ \frac{1 + \sin(\theta - \alpha_j)}{1 - \sin(\theta - \alpha_j)} \right\} \tag{7}$$

   in order to weaken the angular near singularity which arises when $\bar{\mathbf{x}}_s$ is near the edge of the polygon $\bar{S}$. This uses the fact that

$$\frac{d\,\theta}{d\,t} = \frac{1}{\rho_{max}(\theta)} = \frac{\cos(\theta - \alpha_j)}{h_j} \tag{8}$$

6. Use Gauss-Legendre's formula to perform numerical integration in the transformed variables R and t in

$$I = \int_{t(0)}^{t(2\pi)} \frac{dt}{\rho_{max}(\theta)} \int_{R(0)}^{R\{\rho_{max}(\theta)\}} \frac{f J \rho}{r^\alpha} \frac{d\rho}{dR} dR \tag{9}$$

Optimum Radial Variable Transformation

The choice of the radial variable transformation governs the efficiency of the PART method.

For constant planar elements,

$$\rho\, d\rho = r'^\alpha\, dR \quad \text{or} \quad R(\rho) = \int \frac{\rho}{r'^\alpha}\, d\rho \tag{10}$$

where $r' = \sqrt{\rho^2 + d^2}$, is equivalent to analytical integration in the radial variable in equation (6) (only one integration point is required for the exact integration in the radial variable), since $r = r'$.

However, for general curved elements, the log-L_2 transformation:

$$\rho\, d\rho = r'^2\, dR \quad \text{or} \quad R(\rho) = \log \sqrt{\rho^2 + d^2} \tag{11}$$

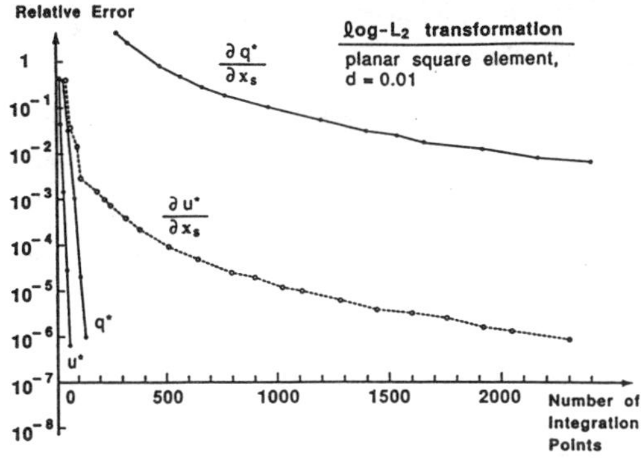

Figure 1: Convergence graph for *log-L_2* transformation

turns out to be more robust and efficient , except for the calculation of the flux, as shown in Fig.1 .

The near optimum radial variable transformation which works efficiently all round, seems to be the log-L_1 transformation:

$$R(\rho) = \log(\rho + d) \tag{12}$$

as shown in Fig.2 .

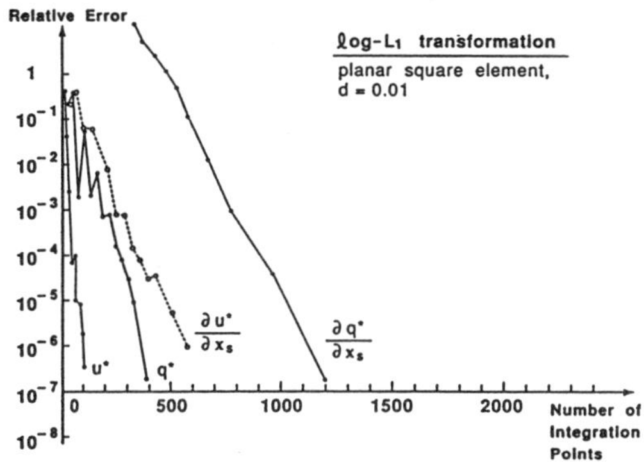

Figure 2: Convergence graph for *log-L_1* transformation

Result of Error Analysis using Complex Function Theory

The essential nature of the integration in the radial variable which appear in the potential integrals $\int_S u^* dS$, $\int_S q^* dS$ and flux integrals $\int_S \partial u^*/\partial \mathbf{x}_s dS$, $\int_S \partial q^*/\partial \mathbf{x}_s dS$ in equations (1) and (3) can be modelled by

$$I_{\alpha,\delta} = \int_0^{\rho_j} \rho^\delta / r'^\alpha d\rho \tag{13}$$

where $\rho_j \equiv \rho_{max}(\theta)$ in equation (6), and $r' \equiv \sqrt{\rho^2 + d^2}$.

Potential calculation of equation (1) gives rise to

- $\alpha = \delta = 1$ for the integration of u^*
- $\alpha = 3, \delta = 1$ for the integration of q^*

Flux calculation of equation (3) gives rise to

- $\alpha = 3$; $\delta = 1, 2$ for the integration of $\partial u^* / \partial \mathbf{x}_s$
- $\alpha = 3, \delta = 1$ and $\alpha = 5$; $\delta = 1, 2$ for the integration of $\partial q^* / \partial \mathbf{x}_s$

Error analysis of the radial integration using the complex contour integral [3,4,5]

$$E_n(f) = \frac{1}{2\pi i} \oint_C \Phi_n(z) f(z) dz \tag{14}$$

for the numerical integration of $I = \int_{-1}^{1} f(z)dz$ and asymptotic expressions of the error characteristic function $\Phi_n(z)$ for $n \rightarrow \infty$ or $| z | \rightarrow \infty$, gives the following estimates for the numerical integration error ε [6]. Here, $D = d/\rho_j$ is equivalent to the source distance relative to the element size and n is the number of radial integration points.

- Identity transformation: $R(\rho) = \rho$

$$\varepsilon \sim D^{\delta - \frac{3}{2}\alpha} \sigma^{-2n} \tag{15}$$

  where $\sigma : 1.04 \sim 1.4$ for $D : 10^{-3} \sim 10^{-1}$.

- log-L_2 transformation : $R(\rho) = \log \sqrt{\rho^2 + d^2}$

  - For δ : *odd*

$$\varepsilon \sim D^{\frac{\delta+1-\alpha}{2}} \left(\frac{\log D}{n} \right)^{2n} \tag{16}$$

  - For δ : *even*

$$\varepsilon \sim (-\log D)^{\frac{\delta+1}{2}} D^{\delta+1-\alpha} n^{-\delta-1} \tag{17}$$

- log-L_1 transformation : $R(\rho) = \log(\rho + d)$

$$\varepsilon \sim (-\log D) D^{\delta+1-\alpha} \sigma^{-2n} \tag{18}$$

  where $\sigma : 1.3 \sim 1.6$ for $D : 10^{-3} \sim 10^{-1}$.

These estimates correspond well with numerical experiment results, and render a firm basis for the optimization of the radial variable transformation : $R(\rho)$.

References

[1] Hayami, K. and Brebbia, C.A. Quadrature Methods for Singular and Nearly Singular Integrals in 3-D Boundary Element Method (Invited Paper), in Boundary Elements X, 1, pp.237-264, Computational Mechanics Publications with Springer-Verlag, 1988.

[2] Hayami, K. A Robust Numerical Integration Method for Three Dimensional Boundary Element Analysis, in Boundary Elements XII, 1 , pp.33-51, Computational Mechanics Publications with Springer-Verlag, 1990.

[3] Barret, W. Convergence Properties of Gaussian Quadrature Formulae, Comput. J., 3, pp.272-277, 1960.

[4] Takahasi,H. and Mori,M. Error Estimation in the Numerical Integration of Analytic Functions, Report of the Computer Centre University of Tokyo, 3, pp.41-108,1970.

[5] Davis, P.J. and Rabinowitz P. Methods of Numerical Integration, Academic Press, 1984.

[6] Hayami, K. (in preparation)

Penalty-Combined Boundary Methods and Applications in Solving Elliptic Problems with Singularities

Z.C. Li, T.D. Bui

Centre de Recherche Informatique de Montréal Inc. 3744 Jean-Brillant, #500, Montréal, Québec H3T 1P1, Canada and Department of Computer Science, Concordia University 1455 de Maisonneuve Blvd. West, Montréal, Québec H3G 1M8, Canada

ABSTRACT

The simplest penalty techniques are proposed for matching the Ritz-Galerkin and finite element methods in solving elliptic problems with multiple singularities. The advantage of this method over the nonconforming combination is its simplicity, without dealing with direct constraints. Error bounds are provided to guide the choice of the penalty constant P_c. These techniques are also applied to the study of potential flow of wind over buildings where there exist multiple singularities and multiple coupling interfaces among different numerical methods, accompanied with error analysis on average wind velocity and average sheer stress along building edges. On the whole, the penalty combined method is beneficial to solving complicated engineering problems.

INTRODUCTION

In order to solve elliptic boundary value problems with singularities, the nonconforming combination has been provided in Li [10,11] by coupling the Ritz-Galerkin method in the subdomains including the singular points, and the finite element method in the remaining subdomains where the solution is smooth enough. Also, a continuity condition for admissible functions is imposed only on the element nodes on the common boundary of two methods. The admissible functions are still discontinuous elsewhere along the entire common boundary. Hence the term *nonconforming* is referred to. Certainly, the techniques for matching different numerical methods are essential to combined methods.

The treatment in [10,11] is rather troublesome, particularly for the problems with multiple singularities and/or multiple coupling-boundaries. We have recently developed the alternative techniques by using additional integrals across the common boundary. The hybrid techniques involving derivative integrands are proposed in [13, 14]. However, the coefficient matrix obtained are not symmetric (or positive definite). The hybrid plus penalty techniques are also given in [12]. Our question here is: could we use only the simplest penalty techniques involving the integral

$$\int_{\Gamma_0} (v^+ - v^-)^2 \, dl \tag{1}$$

for coupling two different admissible functions v^+ and v^- across their common boundary Γ_0? The answer is yes, based on our error analysis and numerical experiments. This is called *the penalty-combined methods*. The methodology and programming of this method is simpler than the methods in [10-14]. Consequently, the penalty-combined methods are beneficial to solving complicated engineering problems.

It is worthy to note that the penalty techniques as Eq.(1) applied to all element boundaries in finite element methods will result in reduced convergence rates (see Ciarlet [5, p. 143]). Also, these penalty techniques have been employed for the exterior boundary of the Dirichlet condition by Babuska [1, 2] and Utku and Carey [21]. This paper will report the application of the penalty techniques to the interface boundary Γ_0.

In the next section, we shall provide the penalty-combined methods, error bounds and optimal convergence rates. In the last section, we give a study on the potential flow of wind over buildings. The penalty-combined method is employed to seek the stream functions, and then wind velocity and sheer stress along the building edges can be evaluated for building engineering and architecture. The surface of buildings always have corners, and the corner points may turn out to be singularity points because near them the wind velocity approaches infinity! This paradox must be cleared out. In this section, we pointed out that the average wind velocity will virtually play an important role. Error analysis has been done for providing the boundedness and error bounds of the average velocity and the average sheer stress on the building roof involving singularities. On the whole, the simplicity and flexibility of this method and the theoretical analysis in this paper are significant for the practical application in solving singularity problems.

PENALTY COMBINED METHODS

Consider the elliptic boundary value problem of two dimensions:

$$-\frac{\partial}{\partial x}(\beta\frac{\partial u}{\partial x}) - \frac{\partial}{\partial y}(\beta\frac{\partial u}{\partial y}) = f, \ \ in \ S, \tag{2}$$

$$u = 0, \ \ on \ \partial S, \tag{3}$$

where S is a polygon, ∂S is its boundary, the functions β and f are sufficiently smooth, and there exists a positive constant β_0 such that

$$\beta = \beta(x, y) \geq \beta_0 \geq 0. \tag{4}$$

The problem, (2) and (3), can be written in the following weak form

$$a(u, v) = f(v), \ \ \text{for } all \ v \in H_0^1(S), \tag{5}$$

where the quadratic functional is

$$a(u, v) = \iint_S \beta(u_x v_x + u_y v_y) \, dS, \tag{6}$$

and the notations are

$$f(v) = \iint_S fv \, dS, \tag{7}$$

$$H_0^1(S) = \left\{ v \mid v, v_x, v_y \in L^2(S) \text{ and } v|_{\partial S} = 0 \right\}. \tag{8}$$

Let the solution domain S be divided into two subdomains S_1 and S_2 by a piecewise straight line Γ_0, i.e., $S = S_1 \cup S_2$. The Ritz-Galerkin method is used in S_2; and the linear finite element method is used in S_1. We choose the admissible functions v as

$$v = \begin{cases} v^+ = \phi + \sum_{l=0}^{L} \tilde{D}_l \phi_l & in\ S_2, \\ v^- = v_1^h & in\ S_1, \end{cases} \tag{9}$$

where $\tilde{D}_l$ are unknown coefficients, ϕ and $\{\phi_l\}$ are known functions, $\{\phi_l\}$ are complete and linearly independent functions in S_2, and v_1^h are piecewise linear interpolation functions on the triangulations of quasi-uniform elements with a maximal boundary length h.

Denote V_h^0 as the above admissible function space satisfying the Dirichlet boundary condition (3). And

$$V_h^0 \not\subset H^1(S), \tag{10}$$

where $H^1(S)$ is the Sobolev space:

$$H^1(S) = \left\{ v \mid v, v_x, v_y \in L^2(S) \right\}. \tag{11}$$

The relation (10) results from:

$$v^+ \neq v^- \quad on\ \Gamma_0, \tag{12}$$

where $v^+ = v|_{S_2}$ and $v^- = v|_{S_1}$.

Define a quadratic functional

$$a_h(u, v) = \iint_{S_1} \beta(u_x v_x + u_y v_y)\, dS + \iint_{S_2} \beta(u_x v_x + u_y v_y)\, dS + \frac{P_c}{h^2} \int_{\Gamma_0} (u^+ - u^-)(v^+ - v^-)\, dl, \tag{13}$$

where P_c is a non-negative, bounded constant independent of L, h, u and v, and h is the maximal boundary length of quasi-uniform elements in S_1. Then the penalty-combined methods are designed to seek and approximate solution, $u_h \in V_h^0$, such that

$$a_h(u_h, v) = f(v), \quad \text{for } all\ v \in V_h^0. \tag{14}$$

Here, the penalty integral, $\frac{P_c}{h^2} \int_{\Gamma_0} (v^+ - v^-)^2 \, dl$, will play the role of matching the Ritz-Galerkin method and the finite element method across their common boundary Γ_0. We call Γ_0 the *coupling boundary*, and P_c the *penalty constant*.

In terms of (13), we define a new norm over V_h^0:

$$\| v \|_H = \left\{ \| v \|_{1,S_1}^2 + \| v \|_{1,S_2}^2 + \frac{1}{h^2} \| v^+ - v^- \|_{0,\Gamma_0}^2 \right\}^{1/2}, \tag{15}$$

where the norm definitions over the Sobolev space $H^m(\Omega)$ are (see [17])

$$\| v \|_{m,\Omega} = \left(\sum_{|\alpha| \leq m} \int_\Omega |D^\alpha v|^2 \, dx \right)^{1/2}, \quad | v |_{m,\Omega} = \left(\sum_{|\alpha| = m} \int_\Omega |D^\alpha v|^2 \, dx \right)^{1/2}. \tag{16}$$

We give here the following theorem without proof.

Theorem 1. There exists a bounded constant C independent of L, h, u and v such that

$$\| u - u_h \|_H \leq C \left\{ \inf_{v \in V_h^0} \| u - v \|_H + h \left| \frac{\partial u}{\partial \nu} \right|_{0,\Gamma_0} \right\}, \tag{17}$$

where ν is the outward normal to ∂S_2, and h is the maximal boundary length of quasi-uniform elements in S_1.

Assume that the solutions on S_1 and Γ_0 (but not in S_2 where there may exist singularities) are smooth enough such that

$$u \in H^2(S_1) \text{ and } u \in H^2(\Gamma_0). \tag{18}$$

On the basis of the Sobolev imbedding theorem [17, p. 64], we have

$$C_0 \left\| \frac{\partial u}{\partial \nu} \right\|_{0,\Gamma_0} \leq \| u \|_{2,S_1} \leq C, \tag{19}$$

with a positive constant C_0. From Theorem 1, error bounds of approximate solutions u_h are:

$$\| u - u_h \|_H \leq C \left\{ \inf_{v \in V_h^0} \| u - v \|_H + h \right\}$$

$$\leq C \left\{ \inf_{v \in V_h^0} \left[\| u - v \|_{1,S_2} + \| u - v \|_{1,S_1} + \frac{1}{h} \| v^+ - v^- \|_{0,\Gamma_0} \right] + h \right\}. \tag{20}$$

Since v^- can be chosen as the piecewise linear interpolation function of u, we can have from (18)

$$\| u - v \|_{1,S_1} = O(h), \tag{21}$$

$$\| u^+ - v^- \|_{0,\Gamma_0} = O(h^2). \tag{22}$$

Then

Corollary 2. Let (18) and the conditions in Theorem 1 hold, then

$$\| u - u_h \|_H \leq C \left\{ \inf_{v \in V_h^0} \| u - v \|_{1,S_2} + h \right\}. \tag{23}$$

Let the true solution

$$u = \phi + \sum_{l=0}^{\infty} D_l \; \phi_l \; in \; S_2 \tag{24}$$

and the remainder

$$\| \sum_{l=L+1}^{\infty} D_l \; \phi_l \|_{1,S_2} \leq Ch. \tag{25}$$

Then an optimal convergence rate

$$\| u - u_h \|_H = O(h) \tag{26}$$

is achieved.

Corollary 2 implies that if the coupling relation between $L + 1$ and h satisfies (25), using only the penalty techniques along Γ_0 can also reach up the optical convergence rate $O(h)$.

For the norm defined in [10,11],

$$\| v \|_h = [\| v \|^2_{1,S_1} + \| v \|^2_{1,S_2}]^{1/2}. \tag{27}$$

Then

$$\| v \|_h \leq \| v \|_H . \tag{28}$$

Therefore, the optimal convergence rate $O(h)$ is also valid for the error norm $\| \cdot \|_h$.

Eq. (14) leads to an algebraic system of discrete equations

$$A\mathbf{x} = \mathbf{b}, \tag{29}$$

where the unknown vector $\mathbf{x}$ consists of $v^-(P_i)$, $P_i \in S_1$ and $\tilde{D}_l$, $\mathbf{b}$ is a known vector, and A is the constant coefficient matrix. We note that the positive definite and symmetric property is granted automatically by the method (14).

In contrast, an additional constraint condition in [10,11]

$$v^+(P_i) = v^-(P_i), \;\; P_i \in \Gamma_0 \tag{30}$$

is imposed on the admissible functions (9). A transformation or substitution of unknowns is needed for obtaining the positive definite and symmetric property of A so that the procedure in [10,11] is rather complicated for coupling multiple boundaries.

In the next section, the penalty techniques will be applied to wind flow over buildings. Of course, the combined methods and their coupling techniques can also applied to other elliptic problems, homogeneous or nonhomogeneous, with singularities as in Grisvard [6] and Grisvard *et al*. [7]. Certain analysis on particular solutions is necessary to provide suitable piecewise functions as long as the expansions of solutions near the singular points are known. Then using the methods and techniques in this paper will provide good approximation of solutions, in particular near singularities.

STUDY ON WIND FLOW OVER BUILDINGS

Since the building often contains multiple corners, there exist multiple singularities in potential flow [8]. Using the penalty techniques is more advantageous for coupling different methods along multiple interfaces. Moreover, the computation of potential flow is the first step of the random vortex method of Chorin [4], and Bui and Oppenheim [3], so that the methods in this paper can also be applied to study on turbulent wind flow. The detailed report will appear later.

1. Potential flow

Consider the potential flow over building:

$$\nabla^2 \psi = 0, \tag{31}$$

$$\frac{\partial \psi}{\partial y} = u, \quad \frac{\partial \psi}{\partial x} = -v, \tag{32}$$

where ψ is the stream function, (u,v) are the components of the wind velocity **u** along the coordinates x and y. For the model building in two dimensions (see Fig. 1), Eqs.(31) and (32) govern ψ on an unbounded domain S. On the ground $y = 0$ and on the surface of the building, we have

$$\mathbf{u}.\mathbf{n} = -\frac{\partial \psi}{\partial S} = 0, \tag{33}$$

where $\mathbf{u} = (u, v)$, **n** is the unit normal to ∂S. Then we can simply let

$$\psi = 0, \tag{34}$$

because adding a constant to ψ does not change u and v. Also, the wind profile is supposed to be known(see Hanson et al.[9])

$$u(y) = \begin{cases} u_\infty \left(\dfrac{y}{y_g}\right)^\alpha, & \text{for } y \leq y_g, \\ u(y) = u_\infty, & \text{for } y \geq y_g, \end{cases} \tag{35}$$

where u_∞ is the velocity at the infinity, and α, y_g are positive constants(see Fig. 1). We then obtain the boundary condition at $x = 0$ (a point far upstream of the building):

$$\psi = \begin{cases} \dfrac{u_\infty}{\alpha+1} \left(\dfrac{y}{y_g}\right)^\alpha y, & \text{for } y \leq y_g, \\ u_\infty y - C_0, & \text{for } y \geq y_g, \end{cases} \tag{36}$$

where the constant $C_0 = \dfrac{\alpha}{\alpha+1} y_g u_\infty$.

We note that in this problem, there exist three angular singular points and the infinity. Naturally, we divide the solution domain S into five subdomains(Fig. 2)

$$S = S_\infty \cup S_h \cup S_1 \cup S_2 \cup S_3, \tag{37}$$

where S_1, S_2 and S_3 are the sectors with the same radius R_0, including the angular points A, B, and C respectively, S_∞ is an unbounded subdomain $(r \geq R^*, 0 \leq \theta \leq \pi/2)$, and S_h is the rest part of S.

In order to apply the combined methods, we have to find asymptotic expansions near the angular points and the infinity. By the method of separation of variables, we can obtain particular solutions on S_∞:

$$\psi_\infty = u_\infty y - \frac{2}{\pi} C_0 \theta + \sum_{n=1}^{L} a_n \left(\frac{r}{R_\infty} \right)^{-2n} \sin 2n\theta, \tag{38}$$

which also satisfy the boundary conditions (33)-(35):

$$\psi = 0 \quad \textit{when } \theta = 0 \text{ and } r \geq R^*, \tag{39}$$

$$\psi = u_\infty y - C_0 \quad \textit{when } \theta = \frac{\pi}{2} \text{ and } r \geq R^*, \tag{40}$$

here we assume $R^* \geq y_g$. In (38), (r, θ) are the polar coordinates with the origin (0,0), a_n unknown coefficients, and R_∞ a scale factor which can be chosen as

$$R_\infty = Max\left\{ y_g, \ \underset{on\ the\ building}{Max\ r} \right\} < R^*. \tag{41}$$

Then the particular solutions are valid for the domain $(r \geq R_\infty, 0 \leq \theta \leq \pi/2)$.

Similarly, the particular solutions $\psi^{(i)}$ on the subdomains $S_i \, (i = 1,2,3)$ can be obtained:

$$\psi_i = \sum_{n=1}^{M_i} b_n^{(i)} \left(\frac{r^{(i)}}{R} \right)^{\sigma_i n} \sin(\sigma_i n \theta^{(i)}), \tag{42}$$

where $(r^{(i)}, \theta^{(i)})$ are the local polar coordinates, $b_n^{(i)}$ are unknown coefficients, and the constants

$$\sigma_i = \frac{\pi}{\theta_M^{(i)}}, \quad \theta_M^{(i)} = 2\pi - \Theta_i, \tag{43}$$

Θ_i are the corner angles of the building. In the denominator of (42), a suitable radius $R(>R_0)$ also serves as a scale factor.

Once the asymptotic solutions on S_1, S_2, S_3 and S_∞ are found, the following penalty-combined method can be easily developed.

2. Penalty-combined method

In this section, we shall give the algorithm of the penalty-combined method. We also derive a quasi-optimal convergence rate of stream-functions, and error bounds for approximate wind velocity and approximate average shear stress on the building roof.

Based on the analysis of asymptotic solutions, the Ritz-Galerkin method is used in $\Omega_i \, (i = 1,2,3)$ and Ω_∞, and the traditional finite element method is used in the subdomain S_h, where $\psi \in H^2(S_h)$. We choose admissible functions as

$$\hat{\psi} = \begin{cases} \psi_\infty = u_\infty y - \dfrac{2}{\pi} C_0 \theta + \sum_{n=1}^{L} a_n \left(\dfrac{r}{R_\infty}\right)^{-2n} \sin 2n\theta \;\; in \;\; S_\infty, \\ \psi_i = \sum_{i=1}^{M_i} b_n^{(i)} \left(\dfrac{r^{(i)}}{R}\right)^{\sigma_i n} \sin(\sigma_i n \theta^{(i)}) \;\; in \;\; S_i, \\ \psi_h^1 \;\; in \;\; S_h, \end{cases} \tag{44}$$

where $i = 1,2,3$, ψ_h^1 are piecewise linear interpolation functions on S_h. Denote the common boundaries $\Gamma_i = S_i \cap S_h$ and $\Gamma_\infty = S_\infty \cap S_h$ with the element nodes $(r_j^{(i)}, \theta_k^{(i)})$ and (r_j, θ_k) respectively.

Since there exist four coupling boundaries, we will choose the penalty-combined method. Denote V_h the space of all functions $\hat{\psi}$ in (44) satisfying (34) and (36), and V_h^0 another space of $\hat{\psi}$ satisfying (34) and $\psi\Big|_{x=0} = 0$. Therefore, the penalty-combined method is to seek an approximate solution $\tilde{\psi}_h \in V_h$ such that

$$B(\tilde{\psi}_h, \psi) = 0, \quad \text{for } all \;\; \psi \in V_h^0, \tag{45}$$

where the bilinear form is

$$B(\psi,\phi) = \iint_{S_\infty} \nabla\psi\nabla\phi dS + \iint_{\hat{S}_h} \nabla\psi\nabla\phi dS + \sum_{i=1}^{3} \iint_{S_i} \nabla\psi\nabla\phi dS$$

$$+ \sum_{i=1}^{3} \frac{P_c}{h^2} \int_{\Gamma_i} (\psi_i - \psi_h^1)(\phi_i - \phi_h^1)\, dl + \frac{P_c}{h^2} \int_{\Gamma_\infty} (\psi_\infty - \psi_h^1)(\phi_\infty - \phi_h^1)\, dl. \tag{46}$$

In (46), $\hat{S}_h$ is the triangulation domain of S_h so that $\hat{S}_h \approx S_h$.

Here exists a difficulty in expressing the infinity condition for the stream function $\hat{\psi}$, because their expansions in (44) have the unboundedness:

$$\psi\big|_{\substack{r\to\infty \\ \theta=\frac{\pi}{2}}} = \infty. \tag{47}$$

However, we can prove

Proposition 3. For $\psi \in V_h$ and $\phi \in V_h^0$, the bilinear forms in (45) are still bounded.

The difficulty (47) can be bypassed away since the useful velocity (32) are computed from the derivatives of ϕ, and since the numerical magnitude involving in the method (45) is still bounded.

In order to give error bounds of u and v, we define a norm

$$\|\psi\|_h = \left\{ \sum_{i=1}^{3} \|\psi\|_{1,S_i}^2 + \|\psi\|_{1,\hat{S}_h}^2 + |\psi|_{1,S_\infty}^2 \right\}^{1/2}, \tag{48}$$

where $\|\psi\|_{1,S_i}$ are the Sobolev norms. It should be noted that on the unbounded domain S_∞, the semi-norm in (48) is defined only with derivatives

$$|\psi|_{1,S_\infty} = \left\{ \iint_{S_\infty} (\psi_x^2 + \psi_y^2) dS \right\}^{1/2}. \tag{49}$$

The norm definition (48) is plausible since only the velocities u and v (i.e., the derivatives of ψ) are our concern in practical application.

Following Li [10,11], we can prove

Theorem 4. Let the radii

$$R_\infty < R^*, \;\; R_0 < R, \tag{50}$$

and the coupling relations satisfy

$$\left(\frac{R_\infty}{R^*}\right)^{2L} \le Ch, \quad \left(\frac{R_0}{R}\right)^{\sigma_k M_k} \le Ch, \quad k = 1, 2, 3, \tag{51}$$

where C is a bounded constant, independent of L, M_k, h and ψ.

Then, a quasi-optimal convergence rate of numerical solutions by the penalty-combined method can be achieved:

$$\| \hat{\psi}_h - \psi \|_h = O(h), \tag{52}$$

where $\hat{\psi}_h$ and ψ are the approximate and exact solutions, and h is the maximal boundary length of triangular elements in S_h.

Denote $\hat{u}_h$ and $\hat{v}_h$ the approximate velocities from $\tilde{\psi}_h$, we have from (32)

Corollary 5. Let conditions in Theorem 4 hold, then there exist optimal convergence rates for $\tilde{u}_h$ and $\tilde{v}_h$ on the unbounded solution domain S:

$$\| \tilde{u}_h - u \|_{0,S} = O(h), \qquad \| \tilde{v}_h - v \|_{0,S} = O(h). \tag{53}$$

where the norm $\| w \|_{0,S} = (\int\int_S w^2 \, dS)^{1/2}$.

Below we consider the singularities of solutions. Suppose that the angles Θ_i of building corners(i.e., convex figuration):

$$0 < \Theta_i < \pi, \tag{54}$$

then following (43), the constants σ_k satisfy:

$$1/2 < \sigma_k < 1. \tag{55}$$

Consequently, the point velocity of wind flow at the corners A,B,C in Fig. 2 will be infinite:

$$|v_k|_{A,B \text{ or } C} = \left| \frac{\partial \psi_k}{r^{(k)} \partial \theta^{(k)}} \right|_{A,B \text{ or } C} = O\left([r^{(k)}]^{\sigma_k - 1} \right) \to \infty. \tag{56}$$

This paradox seems to conflict our intuition. However, the ideal point velocity can never be measured. The velocity (or stress) unboundedness at only a few solitary points may not be serious. In fact, the average wind velocity $\bar{v}$, which is bounded, will play an important role. Hence we define an average velocity of wind along the building edge $\overline{AF}$ (see Fig. 1), containing the singular point A:

$$\bar{v}_{\overline{AF}} = \left\{ \frac{1}{\overline{AF}} \int_{\overline{AF}} v_l^2 dl \right\}^{1/2}, \tag{57}$$

where v_l is the wind velocity along $\overline{AF}$. Then it follows

$$\bar{v}_{\overline{AF}} = \frac{1}{\sqrt{\overline{AF}}} \; |v_l|_{\overline{AF}}, \qquad |v_l|_{\overline{AF}} = \left\{ \int_{\overline{AF}} v_l^2 dl \right\}^{1/2}. \tag{58}$$

The value of $\bar{v}_{\overline{AF}}$ is, indeed, an upper bound

$$v_{\overline{AF}} \leq \bar{v}_{\overline{AF}} \tag{59}$$

of the absolute average velocity $v\,|_{\overline{AF}}$ defined by

$$v_{\overline{AF}} = \frac{1}{\overline{AF}} \int_{\overline{AF}} |v_l| \; dl. \tag{60}$$

Ineq.(59) can be verified by the Schwartz inequality.

Then we have

Lemma 6. Let (50), (54) and (55) hold, then for any $\overline{AF} > 0$, the value $|v|_{\overline{AF}}$ is bounded. Also, suppose

$$0 < \epsilon_0 \leq \overline{AF} \leq R_0, \tag{61}$$

where ϵ_0 is a constant independent of L, M_k and h, then the average velocity $\bar{v}_{\overline{AF}}$ is also bounded.

Based on Lemma 6, we can evaluate the shear stress along building edges (Summers *et al* [19]):

$$\bar{\tau}_{\overline{AF}} = 0.03 \; R_e^{-0.2} \; \rho \; \bar{v}^2_{\overline{AF}}, \tag{62}$$

where R_e is the Reynold number, and ρ is the density of wind flow.

The approximate shear stress is computed by

$$\tilde{\tau}_{\overline{AF}} = 0.03 R_e^{-0.2} \; \rho \; \tilde{v}^2_{\overline{AF}}, \tag{63}$$

where $\tilde{v}_{\overline{AF}}$ is given by

$$\tilde{v}_{\overline{AF}} = \left\{ \frac{1}{\overline{AF}} \int_{\overline{AF}} \tilde{v}_l^2 dl \right\}^{1/2}, \tag{64}$$

$\tilde{v}_l = \dfrac{\partial \tilde{\psi}_i}{r^{(i)} \partial \theta^{(i)}} |_{\theta=0}$, and $\tilde{\psi}_i$ are obtained from the penalty-combined method (45). Then we have

Theorem 7. Let (54), (55), (61) and the conditions in Theorem 4 hold true, then there exist error bounds

$$|\tilde{\tau}_{\overline{AF}} - \bar{\tau}_{\overline{AF}}| = O(h). \tag{65}$$

The above analysis on the average velocity and stress grants a good precision of approximate values near the building covers by our combined methods, bypassing the difficulty (47) and the paradox (56). Above all, the proposed methods and the theoretical analysis in this paper are significant of practical application in solving singularity problems.

3. Numerical Experiments

We consider a model building with different roof configurations by changing the pitch angle β (see Fig. 1). Let $R = 1$, $R_0 = 0.33$, $R^* = 7$, and $R_\infty = \sqrt{17}$. For the boundary condition (35), the constants are given by $\alpha = 0.3$, $u_\infty = 1$ and $y_g = 3$.

Two triangulations in $\hat{S}_h$ are shown in Figs. 3,4. The stream functions $\tilde{\psi}_h$ are computed by the penalty-combined method (45) with $P_c = 10^6$, and the velocity field of wind flow is depicted in $\hat{S}_h$ only. In the subdomains S_i and S_∞, the coefficients are listed in Table 1, from which we obtain the average wind velocity on the building edges in Table 2.

For the coarse partition in Fig. 3, a good match of

$$M_1 = M_2 = M_3 = 4, \quad L = 3$$

is found after trial calculations. Therefore, we can expect a good match of

$$M_1 = M_2 = M_3 = 5, \quad L = 4$$

for the fine partition in Fig. 4, where the maximal boundary length of elements in S_h decreases to its half, based on the coupling relation in [10,11]. Obviously, all the results in the coarse and fine partitions are close to each other.

Finally, we simply take the coarse partition as Fig. 3 to calculate the flow for different angles: $\beta = 0°$, $10°$, $20°$, $40°$ and $60°$. Some velocity fields of wind flow are provided in Figs. 5, and the coefficients and average velocities are listed in Table 2. Then the shear stress can be obtained from (63). We note that when $\beta = 0°$, there exist only two corners of the building. Then the division (37) is substituted by $S = S_\infty \cup S_h \cup S_1 \cup S_3$, and the admissible functions (44) are for $i = 1,3$ only. It seems from Table 2 that the average wind velocity along the building edges will increase at $\overline{AB}$ and $\overline{BC}$ when β increases. This implies that a larger shear stress happens on a sharper roof.

ACKNOWLEDGMENTS

This work was supported in part by the Natural Sciences and Engineering Research Council of Canada, by the Fonds pour la Formation de Chercheurs et l'Aide à la Recherche of Quebéc and by Centre de Recherche Informatique de Montreal, Inc. We also thank George Z. Li for typing this manuscript.

REFERENCES

[1] Babuška, I., *The finite element method for elliptic equation with discontinuous coefficients*, Computing, Vol.5, pp. 207-213, 1970.

[2] Babuška, I., *The finite element method with penalty*, Math. Comput., Vol.27, pp. 221-228, 1973.

[3] Bui,T.D. and Oppenheim, A.K., *Evaluation of wind effects on model buildings by the random vortex method*, Applied Num. Math. Vol.13, pp. 195-207, 1987.

[4] Horin, A.J., *Numerical studies of slightly viscous flow*, J. Fluid Mech. Vol.57, pp. 785-796, 1973.

[5] Ciarlet, P.G., **The Finite Element Method for Elliptic Problem**, Amsterdam, New York, Oxford: North Holland, 1978.

[6] Grisvard, P., **Elliptic Problems in Nonsmooth Domains**, Pitman Advanced Publishing Programs, Boston and London, 1985.

[7] Grisvard, P., Wendland, W. and Whiteman, J.R. (Eds.), **Singularities and Constructive Methods for their Treatment**, Lecture Notes in Mathematics, **1021**, Springer Verlag, 1985.

[8] Habashi, W.G., *The finite element method in the solution of unbounded potential flows*, Inter. J. Numer. Meth. in Eng. Vol.14, pp. 1347-1358, 1979.

[9] Hanson, T., Summers, D.M. and Wilson, C.B., *Numerical modeling of wind flow over buildings in two dimensions*, Inter. J. Numer. Meth. in Fluids, Vol.4, pp. 25-41, 1984.

[10] Li, Z.C., *A nonconforming combined method for solving Laplace's boundary problems with singularities*, Numer, Math. Vol.49, pp. 475-497, 1986.

[11] Li, Z.C., **Numerical Methods for Elliptic Problems with Singularities: Boundary Methods and Nonconforming Combinations**, World Scientific, Singapore, New Jersey, London and Hong Kong, 1990.

[12] Li, Z.C. and Bui, T.D., *A new kind of combinations between the Ritz-Galerkin and finite element methods for singularity problems*, Computing, Vol.40, pp. 475-497, 1988.

[13] Li, Z.C. and Bui, T.D., *The simplified hybrid-combined methods for Laplace's equation with singularities*, J. Comp. and Appl. Math., Vol.29, pp. 171-193, 1990.

[14] Li, Z.C. and Liang, G.P., *On the simplified-combined method*, Math. Comp., Vol.41, pp. 13-25, 1983.

[15] Li, Z.C., Mathon, R. and Sermer, P., *Boundary methods for solving elliptic problems with singularities and interfaces*, **SIAM** J. Numer. Anal., Vol.24, pp. 487-498, 1987.

[16] Motz, A., *Treatment of singularities of partial differential equations by relaxation methods*, Quant. J. Appl. Math. Vol.4, pp. 371-377, 1946.

[17] Sobolev, S.L., **Application of Functional Analysis in Mathematical Physics**, AMS, Providence, R. I., 1963.

[18] Strang, G. and Fix, G.J., **An Analysis of the Finite Element Method** , New Jersey, Prentice-Hall, Englewood Cliffs, 1973.

[19] Summers, D.M., Hanson, T. and Wilson, C.B., *A random vortex simulation of wind-flow over a building*, Inter. J. Numer. Meth. in Fluids, Vol.5, pp. 849-871, 1985.

[20] Thatcher, R.W., *The use of infinite grid refinement at singularities in the solution of Laplace's equation*, Numer. Math., Vol.25, pp. 163-178, 1976.

[21] Utku, M. and Carey, G.F., **Boundary Penalty Techniques**, Comput. Methods Appl. Mech. Eng. Vol.30, pp. 103-118, 1982.

[22] Zienkiewciz, O.C., **The Finite Element Method** (3rd Edition), McGrill-Hill, London 1977.

Table 1. The calculated coefficients in (44) from the penalty-combined methods for $\beta = 40^0$

Coefs.	l	The rough partition in Fig. 3	The fine partition in Fig. 4
	1	0.4108213	0.4170790
	2	0.1529620	0.1576584
$b_l^{(1)}$	3	0.0204069	0.0166925
	4	-0.0182768	-0.0028578
	5	/	-0.0424585
	1	0.7822667	0.7718139
	2	0.0018099	0.0072472
$b_l^{(2)}$	3	-0.0575886	-0.0313106
	4	0.0263113	0.0078526
	5	/	-0.0484961
	1	0.4373974	0.4444181
	2	-0.1517123	-0.1582070
$b_l^{(3)}$	3	0.0241213	0.0138468
	4	-0.0023637	-0.0025032
	5	/	-0.0426734
	1	-1.2292486	-1.2436789
	2	0.8130880	0.9679558
a_l	3	-1.3929712	-1.8918433
	4	/	2.8204021

Table 2. Average wind velocity along the building edges for different angles β

β	$\bar{v}_{\overline{EA}}$	$\bar{v}_{\overline{AB}}$	$\bar{v}_{\overline{BC}}$	$\bar{v}_{\overline{CD}}$
0	0.8275	*	*	0.8075
10	0.7004	1.1956	1.1753	0.6803
20	0.6401	1.1864	1.1704	0.6194
40	0.5197	1.2223	1.2166	0.4962
60	0.3991	1.4327	1.4408	0.3743

* where $\bar{v}_{\overline{AC}}$=1.090.

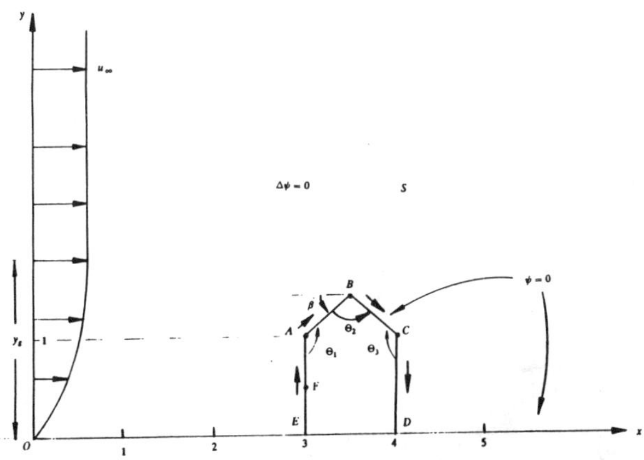

Figure 1. A model building on the earth with an inlet wind

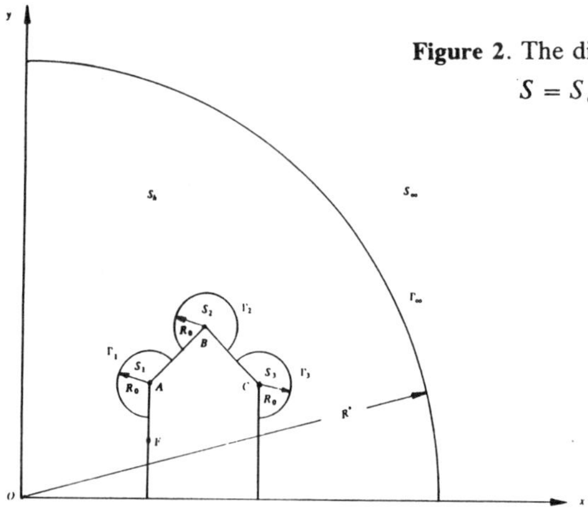

Figure 2. The division of the solution domain

$$S = S_{\infty} \cup S_h \cup S_1 \cup S_2 \cup S_3$$

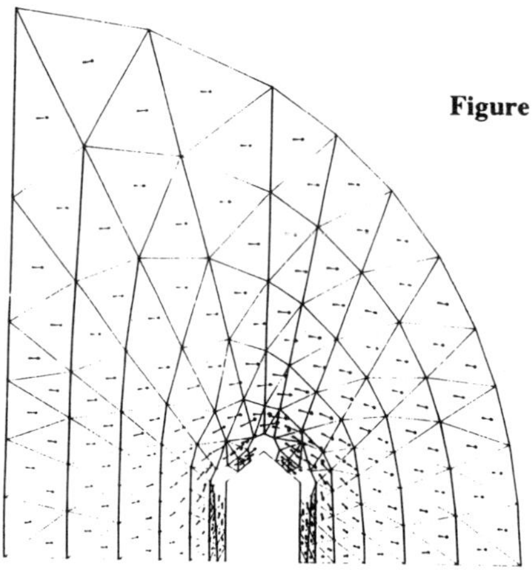

Figure 3. A coarse partition with velocity field for $\beta = 40°$

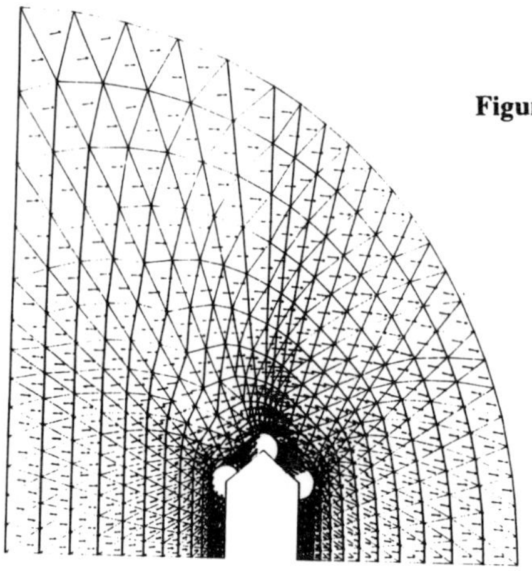

Figure 4. A fine partition with velocity field for $\beta = 40°$

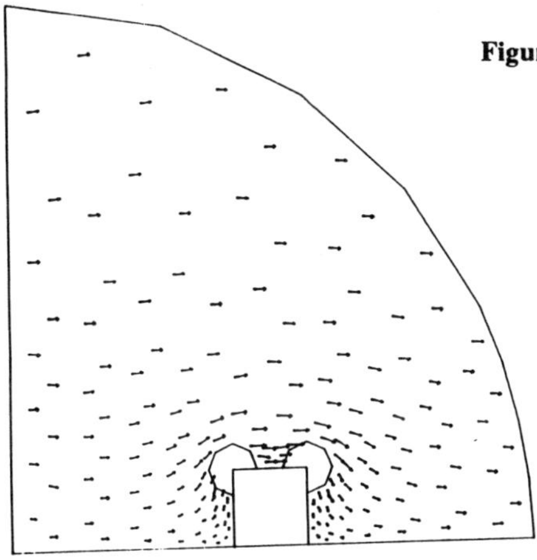

Figure 5. The velocity field in S_h

a) $\beta = 0°$

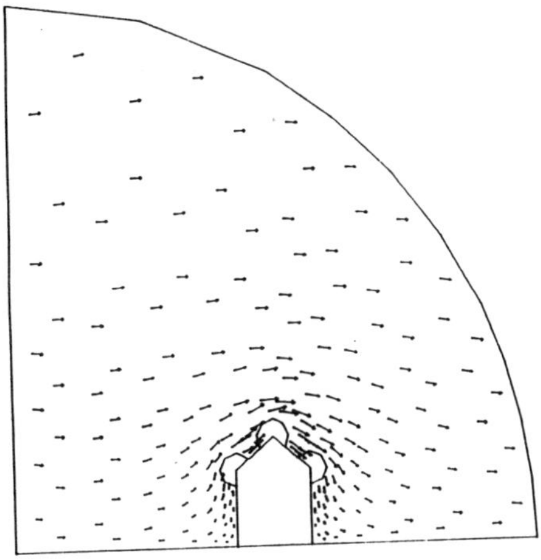

b) $\beta = 40°$

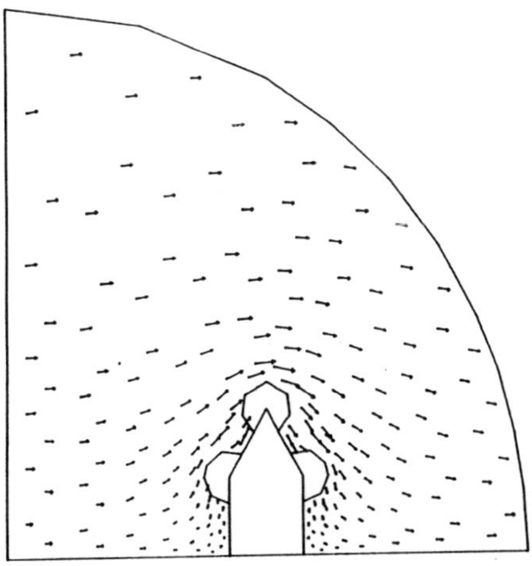

c) $\beta = 60°$

SECTION 6: FRACTURE MECHANICS

Fracture Diagrams for Short Cracks at a Stress Concentration

R.D. Stark, D.J. Cartwright

Department of Mechanical Engineering, Bucknell University, Lewisburg, PA 17837, U.S.A.

ABSTRACT

Fracture diagrams are constructed for one and two cracks at a hole in a finite sheet. The boundary element method is used to produce the weight function for the cracked body. The weight function is integrated to find the stress intensity factors. The stress intensity factors are used together with the weight function to calculate the yield zone length in a strip yield model of the crack. The crack tip opening displacement of the strip yield crack is determined from the known strip yield zone length by using a subsequent boundary element solution. The crack tip opening displacement and the stress intensity factor are used to construct the fracture diagram for symmetrical radial cracks at a hole in a uniformly stressed strip. It is shown that the fracture diagram depends on the ratio of the crack length to the hole radius and that this dependence is more pronounced for smaller cracks and for higher applied stresses.

INTRODUCTION

Fracture of structures due to short cracks at a region of stress concentration may occur at stresses, relative to the materials yield strength, that are too high for analysis using the principles of Linear Elastic Fracture Mechanics (LEFM). This may occur even in high strength materials that would ordinarily be amenable to an LEFM analysis. Under these conditions it is necessary to use a non-linear fracture mechanics (NLFM) approach. One such involves constructing a Fracture Diagram which can be used to assess the damage tolerance of the structure. The crack model of Dugdale [1] and Bilby, Cottrell and Swinden [2] usually forms the basis of the Fracture Diagram. Heald, Spink, and Worthington [3] used this model to predict fracture stresses. The approach of Dowling and Townley [4] was extended by Harrison et.al., [5-8] into a systematic construction of the fracture diagram.

The fracture diagram was compared with a wide range of experimental fracture data [9] and its use extended to include thermal and residual stresses [10]. The method has been the subject of a review [11] and has been used in a wide range of applications [12-21]. In this later work the effect of the structural geometry has been incorporated into the diagram. Structural geometry and non-linear effects of fasteners have also been included in fracture diagrams for stiffened panels typical of those used in aircraft applications [22,23] where it was shown that the geometrical effects of the stiffeners could significantly affect the fracture diagram.

Another structural feature in which the geometry may significantly affect the fracture diagram occurs for short cracks at the edge of a hole. Cracks at regions of stress concentration are of practical importance because they occur frequently in real structures, and can cause a reduction of strength below acceptable levels while they are still smaller than 10% of the hole radius. Holes are introduced in most structures for several reasons, they may be used to connect one structure to another by bolts or rivets, they may be used to lower structural weight, or to allow access for maintenance or service.

In the present work the Boundary Element Method is used to determine Weight Functions for the cracks [24,25]. The stress intensity factor of the crack with no yield zone is determined and the crack opening displacement of the strip yield crack is also determined following the approach [26,27]. These are then used to construct fracture diagrams for one or two radial cracks at a central hole in a uniformly stressed strip.

THEORETICAL FORMULATION

The equation for the fracture diagram can be written in the form [22]

$$K_r = \frac{Y_k \; S_r}{\sqrt{Y_\delta}} \left[\frac{8}{\pi^2} \ln\left\{ \sec\left(\frac{\pi S_r}{2}\right) \right\} \right]^{-\frac{1}{2}} . \qquad (1)$$

where K_r is the ratio of the stress intensity of the crack to the fracture toughness of the material, S_r is the ratio of the applied load (or stress) and the load (or stress) at plastic collapse. The quantities Y_k and Y_δ are the geometry dependent correction factors for the stress intensity in the elastic crack model and the crack opening displacement in the strip yield zone crack model respectively [22]. The fracture diagram [6] is a special case of equation (1) in which Y_k and Y_δ are assumed to be unity which means that the fracture diagram is independent of the geometrical features of the structure. If the effects of geometry are to be included the Y_k and Y_δ must be determined.

The configuration to be analyzed is shown in figure 1a. It consists of a strip of width 2b and length 2h containing a central circular hole of radius R. The ends of the strip are subjected to a uniform traction of magnitude σ. The hole may have either one or two radial cracks on the line of the minimum section y=o. These cracks are either traction free surfaces $R \leq x \leq (R+\ell)$ (elastic crack) or traction free surfaces $R \leq x \leq (R+\ell)$ with uniform tractions of magnitude σ_Y over the surfaces $(R+\ell) \leq x \leq (R+\ell+s)$ (strip yield crack).

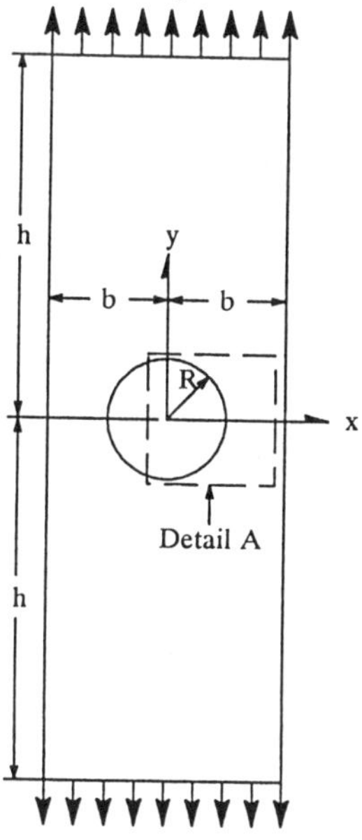

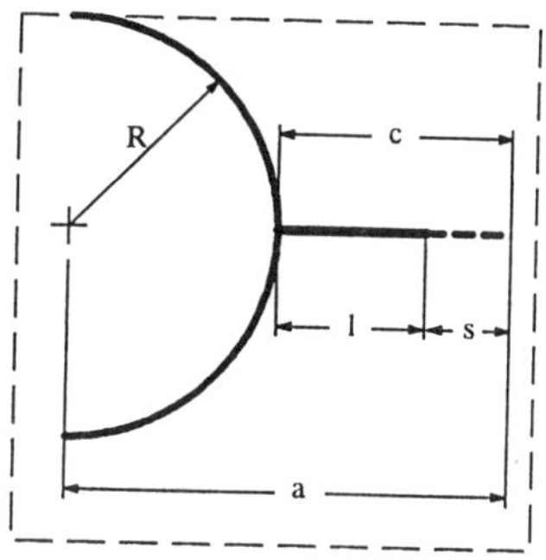

Figure 1 Uniformly Stressed Strip with Radial Cracks at a Central Hole

The stress intensity factor K_I for a cracked body subject to traction boundary conditions can be written [25] as

$$K_I = \frac{E'}{4\sqrt{2\pi}B_0} \int_\Gamma (\bar{t}.\bar{u})d\Gamma, \qquad (2)$$

where E'= E for plane stress or $E' = E/(1-\nu^2)$ for plane strain, $\underline{B_0}$ is the Bueckner field strength, $\bar{t}$ is the traction vector, $\bar{u}$ is the displacement vector, on Γ, the boundary of the body. For the crack without a strip yield zone the stress intensity factor $(K_I)_a$ can be reduced to

$$(K_I)_a = Y_k \, \sigma\sqrt{\pi a} \tag{3}$$

where

$$Y_k = \frac{b}{2\pi a\sqrt{2}} \int_{-1}^{1} U_2(X_1,h)\, dX_1$$

The integrand $U_2(X_1,h)$ is the weight function along $y=h$ which is determined from a boundary element analysis of the configuration in figure 1a when a Bueckner Singular field is applied at the crack tip [25] at $x=R+\ell$. The integration is carried out over the loaded boundary $-1<X_1<1$ where $X_1 = x/b$.

For the strip yield crack which has tractions of magnitude σ_Y a distance s ahead of the tip at $x = (R+\ell)$ the stress intensity factor $(K_I)_b$ can also be determined from equation (1) as

$$(K_I)_b = -\frac{\sigma_Y c}{2\sqrt{2\pi a}} \int_{\ell/c}^{1} \frac{\sum_n A_n X^n}{\sqrt{1 - X^2}}\, dX \tag{4}$$

where $X = x/c$ and the constants A_n are determined from a least squares fit to the weight function after it has been calculated from the boundary element analysis for a Bueckner Singular field at the crack tip at $x = (R+\ell+s)$. The integral in equation (4) was performed analytically using a fourth degree polynomial fit to the weight function. For the strip yield crack the condition that must be satisfied, such that the stresses are bounded in the strip yield zone ahead of the tip is given by

$$(K_I)_a + (K_I)_b = 0 \tag{5}$$

Substituting the expression for the stress intensity factors from equation (3) and (4) gives the relationship between the applied stress ratio σ/σ_Y and the strip yield zone length ratio ℓ/c such that stresses are bounded as

$$\frac{\sigma}{\sigma_Y} = \frac{c}{b}\int_{\ell/c}^{1} \frac{\sum_n A_n X^n}{\sqrt{1 - X^2}}\, dX \Bigg/ \int_{-1}^{1} U_2(X_1,h)\, dX_1. \tag{6}$$

The stress ratio σ/σ_Y is calculated for each ℓ/c ratio using equation (6). The crack opening displacement is calculated following [27] by resolving a boundary integral equation for each pair of ℓ/c and σ/σ_Y values that satisfy equation (6). This solution provides the displacements of the crack surface from which the crack opening displacement δ can be determined at a the coordinate $x = R+\ell$. The geometry dependent factor Y_k is then determined as

$$Y_\delta = \frac{\delta}{\delta_0} , \qquad (7)$$

where $$\delta_0 = \frac{8\sigma_Y a}{\pi E'} \ln \{\sec(\frac{\pi\sigma}{2\sigma_y})\}.$$

is the crack tip opening displacement for a crack of length 2a in a uniformly stressed infinite sheet. The values of Y_k and Y_δ determined from equations (3) and (7) are used in equation (1) to construct the fracture diagrams for one or two cracks at a hole in a uniformly stress strip.

BOUNDARY ELEMENT MODELING

The finite sheet geometry is shown in Figure 1. It is defined by b/R = 2.0 and h/b = 3.0. Because of symmetry in the geometry and the loading, only the top half of the sheet was modeled (Figure 2)

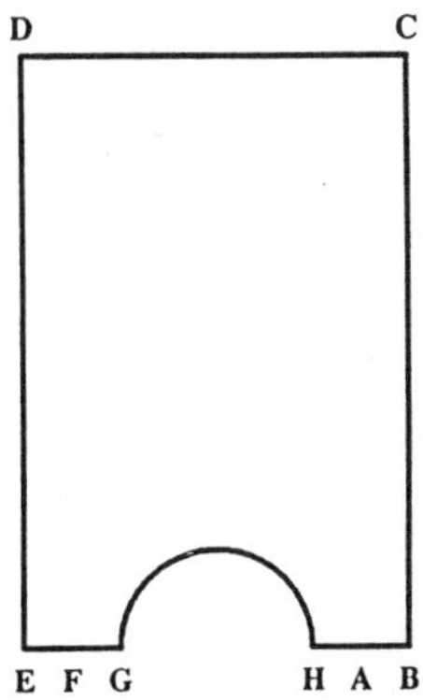

Figure 2 Boundary Element Model of the Finite Sheet

The model was made using quadratic boundary elements. Two conditions were satisfied in meshing the boundary: 1) no element was allowed to be more than twice as long as the element on either side of it and 2) elements meeting at a corner were required to have equal lengths.

The models had between 61 and 107 elements in the mesh. They were made with the minimum number of elements satisfying the above conditions and giving stress intensity correction factors and crack tip opening displacement correction factors within 1-2% of known accurate solutions.

The shortest single crack investigated had an ℓ/R ratio of 0.05. The mesh for that model had 93 elements. The crack, line HA in Figure 2, had 12 elements and line CD had 5 elements. The first two and last two elements along line HA were 5% of the crack length, and the rest of the elements were 10% of the crack length. all of the elements on line CD were 20% of the line length. Line AB had 13 elements and line BC had 10 elements. The elements on lines AB and CD were graded to provide a transition from the short elements at the crack to the larger elements along the top of the sheet. Line DE had 6 elements and line EG had 2 elements. Again these lines were graded to serve as a transition region. Since no information was required on lines DE and EG, they could have larger elements to minimize the number of elements in the mesh. The arc from G to H had 45 elements. The elements on the arc GH were graded from 0.245% of the hole radius near the crack to 31.4% of the hole radius near point G.

The longest single crack has an ℓ/R ratio of 0.95. The boundary element mesh for this model had 61 elements. Line HA had 12 elements, and line CD had 5. Since line AB was so short, it had only one element. Line BC had 10 elements. These elements were graded to provide a transition from the short element on line AB to the longer elements on line BC. Line DE contained 6 elements, and line EG again had 2 elements. Arc GH had 25 elements. They were graded from 3.93% of the hole radius near the crack to 31.4% near point E. For the single crack models, lines CD, DE, and EG were specified for the first model and did not change from model to model. The rest of the lines were remeshed from model to model as the crack length changed.

Similar meshes were used for the double crack with a maximum number of elements of 107 for the shortest crack ($\ell/R = 0.05$) and a maximum of 81 for the largest crack $\ell/R = 0.95$. In the double crack model the half sheet was used in order to avoid reflecting the singular field as would occur if a quarter sheet were used to model this geometry.

COMPARISON WITH KNOWN SOLUTIONS

Shrivakumar and Forman [28] have determined weight functions for one crack at a hole in an infinite sheet. These were used to find normalized stress intensity factors $K_I/(\sigma_Y\sqrt{\pi a})$ due to a traction of magnitude σ_Y over a length s from the tip of the crack. These results are compared with the boundary element results which used the weight function derived from the singular field. From table 1 it can be seen that for both a short c/R = 0.05 and a long crack c/R = 1.0 the results agree with 1%. Even over a wider range differences did not exceed 1.5%.

$\frac{s}{c}$	Weight function BEM		Shivakumar and Forman		% Diff	
	c/R		c/R		c/R	
	0.05	1.0	0.05	1.0	0.05	1.0
1.00	0.235	0.623	0.237	0.618	-0.84	0.81
0.80	0.202	0.559	0.203	0.555	-0.49	0.72
0.60	0.168	0.486	0.169	0.483	-0.59	0.62
0.40	0.132	0.399	0.133	0.397	-0.75	0.5
0.05	0.044	0.143	0.044	0.142	0.00	0.7

Table 1 Comparison of normalized stress intensity factors for one crack at a hole in an infinite sheet

For two cracks at a hole in a uniformity stressed strip the stress intensity factors $K_I/(\sigma\sqrt{\pi a})$ were within 1.6% of those determined by Newman [29] using boundary collocation. These are shown in table 2.

ℓ/R	Weight function BEM	Newman	% Diff
0.05	0.948	0.963	-1.56
0.06	1.02	1.03	-0.97
0.1	1.22	1.23	-0.81
0.2	1.49	1.5	-0.67
0.4	1.8	1.82	-1.10
0.7	2.45	2.48	-1.21
0.8	2.91	2.91	0.00

Table 2 Comparison of Normalized Stress Intensity Factors for Two Cracks at a Hole in a Strip

The crack tip opening displacement correction factors calculated by the boundary element method for a center cracked infinite sheet are shown in Table 3. The infinite sheet was modeled in the boundary element analysis as a large finite sheet, b/a = 20 and h/a = 20.

The crack tip opening displacement was determined by a boundary element stress analysis. The remote stress, σ, was applied along the top of the sheet and the yield stress, σ_Y, was applied on the strip yield zone. The crack tip opening displacement corresponded to the vertical displacement on the crack at the coordinate $x = (R+\ell)$ in the boundary element solution.

$\frac{s}{c}$	$Y_\delta = \frac{\delta}{\delta_0}$ Weight function BEM	Theoretical	% Diff
0.1	1.003	1.000	0.3
0.2	1.003	1.000	0.3
0.4	1.002	1.000	0.2
0.6	1.000	1.000	0.0
0.8	1.002	1.000	0.2
0.9	0.9999	1.000	-0.01

Table 3 Normalized Crack Tip Opening Displacements for a Center Cracked Infinite Sheet

From table 3 it can be seen that the normalized crack tip opening displacements δ/δ_0 are all within 0.3% of the theoretical values. The boundary element crack tip opening displacement correction factors were usually higher than the closed form solution which is consistent with the effect of the finite boundaries which would increase the crack tip opening displacement correction factors above that for an infinite sheet.

The comparisons with the theoretical solutions show that the present method of calculating stress intensity correction factors and crack tip opening displacement correction factors should produce results for one or two cracks at a hole in a finite sheet which are accurate to within 2%.

FRACTURE DIAGRAMS

The fracture diagrams determined from equation (1) for one crack at a hole in a finite sheet are shown in Figure 3 together with the infinite sheet fracture diagram. They are given for ℓ/R ratios from 0.05 to 0.5 for the finite sheet having $b/R = 2.0$ and $h/b = 3.0$ shown in Figure 1. The quantity K_r is the ratio of the stress intensity factor to the plane strain fracture toughness, and S_r is the ratio of the remote stress at failure, σ, to the remote stress to cause net section collapse of the strip at the crack line, σ_0. This was calculated using the upper bound theorem of plasticity for the strip having either one or two cracks at the hole.

In Figure 3, the fracture diagrams for ℓ/R ratios greater than or equal to 0.2 are predominately above the infinite sheet fracture diagram. For these ℓ/R ratios, the infinite sheet fracture diagram is either very close to the curves or slightly conservative and would therefore be a suitable choice as a failure curve. The fracture diagrams for ratios of ℓ/R smaller than 0.2 depart more from the infinite sheet fracture diagram, an effect which is more pronounced at higher stress ratios.

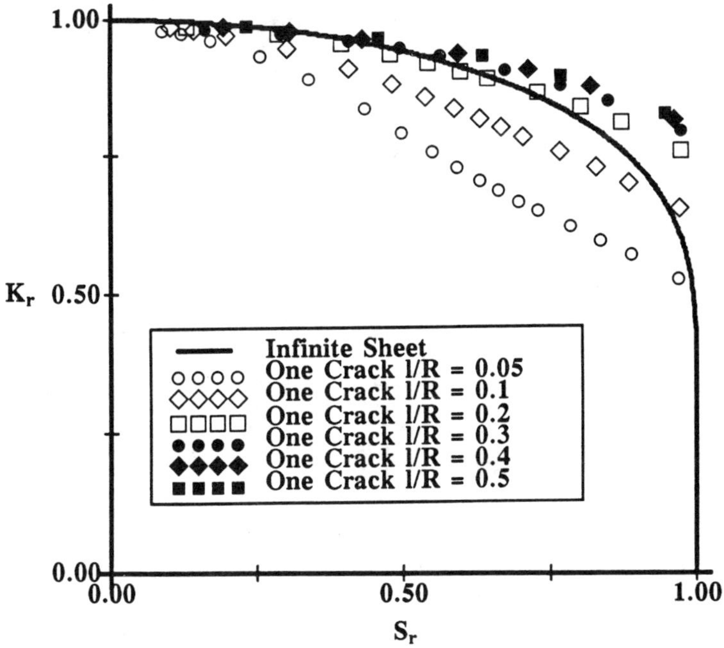

Figure 3 Fracture Diagrams for One Crack at a Hole in a Finite Sheet

The fracture diagrams for the two smaller ratios ℓ/R = 0.05 and 0.1 in Figure 3 are below the infinite sheet fracture diagram. They indicate that the infinite sheet fracture diagram is non-conservative for small ℓ/R ratios for a crack at a hole in a finite sheet and that the stress concentration is effective only for small ℓ/R ratios. All fracture diagrams differ more from the infinite sheet fracture diagram at the higher stress ratios and for small stress ratios, the fracture diagrams approach the infinite sheet fracture diagram.

The fracture diagrams for two equal cracks at a hole in a finite sheet for ℓ/R ratios from 0.05 to 0.5 are in Figure 4. The infinite sheet fracture diagram is the solid line in Figure 4. The fracture diagrams for two cracks at a hole in a finite sheet have the same trends as one crack at a hole in a finite sheet. The fracture diagrams for ℓ/R greater than or equal to 0.2 are close enough to the infinite sheet fracture diagram that it may be used without significant error. The fracture diagrams for ℓ/R = 0.05 and 0.1 for two cracks are below the infinite sheet fracture diagram indicating that the infinite sheet fracture diagram would be unconservative for these ℓ/R ratios. The fracture diagrams for ℓ/R = 0.05 and 0.1 differ from the infinite sheet fracture diagram more at high stress ratios. At small stress ratios, the fracture diagrams approach the infinite sheet fracture diagram.

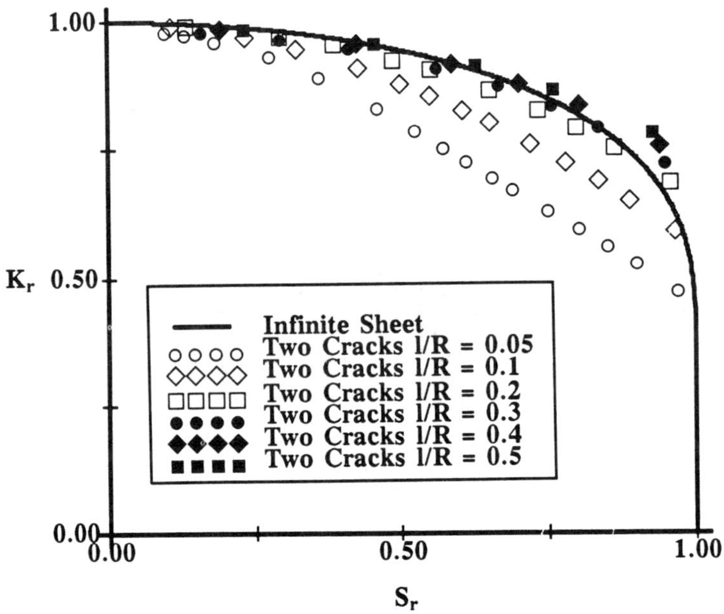

Figure 4 Fracture Diagrams for Two Cracks at a Hole in a Finite Sheet

A comparison is made of the fracture diagrams for two cracks at a hole in a finite sheet (b/R = 2.0 and h/b = 3.0) and, the fracture diagrams for a cracked finite sheet of the same size (h/b = 3.0) with no hole, and the fracture diagram for the infinite sheet, in Figure 5. The two sheets are compared for the smallest and largest ℓ/R ratios. The comparison shows the effect of the hole on the fracture diagram. The fracture diagram for the cracked finite sheet without a hole illustrates the effect of the finite edge boundary only on the fracture diagram.

The fracture diagrams for the two cracks in a finite sheet are close together and are slightly above the infinite sheet fracture diagram at the higher stress ratios. The fracture diagram for $\ell/R = 0.5$ at a hole in an infinite sheet is very

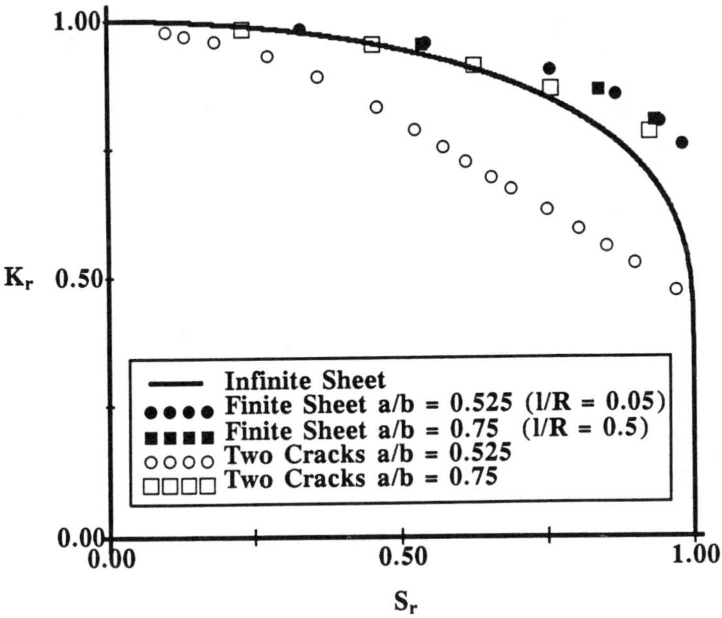

Figure 5 Comparison of Fracture Diagrams for a Cracked Finite Sheet and a Cracked Finite Sheet with a Hole

close to the finite sheet fracture diagrams showing that the hole does not have much influence on the fracture diagram for long cracks. However, for short cracks at a hole there is a significant influence on the fracture diagram. The fracture diagram for short cracks at a hole in a finite sheet differ significantly from the fracture diagram for cracks in a finite sheet for all but small stress ratios. For these ratios the finite width and the hole appear to produce competing effects on the curves in the fracture diagram: the finite width effect raises the curves, while the effect of the hole lowers the curves.

CONCLUSIONS

1. Bueckner singular fields combined with boundary element analysis has been used to determine weight functions and stress intensity factors for one or two cracks at a hole in a finite width strip.

2. The weight functions have been used to determine the crack tip opening displacement of one or two strip yield cracks at

the hole and the stress intensity factors of the crack when the strip yield zone length is zero.

3. Fracture diagrams have been determined for one or two cracks near a region of stress concentration which shows that, for short cracks relative to the hole radius and high stresses relative to the yield stress, neglecting these effects will lead to unconservative estimates fracture stresses.

REFERENCES

1. D. S. Dugdale, (1960) "Yielding of Steel Sheets Containing Slits," Journal of Mechanics and Physics of Solids, 8, pp. 100-104.

2. B.A. Bilby, A. H. Cottrell, and K. H. Swinden, (1963) "The Spread of Plastic Yield from a Notch," Proc. R. Soc. London A, 272, pp. 304-314.

3. P. T. Heald, G. M. Spink, and P. J. Worthington, (1972) "Post Yield Fracture Mechanics," Materials Science and Engineering, 10, pp. 129-138.

4. A. R. Dowling, and C. H. A. Townley, (1975) "The Effect of Defects on Structural Failure: A Two-Criterion Approach," Int. J. Pres. Ves. and Piping, 3, pp. 77-107.

5. R. P. Harrison, K. Lossemore, and I. Milne, (1976) "Assessment of the Integrity of Structure Containing Defects," R/H/R6, Central Electricity Generating Board, Research Division, London.

6. R. P. Harrison, K. Lossemore, and I. Milne, (1981) "Assessment of the Integrity of Structure Containing Defects," R/H/R6 revision 2, Central Electricity Generating Board, Research Division, London.

7. R. P. Harrison, K. Lossemore, and I. Milne, (1981) "Assessment of the Integrity of Structure Containing Defects," R/H/R6 supplement 1, Central Electricity Generating Board, Research Division, London.

8. R. P. Harrison, and I. Milne, (1981) "Assessment of Defects: the GEGB Approach," Phil. Trans. R. Soc. London A, 299, pp. 145-153.

9. G. G. Chell, (1979) "Elastic-Plastic Fracture Mechanics," Developments in Fracture Mechanics, Applied Science Publishers, London, 1, pp. 67-105.

10. G. G. Chell, (1979) "A Procedure for Incorporating Thermal and Residual Stresses into the Concept of a Failure Assessment Diagram, Elastic-Plastic Fracture," ASTM STP 668, J. K. Landes, J. A. Begley, and G. A. Clarke, eds., American Society for Testing and Materials, pp. 581-605.

11. R. deWitt, (1983) "A Review of Generalized Failure Criteria Based on the Plastic Yield Strip Model," Fracture Mechanics: Fourteenth Symposium-Volume I: Theory and Analysis, ASTM STP 791, J. C. Lewis and G. Sines, eds. American Society for Testing and Materials, pp. I-24-I-50.

12. J. M. Bloom, (1981) "An Analytical Assessment of the Effects of Residual Stresses and Fracture Properties on Service Performance of Various Weld Repair Processes," Journal of Pressure Vessel Technology, 103, pp. 373-379.

13. J. M. Bloom, and S. N. Malik, (1982) "Procedure for the Assessment of the Integrity of Nuclear Pressure Vessels and Piping Containing Defects," EPRI Research Project 1237-2, NP-2431.

14. J. M. Bloom, (1983) "A Procedure for the Assessment of the Structural Integrity of Nuclear Pressure Vessels," Journal of Pressure Vessel Technology, 105, pp. 28-34.

15. J. M. Bloom, and S. N. Malik, (1984) "A Failure Assessment Approach for Handling Combined Thermomechanical Loading," Fracture Mechanics: Fifteenth Symposium, ASTM STP 833, R. J. Sanford, ed., American Society for Testing and Materials, Philadelphia, pp. 165-189.

16. J. M. Bloom, and J. L. Hechmer, (1984) "Limits of Linear Elastic Fracture Mechanics," Journal of Pressure Vessel Technology, 106, pp. 196-200.

17. J. M. Bloom, (1985) "Deformation Plasticity Failure Assessment Diagram," Elastic-Plastic Fracture Mechanics Technology, ASTM STP 806, J. C. Newman, Jr., and F. J. Loss, eds., American Society for Testing and Materials, Philadelphia, pp. 114-127.

18. K. K. Yoon, J. M. Bloom, W. A. Pavinich, and H. W. Slager, (1985) "Application of the Failure Assessment Diagram to the Evaluation of Pressure-Temperature Limits for a Pressurized Water Reactor," Journal of Pressure Vessel Technology, 107, pp. 192-196.

19. J. M. Bloom, (1985) "Extensions of the Failure Assessment Diagram Approach-Semi-Elliptical Flaw in Pressurized Cylinder," Journal of Pressure Vessel Technology, 107, pp. 25-29.

20. J. M. Bloom, (1986) "Extensions of the Failure Assessment Diagram Approach-Semi-Elliptical Flaw in Pressurized Cylinder--Part II," Journal of Pressure Vessel Technology, 108, pp. 485-489.

21. J. M. Bloom, (1989) "Simplified Procedures for Handling Self-Equilibrating Secondary Stresses in the Deformation Plasticity Failure Assessment Diagram Approach," Nonlinear Fracture Mechanics: Volume II--Elastic-Plastic Fracture, ASTM STP 995, J. K. Landes, A. Saxena, and J. G. Merkle, eds., American Society for Testing and Materials, Philadelphia, pp. 280-305.

22. T. P. Rich, M. M. Ghassem, and D. J. Cartwright (1985) "Fracture Diagrams for Cracked Stiffened Panels," Engineering Fracture Mechanics, 21, no. 5, pp. 1005-1017.

23. M. Low, and D. J. Cartwright, (1985) "Fracture Diagrams for Stiffened Aircraft Structures: Effects of Material Non-Linearity," Proc. of the 13th Symposium on Aeronautical Fatigue, Pisa, Italy, A. Salvetti and G. Cavallini, eds., EMAS Ltd, UK, pp. 551-583.

24. D. J. Cartwright, and D. P. Rooke, (1985) "An Efficient Boundary Element Model for Calculating Green's Functions in Fracture Mechanics," International Journal of Fracture, 27, pp. R43-R50.

25. M. H. Aliabadi, D. J. Cartwright, and D. P. Rooke, (1989) "Fracture-Mechanics Weight-Functions by the Removal of Singular Fields Using Boundary Element Analysis," International Journal of Fracture, 40, pp. 271-284.

26. D. J. Cartwright, and M. H. Aliabadi, "Boundary Element Analysis of Strip Yield Crack at a Hole," Proceeding of the 10th

27. M. H. Aliabadi, and D. J. Cartwright, "Boundary Element Analysis of Strip Yield Cracks," Engineering Analysis, To be published.

28 V. Shivakumar, and R. G. Forman, (1980) "Green's Function for a Crack Emanating from a Circular Hole in an Infinite Sheet," Internation Journal of Fracture, 16, no. 4, pp. 305-316.

29. J. C. Newman, Jr., (1971) "An Improved Method of Collocation for the Stress Analysis of Cracked Plates with Various shapes," NASA TN D-6376, National Aeronautics and Space Administration.

Cracks Grow in Cement Structures using Laplace Transform and Boundary Element Method

M. Kassem

Department of Engineering Mathematics and Physics, Zagazig Faculty of Engineering

ABSTRACT

The expansion of cracks in a cement structure is treated as a one dimensional moving boundary problem , where Stefan's condition represents the rate of reaction between chlorides and ferrous components inside the structure. The problem is solved using Laplace transform method . The inverse Laplace transform yields a concentration in the form of an error function containing Heaviside functions. This result is similar to the closed form solution , where the position of the moving interface is proportional to the square root of time . This method of solution is tested on a one dimensional problem and compared with the boundary element method . The results obtained are remarkably close to the exact solution.

INDEX : Boundary Element Method, Laplace Transform Method Moving Boundary Element

INTRODUCTION

The problem of creeping of cement due to the interaction of chlorides and sulfates with the impurities present in the cement, was studied by Bazant[1] giving an expectation of the probable distribution of cracks. The size and development of cracks[2] in a cement structure define the critical stage, where diffusion of humidity inside the structure starts[3]. This stage was simulated by an electrochemical cycle[4.5.6] or as a one dimensional mechanical problem[7]. The last approach analyzes the energy released from the cracks into the structure. In this case several theories were developed using the finite element method[8,9,10]. The change of volumes due to the rusting of steels and presence of cracks was also developed through a complete stress analysis[11,12,13]. The chemical reaction between chlorides and ferrous components can also be thought of as a diffusion process through porous material[15,16,17], or as a mixture of both electrochemical and diffusion process[18]. In the present analysis, the state of the structure at the level of the crack is developed as a one dimensional moving problem, using the Laplace transform. The concentration of humidity inside the crack is assumed constant. The position of the interface as well as the concentration at the boundary are defined according to an iterative scheme. The proposed method of solution is compared with the boundary element method of solution[19]. The results obtained are remarkably close to the exact solution.

MATHEMATICAL FORMULATION

The concentration of chloride ions inside the cement and its reaction with ferrous components causes a reduction of homogeneity of the structure, and initiates the appearance of voids. The cement starts to be porous[3,18]. This porosity allows for the diffusion of humidity inside the structure. This phenomena is described by a diffusion equation of the form:

$$\frac{\partial^2 C(x,t)}{\partial x^2} = \frac{1}{D}\frac{\partial C(x,t)}{\partial t} \qquad 0 \leq x < S(t), \quad t > 0 \qquad (1)$$

where $C(x,t)$ is the concentration of humidity inside the fracture. D is the diffusivity coefficient, $S(t)$ is the moving interface. The boundary conditions are:

$$C = C_s \qquad 0 \leq x < S(t),\ t > 0 \qquad (2,a)$$

$$C = C_c \qquad \text{at} \quad x = S(t),\ t > 0 \qquad (2,b)$$

$$\frac{\partial^2 C(x,t)}{\partial x^2} = \frac{\partial S(t)}{\partial t} \qquad \text{at} \quad x = S(t),\ t > 0 \qquad (2,c)$$

and the initial condition

$$C(x,0) = C_c \qquad x \in [0,L] \qquad (2,d)$$

L is the half length of the crack .C_s is the surface concentration ,C_c is the concentration at the moving boundary $S(t)$. The problem is illustrated in Fig.1

NORMALIZATION OF THE PROBLEM

The Stefan problem is normalized according to the following dimensionless quantities:

$$\eta = \frac{x}{L}, \quad \tau = \frac{Dt}{L^2}, \quad \xi(\tau) = S(t)/L, \quad C(\eta,\tau) = \frac{C(x,t)-C_c}{C_s-C_c}$$

where η is the dimensionless spatial variable, L is the fracture length. τ is the dimensionless Fourier number, $\xi(\tau)$ is the location of the interface at a given time. $C(\eta,\tau)$ is the dimensionless concentration of salts inside the crack.

Introducing these quantities into equation 1 and 2 gives:

$$\frac{\partial^2 C(\eta,\xi)}{\partial \eta^2} = \frac{\partial C(\eta,\tau)}{\partial \tau} \qquad 0 \le \eta \le \xi(\tau),\ \tau > 0 \qquad (4)$$

with boundary conditions.

$$\text{at } \eta = 0 \quad \tau > 0 \qquad C(\eta,\tau) = 1 \qquad (5,a)$$
$$\text{at } \eta = \xi(\tau) \ \tau > 0 \qquad C(\eta,\tau) = 0 \qquad (5,b)$$

and

$$\frac{\partial C(\eta,\tau)}{\partial \eta} = \frac{\partial \xi(\tau)}{\partial \tau} \qquad (5,c)$$

The initial condition is

$$C(\eta,0) = 0 \qquad \eta \in [0.L] \qquad (6)$$

This problem was initially solved by Bishri[19] using the boundary element method, and is solved here using the Laplace transform method. Both methods will be compared, with the exact solution given by Crank[20].

LAPLACE TRANSFORMATION METHOD

Application of the Laplace transform to equation 4 gives:

$$SC(\eta,S) - C(\eta.0) = \frac{\partial^2}{\partial \eta^2} C(\eta,S) \qquad (7,a)$$

or

$$C(\eta,S) = A\ e^{-\sqrt{S}\eta} + B\ e^{\sqrt{S}\eta} \qquad (7,b)$$

where A and B coefficients are defined from the boundary conditions (5,a) and (5,b). The result is written as:

$$C(\eta,S) = \frac{e^{\sqrt{S}\eta} - e^{2\sqrt{S}\ \xi(\tau)} \cdot e^{-\sqrt{S}\eta}}{S(1 - e^{2\sqrt{S}\ \xi(\tau)})} \qquad (8)$$

This equation is reduced to a series form by a binomial expansion of the denominator giving.

$$C(\eta,S) = \frac{e^{\sqrt{S}}}{S} [e^{\eta} - e^{2\xi(\tau)-\eta}] . \sum_{n=0}^{\infty} e^{2n\sqrt{S}\xi(\tau)} \qquad (9,a)$$

The summation of the three first is written as:

$$C(\eta,S) = \frac{e^{\sqrt{S}}}{S} [1 + e^{2S\xi(\tau)} + e^{4S\xi(\tau)}][e^{\eta} - e^{2\xi(\tau)-\eta}] \qquad (9,b)$$

The inverse transformation is given by [21]:

$$C(\eta,\tau) = \mathrm{erfc}\left(\frac{1}{2\sqrt{\tau}}\right) + \left[H(\tau - 2\xi(\tau)) + H(\tau - 4\xi(\tau)) + \ldots \right] \left[e^{\eta} - e^{2\xi(\tau)-\eta} \right] \quad (10)$$

The first term is similar to the solution obtained for a diffusion problem. H is the Heaviside step function delayed by the dimensionless quantity $2n.\xi(\tau)$, $n \in [1,\infty]$ This result is written in a closed form as:

$$C(\eta,\tau) = e^{\eta}[1 - e^{2(\xi(\tau)-\eta)}][\mathrm{erfc}\left(\frac{1}{2\sqrt{\tau}}\right) + \sum_{n=1}^{\infty} H(\tau - 2\,\xi(\tau))] \quad (11)$$

The concentration is found to increase with time in a ramp form due to the time wise addition of the Heaviside function. This result is similar to the closed form solution of the problem[20]. For very large time steps as this is the case in cement structure only one term of the Heaviside series is considered. In this case the solution writes as:

$$C(\eta,\tau) = e^{\eta}\,[1 - e^{2(\xi(\tau)-\eta)}]\,[\mathrm{erfc}\left(\frac{1}{2\sqrt{\tau}}\right) + H(\tau - 2\xi(\tau))] \quad (12)$$

NUMERICAL PROCEDURE

The position of the moving interface is defined by a transformation of Stefan's condition, equation (5.c), giving :

$$\xi(t) = \int_0^t \frac{\partial C(\eta,\tau)}{\partial \eta}\, d\tau \quad (13)$$

Where $\xi(t)$ is the position of the moving interface at time t. Equations 12-13 are iteratively solved according to the following steps:

1- The time to the concentration is estimated from equation (12) as:

$$C_k\,(L,t_o) = e^{L}\,H(\tau - 2L) \quad (14)$$

where k is an index for time, L is the crack length.

2- At time $t = t_o + \Delta t$. The approximate positive of the interface is estimated by the following recurrence relation:

$$\xi_{k+1}(\tau) = L + \xi_k\,(\tau) \quad (15a)$$

where[20]

$$\xi_k\,(\tau) = \Delta t^{1/2} \quad (15b)$$

Replacing for $\xi_{k+1}(\tau)$ in equation 12 gives:

$$C_{k+1} = e^{L+\xi_{k+1}} \left[erfc \frac{1}{2\sqrt{\Delta t}} + H(\tau - 2\xi_{k+1}) \right] \qquad (16)$$

3- Evaluate the correct position of the interface from equation (13).

$$\xi_{k+1}(\tau) = (C_{k+1} - C_k)/(\Delta t)^{1/2} \qquad (17)$$

These 3 steps provide a numerical method of solution of the boundary value problem.

APPLICATION

This method is applied to a one dimensional crack illustrated in Fig 1. The problem is solved for different time steps. The results obtained are illustrated in Fig.2.

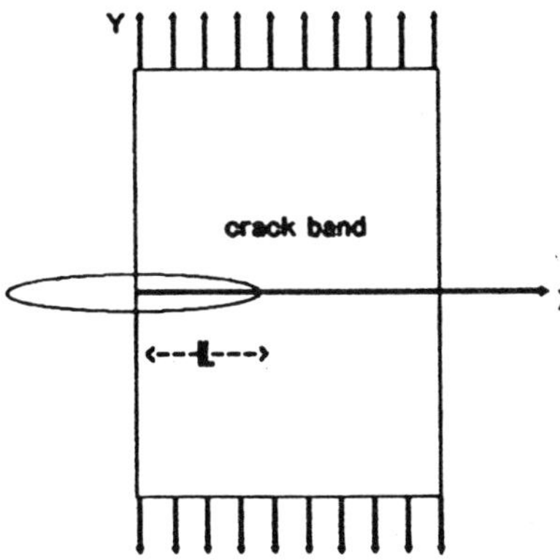

Fig1.Crack inside a cement structure with half length l and moving front

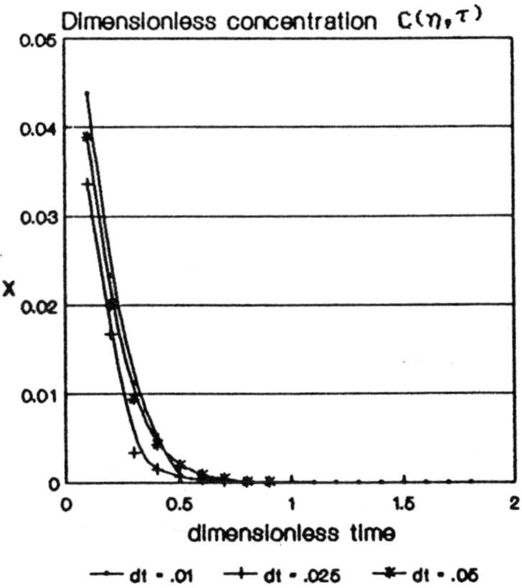

Fig.2 Concentration versus time for different time steps

N

The concentration of humidity at the moving boundary is plotted versus time. Both quantities are dimensionless. It is found that the concentration shows an error function shape diverging with time . This is explained by the interference of the Heaviside function into the solution and is due to the gradual stabilization of chemical reactions between chlorides and impurities , inside the cement structure.

BOUNDARY ELEMENT METHOD

The diffusion problem when solved by the boundary element method [22] reduces equation 4 to:

$$\alpha_i C_i = \int_0^t \int_\Gamma C^*(\eta,\tau)\nabla C(\eta,\tau)\, d\Gamma - \int_\Gamma \nabla C^*(\eta,\tau).C(\eta,\tau)\, d\Gamma - I_\Omega \quad (18)$$

In the forgoing analysis the arguments of C and C^* are maintained whenever confusion may occur and dropped otherwise. α_i is a weighting factor equal to 1/2 on the boundary Γ and to one inside the domain. $C^*(\eta,\tau)$ is the Green solution of the diffusion equation, ∇ is a spatial derivative operator. $C(\eta,\tau)$ is the concentration, previously defined, and I_Ω is written as:

$$I_\Omega = \int_0^t \int_0^{\xi(\tau)} \frac{d(CC^*)}{d\tau}\, d\eta\, d\tau \quad (19)$$

applying Leibniz rule gives:

$$I_\Omega = \int_0^t \left[\frac{\partial}{\partial\tau} \int_0^{\xi(\tau)} CC^*\, d\eta + CC^* \Big|_{\eta=\xi(\tau)} \frac{\partial\xi(\tau)}{\partial\tau} \right] d\tau \quad (20)$$

Substituting for the boundary conditions given in equation 5.b the last term is found to be equal to zero. Inserting this integral in equation 18 gives:

$$d_i C_i = \int_0^t \left[\int_\Gamma C^*\, \nabla C\, d\Gamma - \int_\Gamma \nabla C^*.\, C\, d\Gamma \right] d\tau - \int_\Gamma C^* C\, d\Gamma \quad (21.a)$$

where the last term reduces to :

$$I = \int_\eta C^*\,(\eta,\tau)\, d\eta \quad (21.b)$$

where

$$C^* = \frac{e^{-\eta^2/4\tau}}{\sqrt{4\pi\tau}} \quad (21.c)$$

$$I = \mathrm{erfc}\left(\frac{\eta}{2\sqrt{\tau}}\right) \quad (21.d)$$

Hence equation 21.a reduces to

$$\alpha_i C_i = \int_0^t (C^* \nabla C \mid_{\eta=\xi(\tau)} - C^* \nabla C \mid_{\eta=o}) \, d\tau \qquad (22)$$

This equation is solved numerically with equation 13, according to the numerical scheme previously developed .

RESULTS

The solution of the moving boundary problem stated earlier by the boundary element method is investigated. This problem is chosen as its analytical solution is known[20]. The imposed condition at the surface of the crack is C_s.The numerical solution of diffusion problem is carried for different time steps. The results obtained are displayed in Figure 3. The dimensionless position of the moving front is plotted versus time. The results obtained show a similarity in both the Laplace transform and boundary element method with respect to the exact solution ,with a relative error equal to 6% . The results indicate that maximum errors occur at the first time step and then taper off with time. It was also found that decreasing the time step decreases the error.

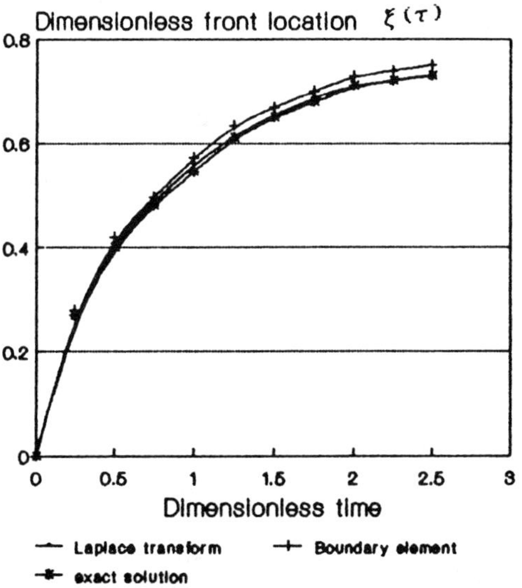

Fig.3 Front location as calculated by the Laplace & Boundary Element method

CONCLUSION

In this work the moving boundary problem is analyzed and solved using an integral Laplace operator and the boundary integral method . Both formulations explicitly couple the unknown front location and the concentration of chlorides at the boundary. In the first method the differential equation is solved in the S plane , and the Inverse Laplace transform yields a complementary function , spreading for different time steps . The position of the moving boundary is obtained by a direct integration of Stefan equation. These two equations are numerically solved, using the closed form solution as an approximation step.The results obtained from this reformulation are compared with the boundary element method of solution. A comparison of these two methods with the exact solution , gives satisfactory results .

NOMENCLATURE

Normal

C : concentration
t : time in sec.
x : space variable
D : diffusinity
L : half length of crack
S(t): position of the moving boundary
H : Heaviside function
C : Green's function
k : index

Greek

η : dimensionless space variable
τ : dimensionless time
ξ : dimensionless position of the boundary
∇ : spatial derivative
Γ : boundary
Ω : domain
α : weighting coefficient

REFERENCES

1- Bazant,Z.P.and Zebich,S. Statistical linear Regression Analysis of Prediction Models for Creep and Shrinkage. **Cement and Concrete Research, Vol.13**, pp.869-876, **1983.**

2- Karihaloo,B.L. and Nallathambi,P.,Size Prediction from Effective Crack Model for Plain Concrete, **Materials and Structures** , **Vol.23** , pp.178-185 **1990.**

3- Bazant,P.Z.and Wang,T.S.Practical Reduction of Cyclic Humidity Effect in Creep and Shrinkage of Concrete , **Material and Structures, Vol.18** , pp.247 -252, **1985.**

4- Hansson,M.C. Comments on Electrochemical Measurements of the Rate of Corrosion of Steel in Concrete **Cement and Concrete Res., Vol.14**,pp.574-584, **1984**

5- Bazant,Z.P.Physical Model for Steel Corrosion in Concrete Sea Structures: Theory. **Proc.of ASCE, JL. Str. Div,**

Vol.105, pp.1137-53, 1979.

6- Bazant,Z.P.Physical Model for Steal Corrosion in Concrete Sea Structures: Applications, Proc.ASCE, Jl. Str. Eng. Div, Vol.105, pp.1156-1166,1979.

7- Bazant,Z.P.and Oh,B.H.Crack Band Theory for Fracture of Concrete,Materials and Structures, Vol.16, pp. 155-157, 1983.

8- Bazant,Z.P.and Chern,J.C.Double Power Logarithmic Law for Concrete Creep. Cement and Research, Vol.14, pp. 793-806 1984

9- Bazant,Z.P.and Cedolini,L.,Finite Element Modelling of Crack Band Propagation, Proc. of ASCE, JL. Str. Eng. Div, pp.69-92 ,1981.

10- Mikami,I. Finite Element Modelling of Crack Band Propagation, Proc. ASCE , J. Str. Eng. Div. pp. 657-60, 1981

11- Erdogan,F.,Gupta,G.D.and Ralwani, M. Interaction Between a Circular Inclusion and an Arbitrary Oriented Crack,Trans. of ASME, J. App. Mech. pp. 1007- 13,1974.

12- Gupta,G.D.and Erdogan,F.The Problem of Edge Cracks in an Infinite Strip,Trans. of ASME, J. Appl. Mech. pp.1001-06, 1971.

13- Martin, P.A. Orthogonal Polynomial Solutions for Pressurized Elliptic Cracks,Q. J . Mech. App. Math., Vol.39, pp.269-287 1986.

14- Martin,P.A.Orthogonal Polynomial Solutions for Elliptic Cracks under Shear Loadings,Q. JL. Mech. App. Math., Vol.39, pp. 519-534, 1986.

15- Vuorinen,J.Applications of Diffusion Theory to Permeability Tests on Concrete:Part 1. Depth of Water Penetration into Concrete and Coefficient of Permeability. Magazine of Concrete Res. Vol.37 , pp. 145-152, 1985.

16- Schrefler,B.A.,Simoni,L.and Majorana,C.E.,A General Model for the Mechanics of Saturated-unsaturated Porous Materials . Materials and Structures, Vol.22, pp. 323-334, 1989.

17- Pullan,A.J.,Boundary Element Solutions of Quasilinearized Time Dependent Infiltration, Appl. Math Model . Vol.12, pp.9-17, 1988.

18- Saad, M. Modelling of Creeping in Cement and Concrete

Structures M.Sc.Thesis, Zagazig Un., 1990.

19- Bishri, A.H.A Boundary Integral Formulation for Moving Interface of two Continua. Int. Conf. of Eng. Math & Physics, Cairo, 1991.

20- Cranck, J. *Free and Moving Boundary Problems*, Clarendon Press, Oxford, 1984.

21- Spiegel, M., *Mathematical Hanbook of formulas and Tables*. Schaum Series. McGraw Hill, N.Y., 1968.

22- Kassem, M. A New Method for a Boundary Integral Solution of a Diffusion Problems. Int. Conf. of Eng. Math & Physics, Cairo, 1991.

Dual Boundary Element Analysis of Pin-Loaded Lugs

A. Portela *, M.H. Aliabadi *, D.P. Rooke **

* *Wessex Institute of Technology, Computational Mechanics Institute, Ashurst, Southampton SO4 2AA, U.K.*

** *Royal Aerospace Establishment, Farnborough, Hants GU14 6TD, U.K.*

Abstract

This paper describes an application of the dual boundary element method to pin-loaded lug problems in linear elastic fracture mechanics analysis. The dual equations of the method are the displacement and the traction boundary integral equations. When the displacement equation is applied on one of the crack surfaces and the traction equation on the other, general mixed-mode crack problems can be solved with a single-region formulation. Both crack surfaces are discretized with discontinuous quadratic boundary elements. The stress intensity factors are evaluated with the J-integral technique. Examples of geometries with embedded cracks are first analyzed to demonstrate the accuracy of the present method. Finally, mixed-mode results are presented for pin-loaded lugs with cracks.

Introduction

Lug-type joints are often used to connect major structural components in the aircraft industry. The cracked-hole lug problem has received much attention because it is a common cause of premature failure of aircraft structures and lacks an exact stress intensity factor solution.

The boundary element method (BEM) is a well established numerical technique in fracture-mechanics analysis, see Aliabadi and Rooke [1]. Its formulation can be based on the classical work theorem, see Portela [2], which leads to a single boundary integral equation. The BEM has been successfully applied to linear elastic problems in domains which contain no degenerated geometries. These degeneracies, either internal or edge surfaces which include no area or volume and across which the displacement field is discontinuous, are defined as mathematical cracks. For symmetric crack problems only one side of the crack need be modelled and a single-region BEM analysis may be used. However, the solution of general crack problems cannot be achieved with the direct application of the BEM, in a single-region analysis, because the coincidence of the crack surfaces gives rise to a singular system of algebraic equations. The equations for a point located at one of the surfaces of the crack are identical to those equations for the point on the opposite surface, with the same coordinates, because the same integral equation is collocated with the same integration path, at both coincident points.

Some special techniques have been devised to overcome this difficulty. Among these, the most important are the subregions method, Blandford, Ingraffea and Liggett [3] and the dual boundary element method, Portela, Aliabadi and Rooke [4]. The subregions method introduces artificial boundaries into the body, which connect the cracks to the boundary, in such a way that the domain is divided into subregions without cracks. The main drawback of this method is that the introduction of artificial boundaries is not unique, and thus cannot be easily implemented into an automatic procedure. In addition, the method generates a larger system of algebraic equations than it is strictly required. Despite these drawbacks, the method of subregions is currently the most widely used technique for crack problems.

The dual boundary element method (DBEM) incorporates two independent boundary integral equations, with the displacement equation applied on one of the crack surfaces and the traction equation on the other. As a consequence, general mixed-mode crack problems can be solved in a single-region formulation; since, although the integration path is still the same for coincident points on the crack surfaces, the respective boundary integral equations are distinct.

The present paper is concerned with an application of the DBEM for solving a linear elastic fracture mechanics pin-loaded lug problem. The dual boundary integral equations are presented and the crack modelling strategy defined. Along the crack, analytic integration is carried out on singular elements. The stress intensity factor is introduced in terms of the J-integral. The accuracy of the present method is shown with numerical results obtained for several internal cracked geometries. Finally, mixed-mode results are presented for the pin-loaded lug problem.

The Dual Boundary Integral Equations

The dual equations, on which the DBEM is based, are the displacement and the traction boundary integral equations. In the absence of body forces and assuming continuity of the displacements at a boundary point $\mathbf{x}'$, the boundary integral representation of the displacement components u_i is given by

$$c_{ij}(\mathbf{x}')u_i(\mathbf{x}') + \int\!\!\!\!\!-_{\Gamma} T_{ij}(\mathbf{x}',\mathbf{x})\, u_j(\mathbf{x})\, d\Gamma(\mathbf{x}) = \int_{\Gamma} U_{ij}(\mathbf{x}',\mathbf{x})\, t_j(\mathbf{x})\, d\Gamma(\mathbf{x}), \qquad (1)$$

where i and j contain cartesian components; $T_{ij}(\mathbf{x}',\mathbf{x})$ and $U_{ij}(\mathbf{x}',\mathbf{x})$ represent the Kelvin traction and displacement fundamental solutions, respectively, at a boundary point $\mathbf{x}$. The distance between the points $\mathbf{x}'$ and $\mathbf{x}$ is denoted by r. The integrals in equation (1) are regular, provided $r \neq 0$. As the distance r tends to zero, the fundamental solutions exhibit singularities; they are a strong singularity of order $\frac{1}{r}$ in T_{ij} and a weak singularity of order $\ln \frac{1}{r}$ in U_{ij}. The symbol $⨍$ stands for the Cauchy principal-value integral, and the coefficient $c_{ij}(\mathbf{x}')$ is given by $\delta_{ij}/2$ for a smooth boundary at the point $\mathbf{x}'$ (δ_{ij} is the Kronecker delta).

The stress boundary integral equations were first derived by Cruse [5]. In the absence of body forces and assuming continuity of both strains and tractions at $\mathbf{x}'$ on a smooth boundary, the stress components σ_{ij} are given by

$$\frac{1}{2}\sigma_{ij}(\mathbf{x}') + \oint\!\!\!\!\!=_{\Gamma} S_{ijk}(\mathbf{x}',\mathbf{x})\, u_k(\mathbf{x})\, d\Gamma(\mathbf{x}) = ⨍_{\Gamma} D_{ijk}(\mathbf{x}',\mathbf{x})\, t_k(\mathbf{x})\, d\Gamma(\mathbf{x}). \qquad (2)$$

In this equation, $S_{ijk}(\mathbf{x}',\mathbf{x})$ and $D_{ijk}(\mathbf{x}',\mathbf{x})$ denote the derivatives of $T_{ij}(\mathbf{x}',\mathbf{x})$ and $U_{ij}(\mathbf{x}',\mathbf{x})$, respectively. The integrals in equation (2) are regular, provided $r \neq 0$. As the distance r tends to zero, S_{ijk} exhibits a hypersingularity of the order $\frac{1}{r^2}$, while D_{ijk} exhibits a strong singularity of the order $\frac{1}{r}$. The symbol $⨎$ stands for the Hadamard principal-value integral. On a smooth boundary, the traction components, t_j, are given by

$$\frac{1}{2}t_j(\mathbf{x}') + n_i(\mathbf{x}')\, ⨎_{\Gamma} S_{ijk}(\mathbf{x}',\mathbf{x})u_k(\mathbf{x})d\Gamma(\mathbf{x}) = n_i(\mathbf{x}')\, ⨍_{\Gamma} D_{ijk}(\mathbf{x}',\mathbf{x})t_k(\mathbf{x})d\Gamma(\mathbf{x}), \quad (3)$$

where n_i denotes the i component of the unit outward normal to the boundary, at the point $\mathbf{x}'$. Equations (1) and (3) constitute the basis of the DBEM, see Portela, Aliabadi and Rooke [4].

On a traction-free crack, the dual integral equations (1) and (3) are simplified; the displacement and the traction equations are given by

$$c_{ij}(\mathbf{x}')u_j(\mathbf{x}') + ⨍_{\Gamma_c} T_{ij}(\mathbf{x}',\mathbf{x})\, u_j(\mathbf{x})\, d\Gamma(\mathbf{x}) = 0 \qquad (4)$$

and

$$n_i(\mathbf{x}')\, ⨎_{\Gamma_c} S_{ijk}(\mathbf{x}',\mathbf{x})\, u_k(\mathbf{x})\, d\Gamma(\mathbf{x}) = 0, \qquad (5)$$

respectively, where Γ_c denotes the crack boundary.

Crack Modelling Strategy

For the sake of efficiency and to keep the simplicity of the standard boundary elements, the present formulation uses discontinuous quadratic elements for the crack modelling, as shown in Figure 1. The general modelling strategy, developed by Portela, Aliabadi and Rooke [4], can be summarized as follows:

- the crack boundaries are modelled with discontinuous quadratic elements, as shown in Figure 1;
- the displacement equation (4) is applied on one of the crack surfaces;
- the traction equation (5) is applied on the opposite surface;

- continuous quadratic elements are used along the remaining boundary of the body, except at an intersection between a crack and an edge, where discontinuous or semi-discontinuous elements are required on the edge in order to avoid a common node at the intersection, as shown in Figure 1.

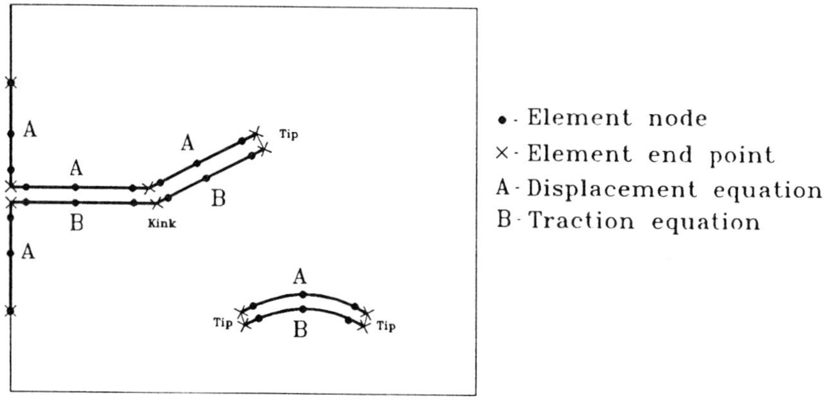

Figure 1: Crack modelling with discontinuous quadratic boundary elements.

This simple strategy is robust and allows the DBEM to effectively model general edge or embedded crack problems; crack tips, crack-edge corners and crack kinks do not require special treatment, since they are not located at nodal points where the collocation is carried out.

In many practical problems the cracks are flat; but cracks do grow along curved paths and are usually modelled as piece-wise flat cracks. For piece-wise flat cracks, all the integrals in equations (4) and (5) are most effectively carried out by direct analytic integration, see Portela, Aliabadi and Rooke [4].

The Stress Intensity Factor Evaluation

The different techniques used in the boundary element method for the evaluation of stress intensity factors are described by Aliabadi and Rooke [1]. In the present paper, the J-integral method was used to obtain the numerical results. The J-integral is an effective method for the determination of stress intensity factors, because the interior elastic field can be accurately determined, with the boundary element method. Consider a cartesian reference system, with the origin at the tip of a traction-free crack, as shown in Figure 2. Rice [6] introduced the path-independent J-integral, which is defined as the rate of energy released per unit of crack translation in the x_1 direction. In the absence of body forces it is given by:

$$J = \int_{\Gamma} (W n_1 - t_j u_{j,1}) \, d\Gamma, \tag{6}$$

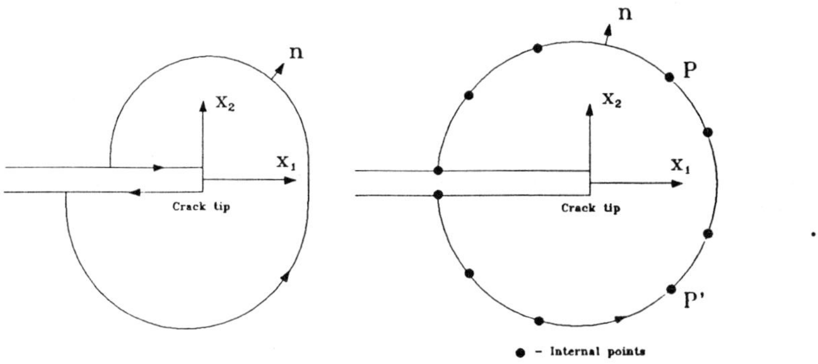

a) General contour b) Circular contour

Figure 2: Coordinate reference system and contour path for J-integral.

where Γ is an arbitrary contour surrounding the crack tip; W is the strain energy density, given by $\frac{1}{2}\sigma_{ij}\varepsilon_{ij}$, where σ_{ij} and ε_{ij} are the stress and strain tensors, respectively; t_j are the traction components, given by $\sigma_{ij}n_i$, where n_i are the components of the unit outward normal to the contour path. The relationship between the J-integral and the stress intensity factors is given by

$$J = \frac{K_I^2 + K_{II}^2}{E'}, \tag{7}$$

where the constant E' is the elasticity modulus E for plane stress conditions and $E' = E/(1-\nu^2)$ for plane strain conditions. A simple procedure, based on the decomposition of the elastic field into symmetric and antisymmetric mode components, can be used to decouple the stress intensity factors of a mixed-mode problem. The integral J is represented by the sum of two integrals as follows:

$$J = J^I + J^{II}, \tag{8}$$

where the superscripts indicate the pertinent deformation mode. For this representation to be possible, it is sufficient to introduce the following decomposition in the elastic fields:

$$\left\{\begin{array}{c} \sigma_{11}^I \\ \sigma_{22}^I \\ \sigma_{12}^I \end{array}\right\} = \frac{1}{2}\left\{\begin{array}{c} \sigma_{11} + \sigma_{11}' \\ \sigma_{22} + \sigma_{22}' \\ \sigma_{12} - \sigma_{12}' \end{array}\right\}, \quad \left\{\begin{array}{c} \sigma_{11}^{II} \\ \sigma_{22}^{II} \\ \sigma_{12}^{II} \end{array}\right\} = \frac{1}{2}\left\{\begin{array}{c} \sigma_{11} - \sigma_{11}' \\ \sigma_{22} - \sigma_{22}' \\ \sigma_{12} + \sigma_{12}' \end{array}\right\} \tag{9}$$

and

$$\left\{\begin{array}{c} u_1^I \\ u_2^I \end{array}\right\} = \frac{1}{2}\left\{\begin{array}{c} u_1 + u_1' \\ u_2 - u_2' \end{array}\right\}, \quad \left\{\begin{array}{c} u_1^{II} \\ u_2^{II} \end{array}\right\} = \frac{1}{2}\left\{\begin{array}{c} u_1 - u_1' \\ u_2 + u_2' \end{array}\right\}, \tag{10}$$

where $\sigma_{ij}'(x_1, x_2) = \sigma_{ij}(x_1, -x_2)$ and $u_i'(x_1, x_2) = u_i(x_1, -x_2)$. Equations (9) and (10) lead to the following decompositions of the elastic field:

$$\sigma_{ij} = \sigma_{ij}^{I} + \sigma_{ij}^{II} \tag{11}$$

and

$$u_i = u_i^{I} + u_i^{II}. \tag{12}$$

When equations (11) and (12) are introduced into equation (6), equation (8) is obtained, with the J-integral components given by

$$J^M = \int_{\Gamma} \left(W^M n_1 - t_j^M u_{j,1}^M \right) d\Gamma, \tag{13}$$

for $M = I$ or $M = II$. Finally, it can be shown that the following relations hold:

$$J^I = \frac{K_I^2}{E'}, \quad J^{II} = \frac{K_{II}^2}{E'}. \tag{14}$$

The implementation of this procedure into the boundary element method is straightforward. A circular contour path, around the crack tip, is defined with a set of internal points, located at symmetrical positions relative to the crack plane, as shown in Figure 2. The two contour points on the crack surfaces are the first and the last points of the path respectively. At these points, it is always verified that $n_1 = -1$ and $n_2 = 0$ and thus, for a traction-free crack, $t_2 = 0$. The integration, along the contour path, can be accomplished with the trapezoidal rule. For the sake of simplicity, only circular paths containing crack nodes were considered in this paper; each path is referred to by a path number which increases as the radius of the contour increases, as shown in Figure 3.

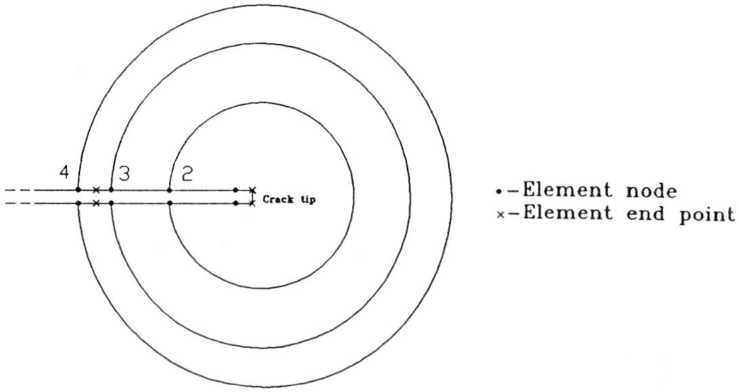

Figure 3: Circular path numbering system for J-integral contours.

Test Results

Consider the analysis of a central slant crack in a rectangular plate, represented in Figure 4. The plate is loaded with a uniform traction $\bar{t} = 1$, applied symmetrically at the ends. The ratio between the height and the width of the

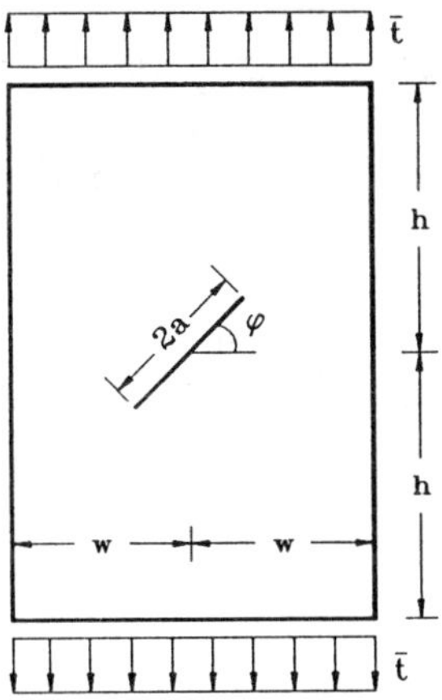

Figure 4: Rectangular plate with a central slant crack ($h/w = 2$, $\theta = 45°$).

plate is given by $h/w = 2$. The crack has the length $2a$ and makes an angle of $\theta = 45°$ with the direction of the traction. Accurate results for this problem have been obtained, see Murakami [8]. To solve this problem with the DBEM, a mesh of 36 quadratic boundary elements was set up, in which 6 discontinuous elements were used on each face of the crack, graded towards the tips, with the ratios 0.25, 0.15 and 0.1. The results obtained are presented in Tables 1 and 2.

a/w	$K_I/(\bar{t}\sqrt{\pi a})$					
	J-Integral Contour Path					Reference
	2	3	4	5	8	[8]
0.2	0.521	0.519	0.521	0.521	0.521	0.518
0.3	0.544	0.542	0.544	0.544	0.544	0.541
0.4	0.575	0.574	0.576	0.576	0.576	0.572
0.5	0.616	0.614	0.617	0.617	0.616	0.612
0.6	0.666	0.665	0.667	0.667	0.666	0.661

Table 1: Mode I stress intensity factors for a central slant crack in a rectangular plate ($h/w = 2$, $\theta = 45°$).

The J-integral computations were carried out with 30 internal points and a trapezoidal integration rule. The results obtained are very accurate; they differ from those of reference [8] by less than 1% and are insensitive to the contour path chosen.

Although discontinuous elements are used to model the crack, its deformed shape is smooth, as shown in Figure 5.

a/w	$K_{II}/(\bar{t}\sqrt{\pi a})$					
	J-Integral Contour Path					Reference
	2	3	4	5	8	[8]
0.2	0.499	0.499	0.501	0.503	0.508	0.507
0.3	0.508	0.508	0.511	0.512	0.517	0.516
0.4	0.521	0.521	0.523	0.525	0.529	0.529
0.5	0.538	0.538	0.541	0.542	0.547	0.546
0.6	0.560	0.560	0.562	0.564	0.569	0.567

Table 2: Mode II stress intensity factors for a central slant crack in a rectangular plate ($h/w = 2$, $\theta = 45°$).

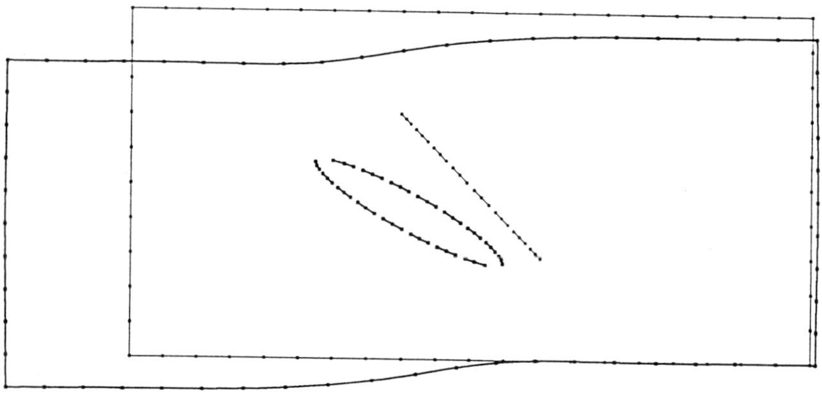

Figure 5: Initial and deformed boundary element meshes of the central slant crack problem.

Pin-Loaded Lugs

Consider the analysis of a straight-shank lug with a cracked hole, represented schematically in Figure 6. The characteristic geometric ratios are $L/W = 3$ and $R/W = 0.4$, where R, L and W are the hole radius, the shaft length and the lug half width, respectively. Five crack lengths with ratios $l/R = 0.1$, 0.3, 0.5, 0.7 and 0.9 with three crack angles of $\varphi = 0$ and ± 15 degrees were considered. Two pin loading cases of normal pressure were considered, a constant distribution of p_0 and a sine distribution of $p_0 \sin\theta$. The bearing loads were assumed to act over half of the hole circumference and symmetrically with respect to the lug symmetry axis. The lug base was constrained with rollers.

A convergence study was carried out with three different meshes (of 40, 48 and 58 boundary elements) refined at the hole boundary, where the stress concentration is high, and at the crack tip, where the stress singularity occurs. For

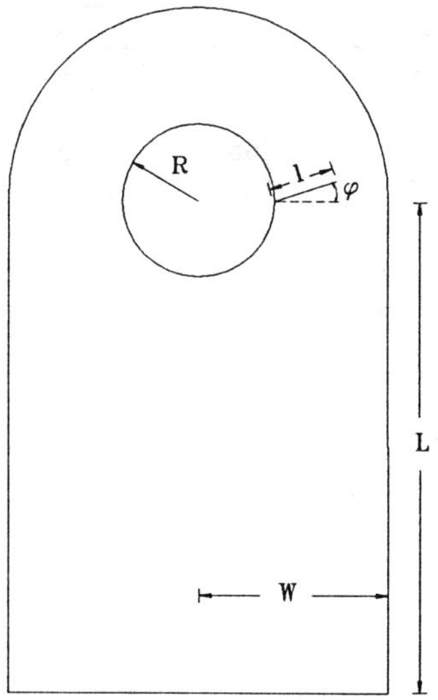

Figure 6: Straight-shank lug geometry.

the largest crack length convergence was achieved with the coarsest mesh, while for the shortest crack length the stress intensity factor solution converged with the mesh with 48 elements. In order to work with the same number of elements for all range of the crack lengths, the most refined mesh was selected for the analysis. Tables 3 and 4 and Figures 7 and 8 present the results obtained for the pin loaded lug problem.

	$K_I/(p_0\sqrt{\pi l})$			$K_{II}/(p_0\sqrt{\pi l})$		
l/R	$\varphi = +15°$	$\varphi = 0°$	$\varphi = -15°$	$\varphi = +15°$	$\varphi = 0°$	$\varphi = -15°$
0.1	1.285	1.522	1.585	0.592	0.282	0.018
0.3	1.065	1.216	1.239	0.464	0.199	0.038
0.5	0.965	1.071	1.066	0.373	0.142	0.063
0.7	0.921	1.002	0.974	0.326	0.104	0.091
0.9	0.922	0.989	0.938	0.305	0.076	0.120
1.1	0.980	1.046	0.960	0.312	0.059	0.157
1.3	1.174	1.265	1.096	0.372	0.054	0.211

Table 3: Stress intensity factors of the pin loaded lug for the uniform distribution pressure p_0.

In all cases, no more than 25 internal points were considered along the path in the J-integral calculations.

	$K_I/(\frac{\pi}{4}p_0\sqrt{\pi l})$			$K_{II}/(\frac{\pi}{4}p_0\sqrt{\pi l})$		
l/R	$\varphi = +15°$	$\varphi = 0°$	$\varphi = -15°$	$\varphi = +15°$	$\varphi = 0°$	$\varphi = -15°$
0.1	1.939	1.999	1.906	0.264	0.024	0.283
0.3	1.489	1.501	1.405	0.165	0.059	0.259
0.5	1.268	1.264	1.155	0.125	0.085	0.265
0.7	1.154	1.139	1.015	0.106	0.107	0.284
0.9	1.106	1.084	0.941	0.096	0.131	0.313
1.1	1.122	1.099	0.924	0.096	0.161	0.361
1.3	1.269	1.265	1.009	0.117	0.215	0.449

Table 4: Stress intensity factors of the pin loaded lug for the distribution pressure $p_0 \sin\theta$.

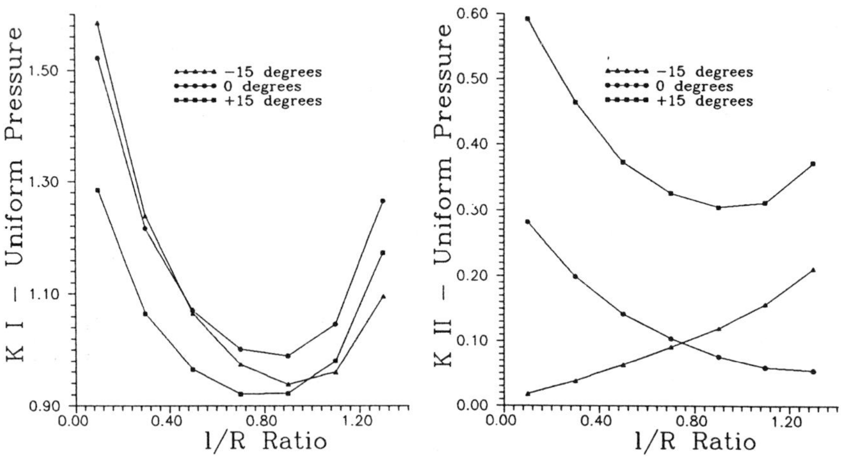

Figure 7: Stress intensity factors of the pin loaded lug for the uniform distribution pressure p_0.

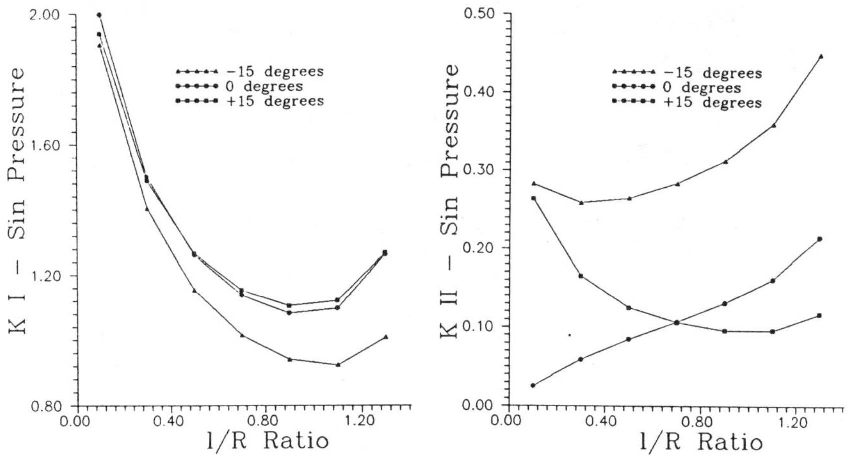

Figure 8: Stress intensity factors of the pin loaded lug for the distribution pressure $p_0 \sin\theta$.

Discussion and Conclusions

The DBEM incorporates two independent boundary integral equations; one is the displacement boundary integral equation and the other is the traction boundary integral equation. When the displacement equation is applied on one of the crack surfaces and the traction equation is applied on the other, general mixed-mode crack problems can be solved in a single-region formulation. Both crack surfaces are discretized with discontinuous quadratic boundary elements, which circumvent the problem of collocation at crack tips, crack kinks and crack-edge corners. For flat boundary elements, the direct analytic integration is the most effective method to deal with the singular integrations. At a crack node, singular integrations occur twice, once on the self-point element and again on the opposite one. This feature, associated with the use of discontinuous quadratic elements, restricts the use of the rigid body condition to evaluate indirectly the diagonal terms at crack nodes. Several test cracked geometries were analyzed with the DBEM and accurate stress intensity factors were always obtained with the J-integral method. New mixed-mode results were obtained for the pin-loaded lug problem.

References

[1] Aliabadi, M.H., Rooke, D.P., Numerical Fracture Mechanics, Computational Mechanical Publications, Southampton, (1990).

[2] Portela, A., Theoretical Basis of Boundary Solutions for Linear Theory of Structures, New Developments in Boundary Element Method, C.A.Brebbia (Editor), Proc. of the Second International Seminar on Recent Advances in Boundary Element Methods, University of Southampton, (1980).

[3] Blandford, G.E., Ingraffea, A.R., Liggett, J.A., Two-Dimensional Stress Intensity Factor Computations Using the Boundary Element Method, *Int. Journ. Num. Meth. Engng.*, 17, 387–404, (1981).

[4] Portela, A., Aliabadi, M.H., Rooke, D.P.,The Dual Boundary Element Method: Effective Implementation for Crack Problems, submited for publication in *Int. Journ. Num. Meth. Engng.*, December (1990).

[5] Cruse, T.A., Mathematical Foundations of the Boundary Integral Equation Method in Solid Mechanics, Report No. AFOSR–TR–77–1002, Pratt and Whitney Aircraft Group, (1977).

[6] Rice, J.R., A Path Independent Integral and the Approximate Analysis of Strain Concentration by Notches and Cracks, Trans. ASME, *J. Appl. Mech.*, 35, 379–386, (1968).

[7] Civelek, M.B., Erdogan, F., Crack Problems for a Rectangular Sheet and an Infinite Strip, *Int. Journ. Frac.*, 19, 139–159, (1982).

[8] Murakami, Y.,Stress Intensity Factors Handbook, Pergamon Press, (1987).

Three-Dimensional Weight Functions for Circular Cracks in Cylindrical Bars

R. Bains *, M.H. Aliabadi *, D.P. Rooke **

* *Wessex Institute of Technology, Southampton, U.K.*

** *Royal Aerospace Establishment, Farnborough, Hants, U.K.*

Abstract

The boundary element method is applied to the determination of three–dimensional weight functions for the penny–shaped crack using the Subtraction of Fundamental Fields (SFF) technique in conjunction with traction–singular, quarter–point elements. A Boussinesq–Papkovitch potential is used to formulate the fields produced by the application of a pair of collinear unit point forces acting at the crack front. At the application of this load the potential generates singularities in the stress and displacement fields of order $\rho^{-\frac{5}{2}}$ and $\rho^{-\frac{3}{2}}$ respectively, where ρ is the distance measured from the point of application of the forces on the crack front. Furthermore, along the entire crack front with the exception of the point of application of the load, there is a standard stress singularity of order $(r-a)^{-\frac{1}{2}}$, where r denotes radial distance in the crack plane and a is the radius of the penny–shaped crack.

Stress intensity factors have been obtained for a penny–shaped crack, centred on the axis of a cylindrical bar. The bar is subjected to uniform tensile loading on its ends. The penny–shaped crack in both infinite and finite domains is considered.

1 Introduction.

The weight function technique was first proposed by Bueckner [1] originally for states of plane strain. It has been extended to three dimensional fields [2, 3, 4]. The weight functions defined by Bueckner are the displacements on the boundary resulting from the application of a pair of point forces on the crack faces, near the crack front, in an elastic body in equilibrium with zero body forces and surface tractions, see Aliabadi and Rooke [5]. The displacements of this field which satisfy Navier's equation have a singularity at the crack front which normally would not be admissible as a physical field since it gives rise to unbounded strain energy of deformation at the load–point. Because of the singular nature of the displacement fields at the point of application of the pair of point forces, Bueckner refers to these fields as "fundamental fields". A fundamental field displays neither body forces nor surface tractions. In contrast any elastic fields arising from applied loadings are referred to as a regular field if the displacements are bounded everywhere and the stresses obey the usual asymtotic relations along the whole crack front.

In this paper the fields arising from the application of a pair of self–equilibrated point forces on a penny–shaped crack, in an infinite elastic solid, are formulated in terms of a Boussinesq–Papkovitch potential. Only mode I fundamental fields are considered. The theoretical formulation presented follows closely that of Sham and Zhou [6], which is based on the work of Bueckner [7]. The fields generated have stress and displacement singularities of order $\rho^{-\frac{5}{2}}$ and $\rho^{-\frac{3}{2}}$ respectively, in the vicinity of the unit point force, where ρ is the distance measured from the point of application on the crack front. Furthermore, along the entire crack front, excluding the point of application of the forces, there is a standard stress singularity of order $(r-a)^{-\frac{1}{2}}$, where r denotes radial distance, in the plane of the crack, measured from the centre of the crack of radius a.

The high–order singularities are removed from the fields using the Subtraction of Fundamental Fields technique described in previous work [8, 9] and the stress singularity of order $(r-a)^{-\frac{1}{2}}$ is modelled using standard traction–singular, quarter–point elements. The resulting weight functions, are used to evaluate stress intensity factors for cracks in loaded bars of circular cross–section. Stress intensity factors have been obtained for cracks in both infinite and finite bodies.

2 Penny–shaped crack.

2.1 Fundamental Field Formulation.

Consider an isotropic, homogeneous elastic body of infinite extent, weakened by a penny–shaped crack. Let a fixed cartesian coordinate system $(x_j,\ j=1,2,3)$ be attached to this configuration such that the coordinate origin is coincident with the centre of the crack of radius a (see figure 1). It is necessary to introduce a cylindrical coordinate system (r,θ,x_3), since this is more suitable for this problem (see figure 1). Let Ω be a crack in the plane $x_3=0$, with Ω^+ and Ω^- used to denote its faces on the upper and lower x_3–halfspace respectively and $\Omega = \Omega^+ \cup \Omega^-$. The periphery of the crack is denoted by $\mho$.

The elastic fields of mode I, generated by the presence of the two collinear unit point forces P, acting in opposite directions on Ω^+ and Ω^-, see figure 1, can be described by a Boussinesq–Papkovitch potential $G(x_1,x_2,x_3)$. Such a potential G is harmonic, i.e.,

$$\nabla^2 G = \sum_{j=1}^{3} G_{,x_j x_j} = 0 \ , \tag{1}$$

where the following convention is used,

$$f_{,\zeta} = \frac{\partial f}{\partial \zeta} \ .$$

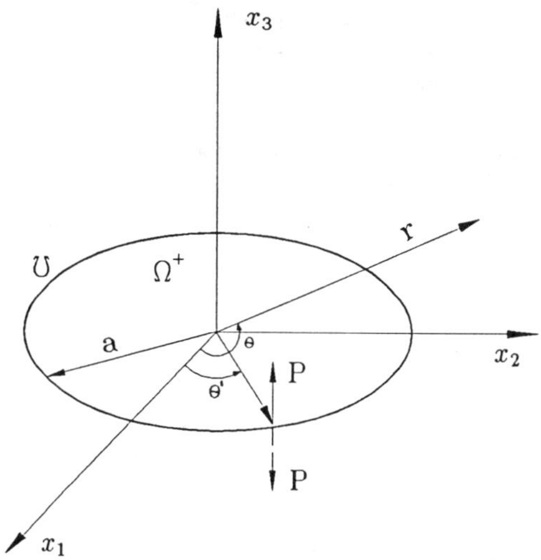

Figure 1: Coordinate systems for the penny–shaped crack configuration.

In cylindrical coordinates (r, θ, x_3), equation (1) becomes

$$\nabla^2 G = G_{,rr} + \frac{G_{,\theta\theta}}{r^2} + \frac{G_{,r}}{r} + G_{,x_3x_3} = 0 \ . \tag{2}$$

This potential gives the displacement fields as

$$u = -x_3 G_{,rx_3} - (1-2\nu)G_{,r} \ , \tag{3}$$

$$v = -x_3 \frac{G_{,\theta x_3}}{r} - (1-2\nu)\frac{G_{,\theta}}{r} \tag{4}$$

and

$$w = -x_3 G_{,x_3x_3} + 2(1-\nu)G_{,x_3} \ ; \tag{5}$$

and the stress fields as

$$\sigma_{rr} = -2\mu \left[(x_3 G_{,rr})_{,x_3} + 2\nu \left(\frac{G_{,\theta\theta}}{r^2} + \frac{G_{,r}}{r} \right) \right] \ , \tag{6}$$

$$\sigma_{\theta\theta} = -2\mu \left[\left(x_3 \frac{G_{,\theta\theta}}{r^2} + x_3 \frac{G_{,r}}{r} \right)_{,x_3} + 2\nu G_{,rr} \right] \ , \tag{7}$$

$$\sigma_{x_3x_3} = -2\mu \left[x_3 G_{,x_3x_3x_3} - G_{,x_3x_3} \right] \ , \tag{8}$$

$$\tau_{\theta x_3} = -2\mu x_3 \frac{G_{,\theta x_3 x_3}}{r} \ , \tag{9}$$

$$\tau_{rx_3} = -2\mu x_3 G_{,rx_3x_3} \tag{10}$$

and

$$\tau_{r\theta} = -2\mu \left[x_3 \left(\frac{G_{,\theta x_3}}{r} \right)_{,r} + (1-2\nu) \left(\frac{G_{,\theta}}{r} \right)_{,r} \right] \ ; \tag{11}$$

where u, v, w are the cylindrical components of the displacement vector and σ_{rr}, $\sigma_{\theta\theta}$, $\sigma_{x_3x_3}$, $\tau_{\theta x_3}$, τ_{rx_3}, $\tau_{r\theta}$ the cylindrical stress components. Material properties are denoted by the shear modulus μ and Poisson's ratio ν. There are no body forces. In order to derive the explicit form of G and its derivatives, it is necessary to introduce oblate spheroidal coordinates s, t, θ (Bueckner [7]). The oblate spheroidal coordinates are convenient for the description of circular cracks. These new coordinates are related to the cylindrical coordinates by

$$r = a \cosh s \cos t \tag{12}$$

and

$$x_3 = a \sinh s \sin t \tag{13}$$

where

$$0 \leq s < \infty \ , \quad -\frac{\pi}{2} \leq t \leq \frac{\pi}{2} \ .$$

Within this range, $s = 0$ on Ω. Also, t is positive for $x_3 > 0$ and negative for $x_3 < 0$; $t = 0$ for $x_3 = 0$ and $r > a$. Bueckner [7] derived a closed form solution for the potential G, namely

$$G = -c_o \left[\frac{1}{\sqrt{q-1}} \log \left(\frac{\sinh s - \sqrt{q-1}}{\sinh s + \sqrt{q-1}} \right) + \frac{\pi}{2} - \arctan\left(\sinh s\right) \right] \ , \tag{14}$$

subject to

$$Im\left(\sqrt{q-1}\right) \geq 0 \tag{15}$$

and

$$-\pi \leq Im(\log(....)) < 0 \ , \tag{16}$$

with

$$q = \frac{r}{a} e^{i(\theta - \theta')} \tag{17}$$

and

$$c_o = -\frac{1}{\sqrt{2}a(1-\nu)} \ ;$$

where $i = \sqrt{-1}$, n is a non negative integer and θ' is the angular location, with respect to x_1, of the point force on $\mho$. Here Im refers to the imaginary part of a complex expression. At this point it must be emphasized that the harmonic potential G represented by equation (14) involves complex variables. Only the real parts denoted by $\Re$ of the final results are implied for various real quantities such as the weight functions. By the appropriate differentiations of G and substitution into the displacement and stress equations the complete solution to the problem of two collinear unit point forces acting on a penny-shaped crack in an infinite elastic body is obtained. The required derivatives of G, for the computation of the fields can be found in [6, 9].

The fundamental fields generated by G have the property that at $\theta = \theta'$ there are singularities of order $\rho^{-\frac{5}{2}}$ and $\rho^{-\frac{3}{2}}$ in the stresses and displacements respectively, where ρ is the distance measured from the point $(a, \theta', 0)$ on the crack front, $\mho$. Furthermore, at all points on $\mho$, excluding the location of the point force (*i.e.*, $\theta = \theta'$), the field generates a standard stress singularity of order

$(r-a)^{-\frac{1}{2}}$. The displacement field is bounded at all points along $\mho$ with the exception of $\theta = \theta'$. Because of the singular nature of the fundamental field (*i.e.*, unbounded stresses and displacements) it gives rise to unbounded strain energy in the vicinity of the point $\theta = \theta'$.

2.2 Model I Stress Intensity Factor.

Bueckner [7] showed that through the application of Betti's reciprocal work theorem, the stress intensity factor $K(\theta')$ due to a regular field can be obtained from knowledge of the fundamental field. For fields in the absence of body forces the stress intensity factor can be expressed as,

$$K(\theta') = \frac{1}{2\pi\sqrt{2\pi}} \int_\Gamma \Re(\underline{u}^*)\cdot \underline{t}^+ d\Gamma \; ; \tag{18}$$

$\Re(\underline{u}^*)$ is the three-dimensional weight function and is proportional to the displacements on the boundary Γ of the elastic body, arising from the point forces at $\theta = \theta'$ on the crack front and $\underline{t}^+$ is the applied traction vector, for which the stress intensity factor is required. The weight functions $\Re(\underline{u}^*)$ are independent of the tractions $\underline{t}^+$, thus the stress intensity factor $K(\theta')$ can be obtained for any loading on Γ once $\Re(\underline{u}^*)$ is known.

3 Subtraction of Fundamental Fields (SFF).

Basically the technique involves modification of the displacement and traction fields, u_j and t_j respectively, before the problem is solved by the boundary element method (BEM). The displacement fields can be written as

$$u_j = (u_j - u_j^{\mathcal{S}}) + u_j^{\mathcal{S}} = u_j^{\mathcal{R}} + u_j^{\mathcal{S}} \tag{19}$$

and the traction fields as

$$t_j = (t_j - t_j^{\mathcal{S}}) + t_j^{\mathcal{S}} = t_j^{\mathcal{R}} + t_j^{\mathcal{S}} \;, \tag{20}$$

where $u_j^{\mathcal{S}}$ and $t_j^{\mathcal{S}}$ are the known singular fields, for a given θ', obtained from equations (3)–(11). Substitution for u_j and t_j into the three-dimensional boundary integral equation gives,

$$C_{ij}(\mathbf{x}')u_j^{\mathcal{R}}(\mathbf{x}') + \int_\Gamma T_{ij}(\mathbf{x},\mathbf{x}')u_j^{\mathcal{R}}(\mathbf{x}')d\Gamma(\mathbf{x}') = \int_\Gamma U_{ij}(\mathbf{x},\mathbf{x}')t_j^{\mathcal{R}}(\mathbf{x}')d\Gamma(\mathbf{x}'). \tag{21}$$

where $\mathbf{x}'$ and $\mathbf{x}$ are points on the surface Γ. The tensors $T_{ij}(\mathbf{x},\mathbf{x}')$ and $U_{ij}(\mathbf{x},\mathbf{x}')$ are known fundamental traction and displacement solutions for a point force in an infinite domain (i.e., Kelvin's solutions). The coefficient C_{ij} can be obtained from rigid body considerations. Only the fields $u_j^{\mathcal{R}}$ and $t_j^{\mathcal{R}}$ appear in equation (21), since $u_j^{\mathcal{S}}$ and $t_j^{\mathcal{S}}$ satisfy the governing equations of elasticity exactly. Equation (21) is now solved, subject to the following new boundary conditions:

$$\overline{u}_j^{\mathcal{R}} = \overline{u}_j - \overline{u}_j^{\mathcal{S}} \tag{22}$$

and

$$\overline{t}_j^{\mathcal{R}} = \overline{t}_j - \overline{t}_j^{S} \quad . \tag{23}$$

After substitution of the new boundary conditions the resulting system of equations can be written as

$$[A]\{\mathbf{b}\} = [C]\{\mathbf{y}\} - [C]\{\mathbf{g}\} \quad . \tag{24}$$

The vector $\{\mathbf{b}\}$ contains the unknowns $u_j^{\mathcal{R}}$ and $t_j^{\mathcal{R}}$; the vector $\{\mathbf{y}\}$ contains the original boundary conditions $\overline{u}_j$, and $\overline{t}_j$. The vector $\{\mathbf{g}\}$ contains the displacements $\overline{u}_j^S$ and the tractions $\overline{t}_j^S$, evaluated at the boundary, while matrices $[A]$ and $[C]$ contain integrals of the Kelvin fundamental solutions T_{ij} and U_{ij}. A rearrangement of equation (24) leads to the following system of equations,

$$[A]\{\mathbf{b}\} = \{\mathbf{f}\} \quad , \tag{25}$$

where

$$\{\mathbf{f}\} = [C]\{\mathbf{y}\} - [C]\{\mathbf{g}\} \quad .$$

This system of equations can be solved for a given value of θ' by the BEM code using standard techniques. Hence, the SFF technique reduces a highly singular set of equations to the more normal singular set that exist in standard crack problems. The subtraction of fundamental fields technique has been used to remove singular terms of order $\rho^{-\frac{3}{2}}$ and $\rho^{-\frac{5}{2}}$ in the displacement and traction fields respectively. The SFF technique is well suited to the removal of the high–order singularities occurring at $\theta = \theta'$. It cannot simultaneously remove the standard stress singularity of order $(r-a)^{-\frac{1}{2}}$, which exists everywhere at $\theta \neq \theta'$ along the crack front. In [8], the additional stress singularity was also encountered and modelled using standard quarter-point displacement elements and quarter-point, traction-singular elements. The coupling of quarter–point elements with the SFF technique will also be implemented in this work. For curved crack fronts that will be encountered in this work, quarter–point elements of constant length l, are placed on either side of the crack front. The quarter–point element length l has been normalised with respect to the radius of the penny–shaped crack. The stress intensity factor at $\theta = \theta'$ can be evaluated from equation (18) for any given loading.

4 Numerical Analysis.

The fundamental field formulation and the SFF technique presented have been incorporated into a standard 3–D boundary element code. To check the validity of the formulation presented, the problem of determining the stress intensity factor for an embedded penny–shaped crack in homogeneous, isotropic, linearly elastic infinite domain under the action of a uniform traction field was considered. In this state of loading the penny–shaped crack will only experience mode I deformation.

The geometry under consideration was a bar of circular cross–section, with the center of the crack lying on the axis of the bar and with the plane of the crack perpendicular to that axis. The radius of the bar will be denoted by R, its height by H. The crack is of radius a. To ensure that the penny–shaped

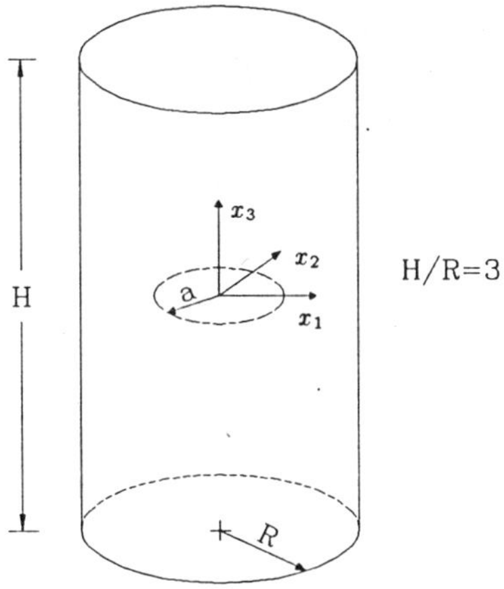

Figure 2: Geometry under analysis.

crack is in an infinite domain (i.e., no interaction between the crack front and the boundaries of the domain), $a/R = 0.1$ and $H/R = 3$ was considered. The basic configuration is shown in figure 2. Due to the symmetry of the problem, only half the geometry above the plane $x_3 = 0$, is considered. For the numerical modelling eight noded isoparametric elements are used throughout this analysis. Normalized stress intensity factors for the half symmetry are obtained according to,

$$\frac{K(\theta')}{K_o} = \frac{1}{2\pi\sigma_o\sqrt{2a}} \int_{\Gamma} \Re(\underline{u}^*)\cdot\underline{t}^+ d\Gamma \qquad (26)$$

where

$$K_o = 2\sigma_o\sqrt{\frac{a}{\pi}} \qquad (27)$$

for the case of uniform tensile loading σ_o, over the ends of the bar. All stress intensity factor results are normalized with respect to K_o. This is the stress intensity factor for an isolated ($R = \infty$) plane penny–shaped crack of radius a in a solid subjected to uniform stress σ_o at infinity. The percentage difference in the normalised stress intensity factor is defined as:

$$\%\text{Difference} = 100 \left\{ \frac{(\frac{K(\theta')}{K_o})_{(\text{KS})} - (\frac{K(\theta')}{K_o})_{(\text{SFF+TSQP})}}{(\frac{K(\theta')}{K_o})_{(\text{KS})}} \right\} \qquad (28)$$

where KS refers to known solution, [10]. The integral presented in equation (26) is evaluated numerically using Gaussian quadrature.

The boundary element mesh depicting the geometry and the fundamental field variation is shown in figure 3. The mesh on the $x_3 = 0$ plane is shown in figure 4. In the x_3–direction the mesh is graded according to the ratio 1:2:3:4:5:6, while ahead of the penny–shaped crack on the intact area the mesh is graded in the radial directions, with ratio 1:2:4:8. This mesh has 120 elements and 1086 degrees

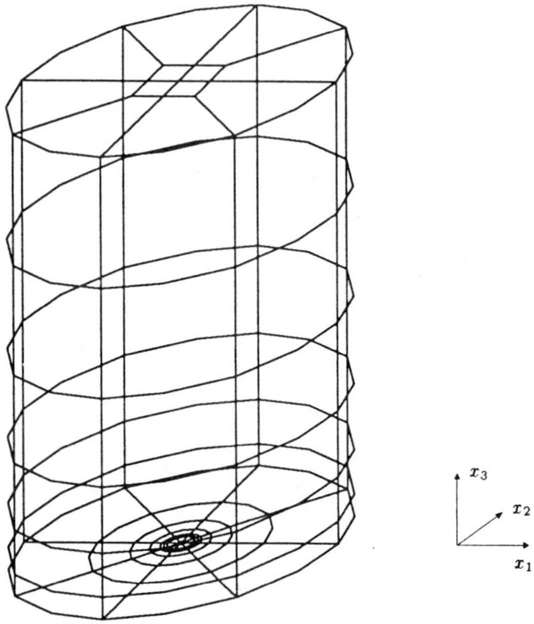

Figure 3: Boundary discretization of the geometry.

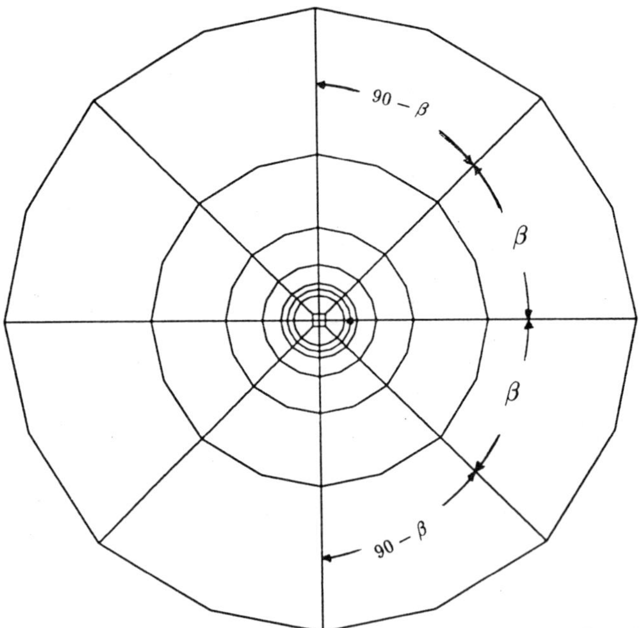

• Site of point force

Figure 4: Crack plane discretization.

of freedom. Furthermore, the mesh contains a ring of traction–singular, quarter–point elements, placed on either side of the penny-shaped crack front. These elements are only used to simulate the traction singularity of order $(r-a)^{-\frac{1}{2}}$. In [8], analysis was conducted to ascertain the dependence of the stress intensity factor on l/a, the ratio of the quarter–point element length to the crack length, for a straight crack front. In the first instance the value of $l/a = 0.2$ obtained in [8] will be used in this work also. Later results will show the dependence of the stress intensity factor on the ratio l/a, when the crack front is circular.

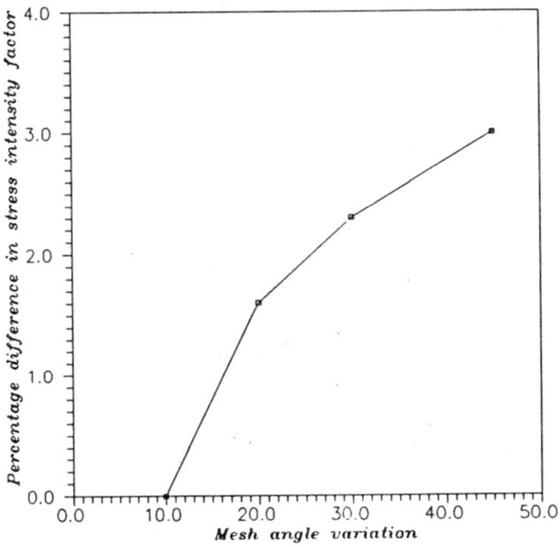

Figure 5: Stress intensity factor variation with mesh angle β.

For the mesh shown in figure 3 and at $\theta' = 0$, a value of 0.97 was obtained for the normalized stress intensity factor. This value differs from the exact theoretical solution as presented in [10] by 3%. The first radial grid lines either side of the line through the forces make angle β with that line. The accuracy of the results depended on β; decreasing the angle β (shown in figure 4) improved the accuracy. The actual dependence of the stress intensity factor on mesh angle β is shown in figure 5. It must be emphasized that all the results obtained in figure 5 have the same number of elements and degrees of freedom as the general mesh shown in figure 3. The final mesh, which is shown in figure 6, is for $\beta = 10°$. A 3% improvement in the results has been obtained by reducing the mesh angle β, from 45° to 10°.

To investigate the effect of a penny-shaped crack in a finite size bar, the radius R was decreased, still maintaining $H/R = 3$ and crack radius a, constant. For this analysis the mesh distribution on the surface $x_3 > 0$ and on the end of the bar were kept the same as that in figure 6. The following ratios of 0.1,0.33,0.53 and 0.75 for a/R were considered. Results showing the stress intensity factor variation as a function of a/R are presented in figure 7. For a/R equal to 0.1 and 0.33, both meshes contained 120 elements 1086 degrees of freedom. The other two ratios of 0.53 and 0.75 contained 104 elements and 942 degrees of freedom. The reason why the total number of elements are decreasing with increasing a/R ratio is that the ligament ahead of the crack front decreases as $R \to a$; thus if the

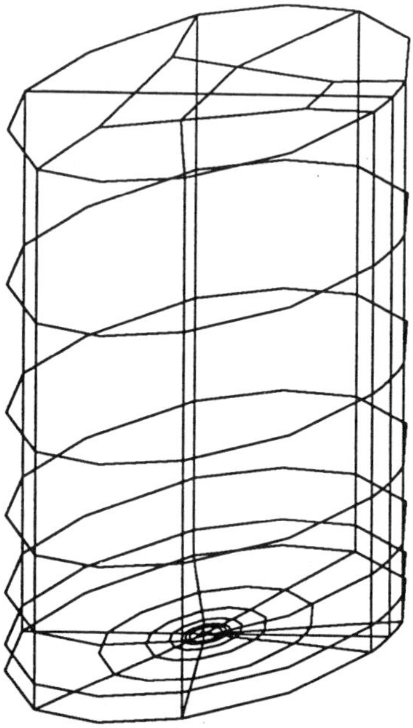

Figure 6: Mesh with $\beta = 10°$.

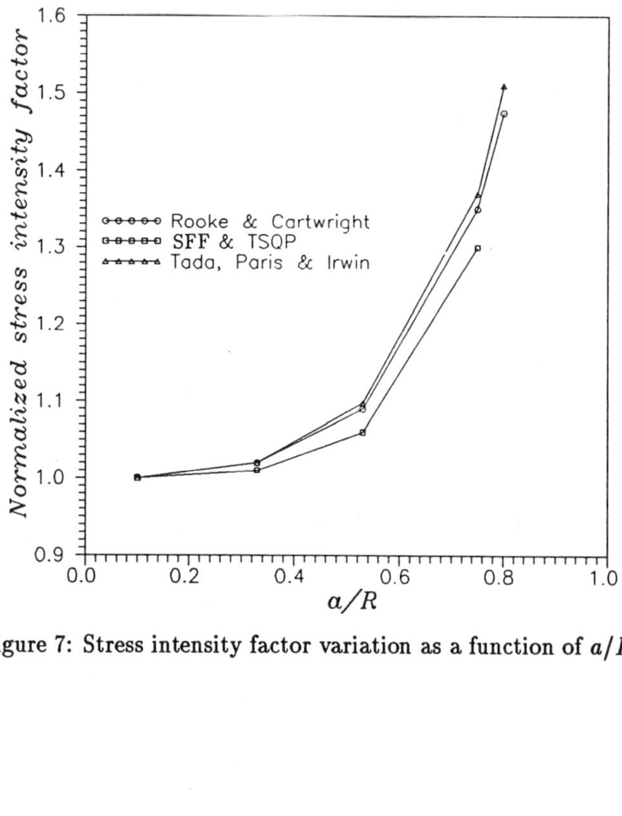

Figure 7: Stress intensity factor variation as a function of a/R

number of elements on the intact area is maintained constant, then the element length to width ratio becomes large. Furthermore, as $R \to a$, it was found that varying the discretization of the crack face produced virtually no change in the stress intensity factor. Results obtained from two independent techniques, namely integral transforms [10] and asymptotic approximation [11] are also plotted in figure 7. In [10] and [11] the authors state the accuracies of their results to

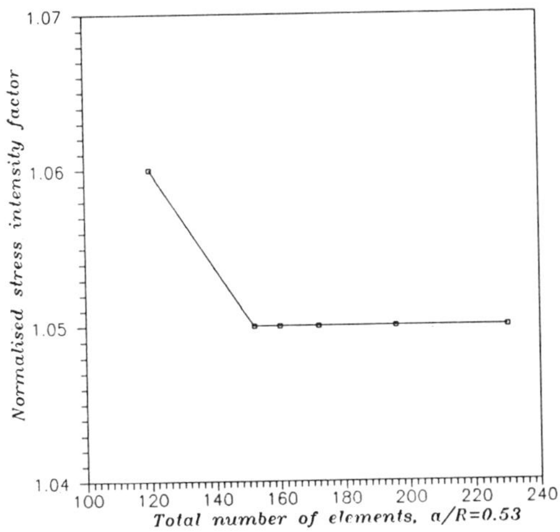

Figure 8: Mesh convergence for $a/R = 0.53$.

better than 1%. At $a/R = 0.1$, all the results presented are in agreement with each other and the exact solution. For other a/R values the results presented are within bands of 1%,4% and 5% for ratios of 0.33,0.53 and 0.75 respectively. Mesh convergence studies for $a/R = 0.53$, showing the variation of the normalised stress intensity factor with total number of elements, is presented in figure 8. This figure shows that after 148 elements the stress intensity factor attains a constant value. At a/R ratio of 0.75 and 0.8 stress intensity factor results presented by [10] and [11] differ by 1.5% and 2.6% respectively. Overall the results have the same pattern; with increasing a/R ratio the interaction between the crack front and the external surface boundaries becomes more dominant and the stress intensity factor increases.

All the results obtained so far have been generated on the assumption that the optimum value of l/a, namely 0.2, obtained from the straight crack–front analysis [8], will also give the best results, when the crack front is circular. To check the validity of this assumption, stress intensity factor values were obtained for l/a ratios equal to $0.05, 0.1, 0.2$, and 0.3, with $\theta' = 0°$, $a/R = 0.5$ and $H/R = 3$. Meshes with the first two l/a ratios consisted of 104 elements with 942 degrees of freedom, while meshes with the latter two ratios had 112 elements and 1014 degrees of freedom. With decreasing l/a ratio, extra rings of elements were added on the intact area to maintain radial element to element ratios of 1:2. Results showing the percentage difference in the normalized stress intensity factor, relative to [10], for different l/a ratios are presented in figure 9. This figure shows that over all the range of l/a considered there is only a 1% variation in the

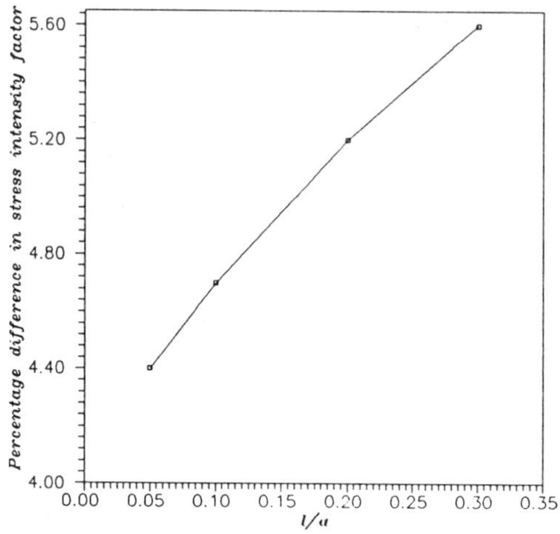

Figure 9: Stress intensity factor variation as a function of l/a

stress intensity factor. The results of figure 9 have been obtained with element distributions similar to that shown in figure 3.

5 Conclusions.

Fracture mechanics weight functions for penny-shaped cracks have been obtained efficiently, through the coupling of the subtraction of fundamental fields with traction-singular, quarter-point elements. Stress intensity factors obtained from these weight functions, for the penny-shaped crack embedded in an infinite solid, are in agreement with those presented in the literature. For cracks in finite geometries, the interaction between the crack front and the boundaries becomes important. Results presented to study this, follow the same pattern as those given in the literature but are significantly lower as the crack front approaches the boundaries. In this study stress intensity factors have been found to be insensitive to variations in the traction-singular, quarter-point element length.

6 Acknowledgement.

This work has been carried out with the support of Procurement Executive, Ministry of Defence, UK.

References

[1] Bueckner, H.F., A novel principle for the computation of stress intensity factors, *Zeitschrift fur angewandte mathematik und mechanik (ZAMM)*, Vol. 46, pp. 529–545, (1970).

[2] Bueckner, H.F., Field singularities and related integral representations, Chapter 5 in Methods of Analysis and Solutions of Crack Problems, edited by G.C. Sih, Noordhoff International Publishing, (1973).

[3] Rice, J.R., Some remarks on elastic crack tip stress fields, *Int. J. Solids Structures*, Vol. 8, pp. 751–758, (1972).

[4] Rice, J.R., Weight function theory for three-dimensional elastic crack analysis, Fracture Mechanics: Perspectives and Directions (Twentieth Symposium), ASTM STP 1020, (Edited by Wei, R.P., and Gangloff, R.P.), American Society for Testing and Materials, Philadelphia, pp. 29–57, (1989).

[5] Aliabadi, M. H., and Rooke, D. P., Numerical Fracture Mechanics. Computational Mechanics Publications and Kluwer Academic Publishers, (1991).

[6] Sham, T.-L. and Zhou, Y., Computation of three-dimensional weight functions for circular and elliptic cracks, *Int. J. Fract.*, 41, pp. 51–75, (1989).

[7] Bueckner, H.F., Weight functions and fundamental fields for the penny-shaped and half-plane crack in three-space. *Int. J. Solids Structures*, 23, No. 1, pp. 59–93, (1987).

[8] Bains, R.S., Weight functions for three-dimensional crack problems, Report No. 2, Research Agreement No. 2235/01 XR/ST, Computational Mechanics Institute, Wessex Institute of Technology, March (1990).

[9] Bains, R.S., Weight functions for the penny-shaped and elliptical cracks using SFF technique, Report No. 3, Research Agreement No. 2235/01 XR/ST, Computational Mechanics Institute, Wessex Institute of Technology, March (1991).

[10] Rooke, D.P., and Cartwright, D.J., Compendium of Stress Intensity Factors, Procurement Executive, Ministry of Defence, HMSO, (1976).

[11] Tada, H., Paris, P.C., and Irwin, G.C., The Stress Analysis of Cracks Handbook, Second Edition, Paris Production Incorporation and Del Research Corporation, (1985).

Stress Analysis of Cyclic Symmetric Structures by the New Type Boundary Integral Equation

J.F. Xie, Y.L. Wu

Institute of Mechanics, Chinese Academy of Sciences, Beijing 100080, People's Republic of China

ABSTRACT

According to the idea in Ref. [1], a new type of boundary integral equation for cyclic symmetric structures including two types of boundary condition has been deduced. Its boundary element formulation is presented. Numerical result for a rotating hollow disk is given to demonstrate the accuracy of the new type of boundary integral equation.

INTRODUCTION

The boundary element method based on the boundary integral equations has been a powerful computational tool for structure analysis [2] since 1967. It was initiated by Rizzo [3] and Cruse [4] who used the method of weighted residuals. Recently, Hu [1] proposed a new type of boundary integral equation for theory of elasticity to solve the problems of elasticity. This new type of boundary integral equation is different from that type which was given by Rizzo and Cruse. It is generated by using two-state conservation integral of elasticity. It expresses the domain stresses of elastic body with the displacements and tractions on the boundary. The boundary integral equation by the above method is of the second kind integral equation on boundary where boundary displacements are given. This result complement Rizzo type equation with each other [5]. In this paper, we applied the new type of boundary integral equation proposed in Ref. [1] into cyclic symmetric structures. The boundary element formulation and two types of boundary conditions are presented in detail. Numerical example of a rotating hollow disk is given to demonstrate the accuracy of the new type of boundary integral equation.

TWO-STATE CONSERVATION INTEGRAL INCLUDING DISTRIBUTED LOADS

To consider the plane problems of elastic body, we choose Cartesian coordinate. The components of displacements,strains,stresses, distributed loads in the region and tractions on the boundary are u_i, ε_{ij},σ_{ij},f_i and p_i respectively.

Basic equations of elasticity are:

$$\varepsilon_{ij} = \frac{1}{2}(u_{i,j} + u_{j,i}) \tag{1.a}$$

$$\sigma_{ij} = A_{ijlm}\varepsilon_{lm} \tag{1.b}$$

$$\sigma_{ij,j} + f_i = 0 \tag{1.c}$$

For an isotropic elastic material, A_{ijlm} is a constant.

The equilibrium condition when appied to the boundary B of the domain Ω yields

$$p_i = \sigma_{ij}n_j \tag{2}$$

in which n_j represents the direction cosins of the outside normal of B.

The strain energy density $U(\varepsilon)$ is

$$U(\varepsilon) = \frac{1}{2} A_{ijlm} \varepsilon_{ij} \varepsilon_{lm} \tag{3}$$

If u_i,ε_{ij},σ_{ij},f_i and p_i satisfy Eq. (1), they consitute an exact elasticity state.

Differentiating such a state with respect to x_1 and x_2 respectively, we get other two exact states:

$$\begin{array}{lll} u_i^k = u_{i,k} \quad , & \varepsilon_{ij}^k = \varepsilon_{ij,k} \quad , & \sigma_{ij}^k = \sigma_{ij,k} \\ f_i^k = f_{i,k} \quad , & p_i^k = n_j \sigma_{ij}^k = n_j \sigma_{ij,k} (k = 1, 2) & \end{array} \tag{4}$$

In here, please notice that p_i^k is gotten by using equilibrium condition, not by differentiating p_i with respect to x_1 and x_2.

Substituting an exact state and one of its generated exact states in Eq. (4) into virtual work principle, we have

$$\int_\Omega A_{ijlm} \varepsilon_{ij} \varepsilon_{lm}^k \, d\Omega = \int_\Omega f_i u_i^k \, d\Omega + \int_B p_i u_i^k \, dB \tag{5}$$

From Eq. (3) and Eq. (4), the left-hand of Eq. (5) is

$$\int_\Omega A_{ijlm} \varepsilon_{ij} \varepsilon_{lm}^k \, d\Omega = \int_\Omega \left[U(\varepsilon) \right]_{,k} \, d\Omega = \int_B n_k U(\varepsilon) \, dB \tag{6}$$

Substituting Eq. (6) into Eq. (5), we get

$$\int_\Omega f_i u_{i,k} \, d\Omega = \int_B \left[n_k U(\varepsilon) - p_i u_{i,k} \right] dB \tag{7}$$

Eq. (7) in single-state conservation integral.

Suppose $u_i^{(1)}$,$\varepsilon_{ij}^{(1)}$,$\sigma_{ij}^{(1)}$, $f_i^{(1)}$,$p_i^{(1)}$ and $u_i^{(2)}$,$\varepsilon_{ij}^{(2)}$, $\sigma_{ij}^{(2)}$,$f_i^{(2)}$,$p_i^{(2)}$ are two exact elasticity states, their linear combination

$$\begin{array}{lll} u_i^{(0)} = u_i^{(1)} + a u_i^{(2)} \quad , & \varepsilon_{ij}^{(0)} = \varepsilon_{ij}^{(1)} + a \varepsilon_{ij}^{(2)} \quad , & \sigma_{ij}^{(0)} = \sigma_{ij}^{(1)} + a \sigma_{ij}^{(2)} \\ f_i^{(0)} = f_i^{(1)} + a f_i^{(2)} \quad , & p_i^{(0)} = p_i^{(1)} + a p_i^{(2)} & \end{array} \tag{8}$$

is also an exact state, where a is an arbitrary constant. Substituting Eq. (8) into Eq. (7), we have

$$\int_\Omega f_i^{(0)} u_{i,k}^{(0)} \, d\Omega = \int_B \left[n_k U(\varepsilon^{(0)}) - p_i^{(0)} u_{i,k}^{(0)} \right] dB \tag{9}$$

Because Eq. (9) is an identity with respect to a, the coefficients of the terms linear in a on the two sides of this equation must be equal, we obtain

$$\begin{aligned} &\int_\Omega \left[f_i^{(1)} u_{i,k}^{(2)} + f_i^{(2)} u_{i,k}^{(1)} \right] d\Omega \\ &= \int_B \left[n_k A_{ijlm} \varepsilon_{ij}^{(1)} \varepsilon_{lm}^{(2)} - p_i^{(1)} u_{i,k}^{(2)} - p_i^{(2)} u_{i,k}^{(1)} \right] dB \end{aligned} \tag{10}$$

This equation is an identity satisfied by two exact elasticity states. It is called the two-state conservation integral.

BOUNDARY INTEGRAL OF DOMAIN DISPLACEMENTS AND STRESSES BASED ON THE TWO-STATE CONSERVATION INTEGRAL

For plane strain applications where the fundamental solution corresponds to unit point loads applied within an infinite plane (Kelvin), the components of the fundamental displacements and tractions are given by

$$u_i^h = \frac{1}{8\pi G(1-\nu)}[(3-4\nu)\ln\frac{1}{r}\delta_{ih} + r_{,i}r_{,h}] \tag{11}$$

$$p_i^h = -\frac{1}{4\pi(1-\nu)r}\left\{\frac{\partial r}{\partial n}[(1-2\nu)\delta_{ih} + 2r_{,i}r_{,h}] - (1-2\nu)[r_{,i}n_h - r_{,h}n_i]\right\} \tag{12}$$

where $r = r(\xi, x)$ represents the distance between the load point ξ and the field point x and its derivatives are taken with reference to the coordiates.Applying Betti's reciprocal theorem of work to an unknown elasticity state and the fundamental solution, we have

$$\int_\Omega f_i^h u_i \, d\Omega + \int_B p_i^h u_i \, dB = \int_\Omega f_i u_i^h \, d\Omega + \int_B p_i u_i^h \, dB \tag{13}$$

So

$$u_h(\xi) = \int_B (p_i u_i^h - p_i^h u_i) \, dB + \int_\Omega f_i u_i^h \, d\Omega \tag{14}$$

in which ξ is a point in the domain Ω, and the domain integral of body forces can be suitably transformed into a surface integral.

Eq. (14) is the expression of boundary integral of displacements. In order to obtain the expressions of domain stresses, we first have to get the derivative of domain displacements. We choose two states of the conservation integral as the outstanding state and fundamental solution respectively.

$$\int_\Omega (f_i u_{i,k}^h + f_i^h u_{i,k}) \, d\Omega = \int_B (n_k A_{ijlm} \varepsilon_{ij} \varepsilon_{lm}^h - p_i u_{i,k}^h - p_i^h u_{i,k}) \, dB \tag{15}$$

$$u_{h,k}(\xi) = \int_B (n_k A_{ijlm} \varepsilon_{ij} \varepsilon_{lm}^h - p_i u_{i,k}^h - p_i^h u_{i,k}) \, dB - \int_\Omega f_i u_{i,k}^h \, d\Omega \tag{16}$$

Hereby we obtain the expressions of domain displacements derivative represented by boundary integral. Eq. (16) is the main formula of the new type of boundary integral equation.

For the plane problems of elasticity, $f_i = 0$ is assumed, we have further arrangement to Eq. (16),

$$u_{h,1}(\xi) = \int_B (\sigma_{12}^h u_{1,s} + \sigma_{22}^h u_{2,s} - p_i u_{i,1}^h) \, dB \tag{17}$$

$$u_{h,2}(\xi) = \int_B (-\sigma_{11}^h u_{1,s} - \sigma_{12}^h u_{2,s} - p_i u_{i,2}^h) \, dB \tag{18}$$

Substituting Eq. (17) and Eq. (18) into the constitutive equations of elasticity, we have the expressions of boundary integral of domain stress. For the plane strain problems of isotropic elastic body, the domain stresses are

$$\sigma_{11}(\xi) = \frac{(1-\nu)E}{(1+\nu)(1-2\nu)} \int_B [u_{1,s}(\sigma^1_{12} - \frac{\nu}{1-\nu}\sigma^2_{11}) + u_{2,s}(\sigma^1_{22} - \frac{\nu}{1-\nu}\sigma^2_{12})$$
$$- p_i(u^1_{i,1} + \frac{\nu}{1-\nu}u^2_{i,2})]\, dB \tag{19}$$

$$\sigma_{22}(\xi) = \frac{(1-\nu)E}{(1+\nu)(1-2\nu)} \int_B [u_{1,s}(\sigma^2_{11} - \frac{\nu}{1-\nu}\sigma^1_{12}) + u_{2,s}(\sigma^2_{12} - \frac{\nu}{1-\nu}\sigma^1_{22})$$
$$- p_i(u^2_{i,2} + \frac{\nu}{1-\nu}u^1_{i,1})]\, dB \tag{20}$$

$$\sigma_{12}(\xi) = \frac{E}{4(1+\nu)} \int_B [u_{1,s}(\sigma^2_{12} - \sigma^1_{11}) + u_{2,s}(\sigma^2_{22} - \sigma^1_{12})$$
$$- p_i(u^2_{i,1} + u^1_{i,2})]\, dB \tag{21}$$

in which E is the modulus of elasticity, ν is the Poisson ratio.

An important remark is now due : all the expressions presented here are assumed valid for plain strain problems.

BOUNDARY INTEGRAL EQUATION OF ISOTROPIC ELASTIC BODY

When the point $\xi \in \Omega$ in Eq. (17) and Eq. (18) approaches boundary B, the boundary integral in these formulae become singular at ξ. So we need to expand the formulae to the boundary.

Suppose (ξ_1, ξ_2) is a point on the boundary B, where the boundary curve has continuous tangent. Let B_ε be a half circle with radius ε, centred at (ξ_1, ξ_2). Let Ω_ε be the half curve region surrounded by B_ε, as shown in Fig. 1. Apply Eq. (10) on Ω_ε with one outstanding state and another being the fundamental solution of elasticity body. Since in Ω_ε, $f_i^{(1)} = f_i^{(2)} = 0$, so

$$\int_\Omega \left[f_i u^h_{i,k} + f^h_i u_{i,k}\right] d\Omega + \int_{\Omega_\varepsilon} 0\, d\Omega$$
$$= \int_{B-B_\varepsilon} \left[n_k A_{ijlm}\varepsilon_{ij}\varepsilon^h_{lm} - p_i u^h_{i,k} - p^h_i u_{i,k}\right] dB$$
$$+ \int_{B_\varepsilon} \left[n_k A_{ijlm}\varepsilon_{ij}\varepsilon^h_{lm} - p_i u^h_{i,k} - p^h_i u_{i,k}\right] dB \tag{22}$$

On the boundary, substituting Eq. (11) into the integral in Eq. (22) and letting $\varepsilon \to 0$, we have

$$\frac{1}{2}u_{h,k}(\xi) = \int_{\dot{B}} (n_k A_{ijlm}\varepsilon_{ij}\varepsilon^h_{lm} - p_i u^h_{i,k} - p^h_i u_{i,k})\, dB \tag{23}$$

where $\dot{B}$ represents the integral in the sense of Cauchy principal value.
Which can be further written

$$\frac{1}{2}u_{h,1}(\xi) = \int_{\dot{B}} (\sigma^h_{12}u_{1,s} + \sigma^h_{22}u_{2,s} - p_i u^h_{i,1})\, dB \tag{24}$$

and

$$\frac{1}{2}u_{h,2}(\xi) = \int_{\dot{B}} (-\sigma_{11}^h u_{1,s} - \sigma_{12}^h u_{2,s} - p_i u_{i,2}^h)\, dB \tag{25}$$

Substituting Eq. (24) and Eq. (25) into the constitutive equations, we obtain

$$\begin{aligned}\frac{1}{2}\sigma_{11}(\xi) = \int_{\dot{B}} \{&\frac{(1-\nu)E}{(1+\nu)(1-2\nu)}[u_{1,s}(\sigma_{12}^1 - \frac{\nu}{1-\nu}\sigma_{11}^2)\\ &+ u_{2,s}(\sigma_{22}^1 - \frac{\nu}{1-\nu}\sigma_{12}^2)] - p_i\sigma_{11}^i\}\, dB\end{aligned} \tag{26}$$

$$\begin{aligned}\frac{1}{2}\sigma_{22}(\xi) = \int_{\dot{B}} \{&\frac{(1-\nu)E}{(1+\nu)(1-2\nu)}[u_{1,s}(\sigma_{11}^2 - \frac{\nu}{1-\nu}\sigma_{12}^1)\\ &+ u_{2,s}(\sigma_{12}^2 - \frac{\nu}{1-\nu}\sigma_{22}^1)] - p_i\sigma_{22}^i\}\, dB\end{aligned} \tag{27}$$

$$\frac{1}{2}\sigma_{12}(\xi) = \int_{\dot{B}} \{\frac{E}{4(1+\nu)}[u_{1,s}(\sigma_{12}^2 - \sigma_{11}^1) + u_{2,s}(\sigma_{22}^2 - \sigma_{12}^1)] - p_i\sigma_{12}^i\}\, dB \tag{28}$$

Let n_1, n_2 are the direction cosines of the outward normal of the boundary curve on (ξ_1, ξ_2). Applying boundary condition, we have

$$\begin{aligned}\frac{1}{2}p_1(\xi) &= n_1 \cdot \frac{1}{2}\sigma_{11}(\xi) + n_2 \cdot \frac{1}{2}\sigma_{12}(\xi)\\ &= \int_{\dot{B}} \{u_{1,s}[n_1\frac{(1-\nu)E}{(1+\nu)(1-2\nu)}(\sigma_{12}^1 - \frac{\nu}{1-\nu}\sigma_{11}^2) + n_2\frac{E}{4(1+\nu)}(\sigma_{12}^2 - \sigma_{11}^1)]\\ &+ u_{2,s}[n_1\frac{(1-\nu)E}{(1+\nu)(1-2\nu)}(\sigma_{22}^1 - \frac{\nu}{1-\nu}\sigma_{12}^2) + n_2\frac{E}{4(1+\nu)}(\sigma_{22}^2 - \sigma_{12}^1)]\\ &- p_i p_1^i\}\, dB\end{aligned} \tag{29}$$

$$\begin{aligned}\frac{1}{2}p_2(\xi) &= n_1 \cdot \frac{1}{2}\sigma_{12}(\xi) + n_2 \cdot \frac{1}{2}\sigma_{22}(\xi)\\ &= \int_{\dot{B}} \{u_{1,s}[n_1\frac{E}{4(1+\nu)}(\sigma_{12}^2 - \sigma_{11}^1) + n_2\frac{(1-\nu)E}{(1+\nu)(1-2\nu)}(\sigma_{11}^2 - \frac{\nu}{1-\nu}\sigma_{12}^1)]\\ &+ u_{2,s}[n_1\frac{E}{4(1+\nu)}(\sigma_{22}^2 - \sigma_{12}^1) + n_2\frac{(1-\nu)E}{(1+\nu)(1-2\nu)}(\sigma_{12}^2 - \frac{\nu}{1-\nu}\sigma_{22}^1)]\\ &- p_i p_2^i\}\, dB\end{aligned} \tag{30}$$

where

$$p_1^i = n_1\sigma_{11}^i + n_2\sigma_{12}^i \quad , \quad p_2^i = n_1\sigma_{12}^i + n_2\sigma_{22}^i \tag{31}$$

Eq. (29) and Eq. (30) are the final form of the new type of boundary integral equations in plane elasticity.

BOUNDARY DISPLACEMENT DERIVATIVE

In this section, we discuss how to express $u_{1,s}$ and $u_{2,s}$ in Eq. (29) and Eq. (30) by using u_i and p_i on boundary.

Boundary tractions are

$$p_1 = n_1\sigma_{11} + n_2\sigma_{12} \quad , \quad p_2 = n_1\sigma_{12} + n_2\sigma_{22} \tag{32}$$

For the plain strain problems of elasticity, the stresses are

$$\sigma_{11} = \frac{(1-\nu)E}{(1+\nu)(1-2\nu)}(u_{1,1} + \frac{\nu}{1-\nu}u_{2,2}) \tag{33.a}$$

$$\sigma_{22} = \frac{(1-\nu)E}{(1+\nu)(1-2\nu)}(u_{2,2} + \frac{\nu}{1-\nu}u_{1,1}) \tag{33.b}$$

$$\sigma_{12} = \frac{E}{2(1+\nu)}(u_{1,2} + u_{2,1}) \tag{33.c}$$

Since the boundary is divided into several elements. In each element we have (Fig. 2)

$$\frac{\partial u_i}{\partial \eta_k} = \frac{\partial u_i}{\partial x_m} \cdot \frac{\partial x_m}{\partial \eta_k} \tag{34}$$

where η_k is the local coordinate on one element.

To linear element, displacement u_i and coodinate x_m of arbitray point on boundary can be represented as

$$u_i = \frac{1}{2}(u_i^{j+1} + u_i^j) + \frac{1}{2}(u_i^{j+1} - u_i^j)\eta_2 \tag{35.a}$$

$$x_m = \frac{1}{2}(x_m^{j+1} + x_m^j) + \frac{1}{2}(x_m^{j+1} - x_m^j)\eta_2 \tag{35.b}$$

respectively where j and $j+1$ are the nodal numbers of element.

Since Eq. (35) has nothing to do with local coordinate η_1, there are

$$\frac{\partial u_i}{\partial \eta_2} = \frac{1}{2}(u_i^{j+1} - u_i^j) \quad , \quad \frac{\partial x_m}{\partial \eta_2} = \frac{1}{2}(x_m^{j+1} - x_m^j) \tag{36}$$

The displacement derivative can be represented as

$$u_{1,1}(x_1^{j+1} - x_1^j) + u_{1,2}(x_2^{j+1} - x_2^j) = u_1^{j+1} - u_1^j \tag{37.a}$$

$$u_{2,1}(x_1^{j+1} - x_1^j) + u_{2,2}(x_2^{j+1} - x_2^j) = u_2^{j+1} - u_2^j \tag{37.b}$$

Solving simultaneous equations Eq. (32) and Eq. (37), substituting into prescribed conditions, we will obtain the displacement derivative $u_{i,m}$.

Since

$$u_{1,s} = n_1u_{1,1} + n_2u_{1,2} \quad , \quad u_{2,s} = n_1u_{2,1} + n_2u_{2,2} \tag{38}$$

So we can get $u_{1,s}$ and $u_{2,s}$ represented by using boundary displacement u_i and traction p_i.

ROTATING BODY FORCE AND THE BOUNDARY INTEGRAL EQUATIONS INCLUDING BODY FORCES

When $f_i \neq 0$, the boundary integral equation has one more item than the equation which f_i is equal to zero. This item is the body integral of body forces.

$$B_{hk} = \int_\Omega f_i u_{i,k}^h \, d\Omega \tag{39}$$

Because B_{hk} is domain integral, the calculation of B_{hk} requires the domain to be divided into internal cells. It would lose the advantage of boundary element method.

For rotational problem, the body force is centrifugal load. If the axis of rotation passes through the origin of the coordinate system, the problem is equivalent to a prescribed body force of the form

$$f_i = \rho\omega^2 x_i \tag{40}$$

in which ω is the angular velocity, ρ is the material density.

Let us call G_{ij} the Garlerkin tensor which is related to the fundamental solution u_i^h by the following expression :

$$u_i^h = G_{ih,jj} - \frac{G_{ij,jh}}{2(1-\nu)}$$

For two-dimension plane strain state

$$G_{ih} = \frac{1}{8\pi G} r^2 \ln\frac{1}{r}\delta_{ih}$$

So

$$u_i^h = \frac{1}{8\pi G(1-\nu)}\left[(3-4\nu)\ln\frac{1}{r}\delta_{ih} - \frac{7-8\nu}{2}\delta_{ih} + r_{,i}r_{,h}\right] \tag{41}$$

the difference of Eq. (41) and Eq. (11) simply corresponds to a rigid-body translation.

The above expressions Eq. (40) and Eq. (41) when be substituted into Eq. (39) lead to the following boundary integral

$$\begin{aligned} B_{hk} &= \int_\Omega \rho\omega^2 x_i u_{i,k}^h \, d\Omega = \rho\omega^2 \int_\Omega \left\{ [x_i u_i^h]_{,k} - u_k^h \right\} d\Omega \\ &= \frac{\rho\omega^2}{8\pi G(1-\nu)} \int_B \{x_i n_k[(3-4\nu)ln\frac{1}{r}\delta_{ih} - \frac{7-8\nu}{2}\delta_{ih} + r_{,i}r_{,h}] \\ &- (1-\nu)(2rln\frac{1}{r} - r)[\delta_{kh}r_{,j}n_j - \frac{1}{2(1-\nu)}r_{,h}n_k]\}\, dB \end{aligned} \tag{42}$$

Therefore, the displacement derivatives are

$$u_{h,1}(\xi) = \int_B (\sigma_{12}^h u_{1,s} + \sigma_{22}^h u_{2,s} - p_i u_{i,1}^h)\, dB - B_{h1} \tag{43.a}$$

$$u_{h,2}(\xi) = \int_B (-\sigma_{11}^h u_{1,s} - \sigma_{12}^h u_{2,s} - p_i u_{i,2}^h)\, dB - B_{h2} \tag{43.b}$$

When the point ξ approaches boundary B,

$$u_{h,k}(\xi) \rightarrow \frac{1}{2} u_{h,k}(\xi)$$

We can prove that the integral of B_{hk} on B_ϵ is also zero. But the integral limit is changed to $\dot{B}$. Now the boundary integral including body forces is

$$\frac{1}{2} u_{h,1}(\xi) = \int_{\dot{B}} (\sigma_{12} u_{1,s}^h + \sigma_{22} u_{2,s}^h - p_i^h u_{i,1})\, dB - B_{h1} \tag{44.a}$$

$$\frac{1}{2}u_{h,2}(\xi) = \int_{\dot{B}} (-\sigma_{11}u^h_{1,s} - \sigma_{12}u^h_{2,s} - p^h_i u_{i,2})\, dB - B_{h2} \tag{44.b}$$

Substituting Eq. (44) into the constitution equations, we have the boundary integral of stresses :

$$\begin{aligned}
&\frac{1}{2}\sigma_{11}(\xi) \\
&= \int_{\dot{B}} \{\frac{(1-\nu)E}{(1+\nu)(1-2\nu)}[u_{1,s}(\sigma^1_{12} - \frac{\nu}{1-\nu}\sigma^2_{11}) + u_{2,s}(\sigma^1_{22} - \frac{\nu}{1-\nu}\sigma^2_{12})] \\
&- p_i\sigma^i_{11} - \frac{(1-\nu)E}{(1+\nu)(1-2\nu)}(B_{11} + \frac{\nu}{1-\nu}B_{22})\}\, dB
\end{aligned} \tag{45}$$

$$\begin{aligned}
&\frac{1}{2}\sigma_{22}(\xi) \\
&= \int_{\dot{B}} \{\frac{(1-\nu)E}{(1+\nu)(1-2\nu)}[u_{1,s}(\sigma^2_{11} - \frac{\nu}{1-\nu}\sigma^1_{12}) + u_{2,s}(\sigma^2_{12} - \frac{\nu}{1-\nu}\sigma^1_{22})] \\
&- p_i\sigma^i_{22} - \frac{(1-\nu)E}{(1+\nu)(1-2\nu)}(B_{22} + \frac{\nu}{1-\nu}B_{11})\}\, dB
\end{aligned} \tag{46}$$

$$\begin{aligned}
\frac{1}{2}\sigma_{12}(\xi) = \int_{\dot{B}} \{&\frac{E}{4(1+\nu)}[u_{1,s}(\sigma^2_{12} - \sigma^1_{11}) + u_{2,s}(\sigma^2_{22} - \sigma^1_{12})] \\
&- p_i\sigma^i_{12} - \frac{E}{4(1+\nu)}(B_{12} + B_{21})\}\, dB
\end{aligned} \tag{47}$$

So

$$\begin{aligned}
\frac{1}{2}p_1(\xi) &= n_1 \cdot \frac{1}{2}\sigma_{11}(\xi) + n_2 \cdot \frac{1}{2}\sigma_{12}(\xi) \\
&= \int_{\dot{B}} \{u_{1,s}[n_1\frac{(1-\nu)E}{(1+\nu)(1-2\nu)}(\sigma^1_{12} - \frac{\nu}{1-\nu}\sigma^2_{11}) + n_2\frac{E}{4(1+\nu)}(\sigma^2_{12} - \sigma^1_{11})] \\
&+ u_{2,s}[n_1\frac{(1-\nu)E}{(1+\nu)(1-2\nu)}(\sigma^1_{22} - \frac{\nu}{1-\nu}\sigma^2_{12}) + n_2\frac{E}{4(1+\nu)}(\sigma^2_{22} - \sigma^1_{12})] - p_i p^i_1 \\
&- [n_1\frac{(1-\nu)E}{(1+\nu)(1-2\nu)}(B_{11} + \frac{\nu}{1-\nu}B_{22}) + n_2\frac{E}{4(1+\nu)}(B_{12} + B_{21})]\}\, dB
\end{aligned} \tag{48}$$

$$\begin{aligned}
\frac{1}{2}p_2(\xi) &= n_1 \cdot \frac{1}{2}\sigma_{12}(\xi) + n_2 \cdot \frac{1}{2}\sigma_{22}(\xi) \\
&= \int_{\dot{B}} \{u_{1,s}[n_1\frac{E}{4(1+\nu)}(\sigma^2_{12} - \sigma^1_{11}) + n_2\frac{(1-\nu)E}{(1+\nu)(1-2\nu)}(\sigma^2_{11} - \frac{\nu}{1-\nu}\sigma^1_{12})] \\
&+ u_{2,s}[n_1\frac{E}{4(1+\nu)}(\sigma^2_{22} - \sigma^1_{12}) + n_2\frac{(1-\nu)E}{(1+\nu)(1-2\nu)}(\sigma^2_{12} - \frac{\nu}{1-\nu}\sigma^1_{22})] - p_i p^i_2 \\
&- [n_1\frac{E}{4(1+\nu)}(B_{12} + B_{21}) + n_2\frac{(1-\nu)E}{(1+\nu)(1-2\nu)}(B_{22} + \frac{\nu}{1-\nu}B_{11})]\}\, dB
\end{aligned} \tag{49}$$

The above two equations (48) and (49) are the boundary integral equations including rotational loads, in which $u_{1,s}$ and $u_{2,s}$ consist of boundary components u_i and p_i, σ_{ij}^h can be deduced from the fundamental solutions of elasticity. Dividing whole boundary into several elements and applying Eq. (48) and Eq. (49) on each element, we can deduce matrix equations including boundary displacements and tractions. Substituting the prescribed boundary conditions, seprating the known and unknown varibles on each side of the equations, solving the equtions, we can have the unknown boundary displacements and tractions. Once that all the nodal values of boundary displacements and tractions are calculated, the values of displacements and stresses can be computed at any internal points of domain by simply using Eq. (14) and Eq. (19) to Eq. (21) respectively. So the whole problem can be solved.

TWO TYPES OF BOUNDARY CONDITIONS

When analysing a body whose geometry is the same from one sector to another around a given axis, and when loading is symmetric force such as centrifugal force or torsion, it is possible to restrain the analysis of the behaviour of the total structure to the only one sector. The displacements and stresses of the substructure are symmetric or anti-symmetric to the given axis of the structure to the other substrucetures. We call this cyclic symmetric problem. For example, when a disk is forced by certrifugal force, the displacements and stresses are symmetric to the rotational axis, it is only to calculate one sector of the disk to obtain the displacemetnts and stresses of the whole disk. We use the inclined boundary conditions to solve the problem. When torsion affects the disk, the displacements and stresses are anti-symmetric to the center axis of the body. We calculate this disk by using the equal boundary conditions.

Now we deal with the two types of boundary conditions respectively.

Firstly, we consider the inclined boundary conditions. If we choose one sector of the whole structure to calculate, we have to give the boundary conditions of this sector. It is easy to give out the inclined boundary conditions under the polar coordinates (Fig. 3):

$$u_n^i = 0, \quad u_n^j = 0 \tag{50}$$

which i, j are two points on the boundary of the sector and $r_i = r_j$.

Since we always set up the boundary element equations on Cartesian coordinate system, we need a coordinate transformation:

$$\begin{Bmatrix} u_1 \\ u_2 \end{Bmatrix} = \begin{bmatrix} \cos\theta & -\sin\theta \\ \sin\theta & \cos\theta \end{bmatrix} \begin{Bmatrix} u_t \\ u_n \end{Bmatrix} \tag{51}$$

which u_1, u_2 are displacements on Cartesian coordinate, u_t, u_n are tangential and normal displacements of the point i respectively, θ is the angle of the axis x_1 to the tangent of the point i. Since u_1 and u_2 are dependent now, we must use u_t and u_n as the independent variables. After we substitute $u_n = 0$ into the equtions, we can solve this problem.

Secondly, we discuss the equal boundary conditions. The equal boundary conditions mean that the components of the displacements and tractions on the boundary B_1 and B_2 of one sector of the structure under the polar coordinate system are equal respectively. If i is a point on B_1 and j is a point on B_2, when $r_i = r_j$,

$$u_r^i = u_r^j \quad , \quad u_\theta^i = u_\theta^j \quad , \quad p_r^i = p_r^j \quad , \quad p_\theta^i = p_\theta^j \tag{52}$$

Since we use the tangential and normal components of the displacements and stresses on the local coordinate, we should calculate the transformation of the two types.

For point i

$$u_\theta^i = -u_n^i \quad , \quad u_r^i = u_t^i \tag{53.a}$$

For point j

$$u_\theta^j = u_n^j \quad , \quad u_r^j = -u_t^j \tag{53.b}$$

Since

$$u_r^i = u_r^j \quad , \quad u_\theta^i = u_\theta^j$$

So

$$u_t^j = -u_t^i \quad , \quad u_n^j = -u_n^i \tag{54}$$

We can also get the similar relationship for the tractions.

Now we dispose the above boundary conditions by reducing the coefficient matrices. Supposing for each point on the boundary, the boundary element equations can be expressed as

$$\begin{bmatrix} H_{11} & H_{12} \\ H_{21} & H_{22} \end{bmatrix} \begin{Bmatrix} u_1 \\ u_2 \end{Bmatrix} = \begin{bmatrix} G_{11} & G_{12} \\ G_{21} & G_{22} \end{bmatrix} \begin{Bmatrix} p_1 \\ p_2 \end{Bmatrix} \tag{55}$$

in which u_1, u_2 and p_1, p_2 are components of displacements and tractions respectively. H_{lk} and G_{lk} are the coefficient matrices.

For point i:

$$\begin{bmatrix} H_{11} & H_{12} \\ H_{21} & H_{22} \end{bmatrix}^i \begin{Bmatrix} u_1 \\ u_2 \end{Bmatrix}^i = \begin{bmatrix} G_{11} & G_{12} \\ G_{21} & G_{22} \end{bmatrix}^i \begin{Bmatrix} p_1 \\ p_2 \end{Bmatrix}^i \tag{56.a}$$

For point j:

$$\begin{bmatrix} H_{11} & H_{12} \\ H_{21} & H_{22} \end{bmatrix}^j \begin{Bmatrix} u_1 \\ u_2 \end{Bmatrix}^j = \begin{bmatrix} G_{11} & G_{12} \\ G_{21} & G_{22} \end{bmatrix}^j \begin{Bmatrix} p_1 \\ p_2 \end{Bmatrix}^j \tag{56.b}$$

Since

$$\begin{Bmatrix} u_1 \\ u_2 \end{Bmatrix}^i = \begin{bmatrix} \cos\theta_i & -\sin\theta_i \\ \sin\theta_i & \cos\theta_i \end{bmatrix} \begin{Bmatrix} u_t \\ u_n \end{Bmatrix}^i$$

and

$$\begin{Bmatrix} u_1 \\ u_2 \end{Bmatrix}^j = \begin{bmatrix} \cos\theta_j & -\sin\theta_j \\ \sin\theta_j & \cos\theta_j \end{bmatrix} \begin{Bmatrix} u_t \\ u_n \end{Bmatrix}^j$$

So

$$\begin{Bmatrix} u_1 \\ u_2 \end{Bmatrix}^j = \begin{bmatrix} \cos\theta_j & -\sin\theta_j \\ \sin\theta_j & \cos\theta_j \end{bmatrix} \begin{bmatrix} -1 & 0 \\ 0 & -1 \end{bmatrix} \begin{Bmatrix} u_t \\ u_n \end{Bmatrix}^i \tag{57}$$

Substituting H_{lk} and G_{lk} of point j into those of point i respectively, the reducing H_{lk} and G_{lk} of point i will be

$$\begin{aligned} \begin{bmatrix} \bar{H}_{11} & \bar{H}_{12} \\ \bar{H}_{21} & \bar{H}_{22} \end{bmatrix}^i &= \begin{bmatrix} H_{11} & H_{12} \\ H_{21} & H_{22} \end{bmatrix}^i \begin{bmatrix} \cos\theta_i & -\sin\theta_i \\ \sin\theta_i & \cos\theta_i \end{bmatrix} \\ &+ \begin{bmatrix} H_{11} & H_{12} \\ H_{21} & H_{22} \end{bmatrix}^j \begin{bmatrix} \cos\theta_j & -\sin\theta_j \\ \sin\theta_j & \cos\theta_j \end{bmatrix} \begin{bmatrix} -1 & 0 \\ 0 & -1 \end{bmatrix} \end{aligned} \tag{58}$$

$$\begin{aligned} \begin{bmatrix} \bar{G}_{11} & \bar{G}_{12} \\ \bar{G}_{21} & \bar{G}_{22} \end{bmatrix}^i &= \begin{bmatrix} G_{11} & G_{12} \\ G_{21} & G_{22} \end{bmatrix}^i \begin{bmatrix} \cos\theta_i & -\sin\theta_i \\ \sin\theta_i & \cos\theta_i \end{bmatrix} \\ &+ \begin{bmatrix} G_{11} & G_{12} \\ G_{21} & G_{22} \end{bmatrix}^j \begin{bmatrix} \cos\theta_j & -\sin\theta_j \\ \sin\theta_j & \cos\theta_j \end{bmatrix} \begin{bmatrix} -1 & 0 \\ 0 & -1 \end{bmatrix} \end{aligned} \tag{59}$$

The boundary element equations are

$$\begin{bmatrix} \bar{H}_{11} & \bar{H}_{12} \\ \bar{H}_{21} & \bar{H}_{22} \end{bmatrix}^i \begin{Bmatrix} u_t \\ u_n \end{Bmatrix}^i = \begin{bmatrix} \bar{G}_{11} & \bar{G}_{12} \\ \bar{G}_{21} & \bar{G}_{22} \end{bmatrix}^i \begin{Bmatrix} p_t \\ p_n \end{Bmatrix}^i \tag{60}$$

If we divide the boundary into n segments, The node number is n, too. The number of the equal boundary conditions is m. So the number of reduced equations will be $2(n-m)$. Solving these equations, we will get the unknown components of the displacements and stresses on the boundary.

NUMERICAL RESULTS

Calculating a hollow disk as shown in Fig. 4, we only deal with a 30° sector of the disk with the inclined boundary condition. The shape parameters are : θ=30°, r_a=10mm, r_b=100mm; the physical parameters are given in the tables. The exact solutions of displacements and stresses can be found in every text book of Elasticity Mechanics.

Table 1 and Table 2 show the relative errors of displacement u_r and stress σ_θ of the rotating hollow disk respectively.

$$E = 16000Mpa, \nu = 0.3, \rho = 8.01g/cm^3, \omega = 58000r.p.m.$$

Nodal number of Element A = 40 , Nodal number of Element B = 52

Unit : R - mm , θ - Degree

Table 1. Relative Errors of Displacement u_r

R , θ	55,0	55,30	100,15	10,15	55,15
Element A	3.15	3.16	3.23	3.41	2.84
Element B	2.92	2.92	3.05	3.32	2.75

Table 2. Relative Errors of Stress σ_θ

R , θ	55,0	55,30	100,15	10,15	55,15
Element A	3.14	2.46	4.47	3.31	5.74
Element B	2.82	2.25	4.02	3.08	4.83

CONCLUSIONS

From the above numerical results, we can see the errors decrease when the nodal numbers of elements increase. The boundary conditions are reasonable to the computation. The new type of boundary integral equations provide better numerical results on the case that boundary displacements are given than the ones on the case that the boundary tractions are given. The new type of boundary integral equations have great potentialities. It is worth extending the applicable regions of the new method.

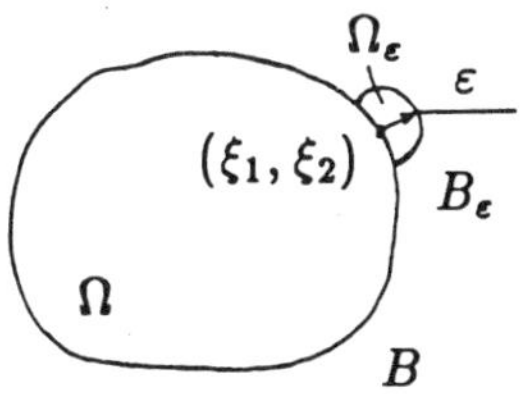

Fig. 1

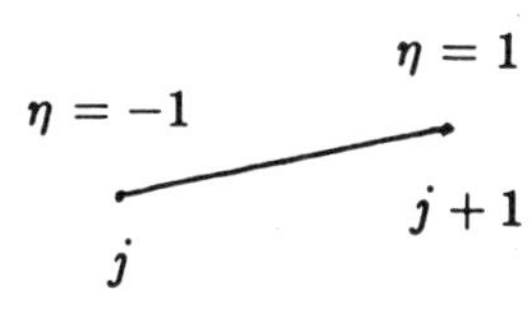

Fig. 2

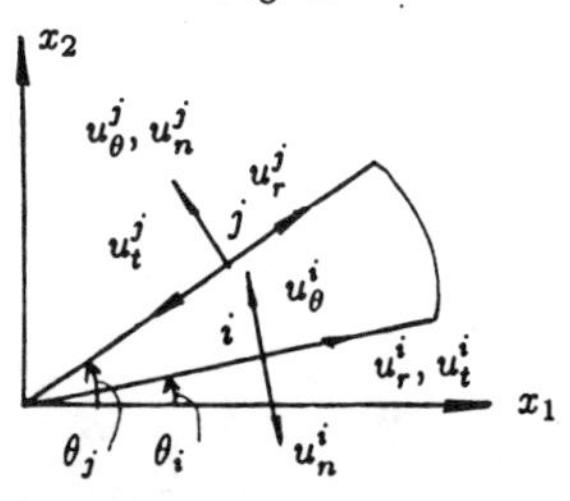

Fig. 3

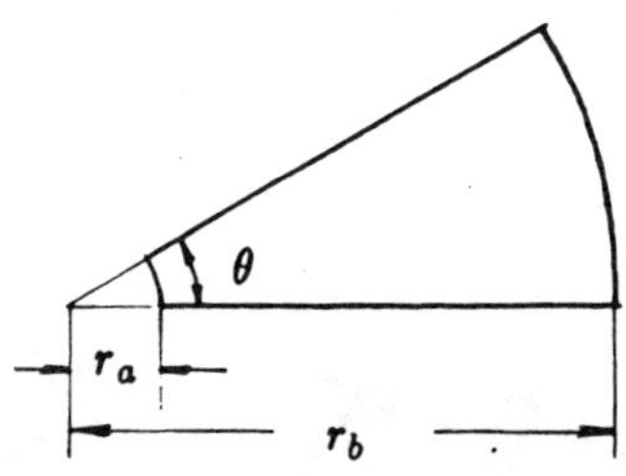

Fig. 4

REFERENCES

[1]. Hu Haichang, *A New Type of Boundary Integral Equation in Elasticity*, Scientia Sinica(Science in China), Series A , **Vol. XXX, No. 4** (April, 1987), pp. 385-390.
[2]. Brebbia, C.A. & Connor, J.J., *Advances in Boundary Elements*, Proceddings of the Eleventh International Conference on Boundary Element Methods, Springer-Verlag (1989),.
[3]. Rizzo, F.J., *An Integral Equation Approch to Boundary Value Problems of Classical Elastostatics*, Quarterly applied Math. , **Vol. 25, No. 1** (1967), pp. 83-95.
[4]. Cruse, T.A., *Numerical Solution in Three Dimensional Elastostatics*, Inter. Jour. of Solid Structure , **Vol. 5, No. 2** (1969), pp. 1259-1274.
[5]. Shen Xinzhuo & Hu Haichang, *A New Type of Boundary Integral Equation in Plane Problem of Elasticity*, Proceedings of The 2nd China-Japan Symposium on Boundary Element Methods, Tsinghua University Press, Beijing (Oct., 1988), pp. 281-288.

AUTHORS' INDEX

Adey R.A. 253
Aliabadi M.H. 381, 393
Alujevic A. 3
Anza J.J. 279

Bains R. 393
Bamps N. 89
Bartels F. 147
Baynham J. 253
Beswick G.E. 321
Bonnet M. 185
Brebbia C.A. 201, 295
Brown D.K. 29
Bui T.D. 339

Cabral J.J.S.P. 135
Carrer J.A.M. 265
Carte I.N. 51
Cartwright D.J. 357
Castro L.M. 215

DeFigueiredo T. 295
Delincé F. 77, 89
De Mey G. 39
De Smet S. 39
Driscart M. 39
Dular P. 77, 89

Genon A. 77, 89

Hayami K. 335
Hedia H. 77
Homm A. 147
Honma T. 65

Igarashi H. 65

Jiang Y.S. 227
Jin X. 29

Karamanoglu M. 321
Kassem M. 371
Kost A. 101

Legros W. 77, 89
Li Z.C. 339
Lin F. 161
Longo A. 279

Mansur W.J. 117
Mason J.C. 309

Neves A.C. 201
Nicolet A. 77, 89
Niku S.M. 253

Parreira P. 215
Pinho J. 215
Portela A. 381

Ramalho M.A. 173
Ribeiro G.O. 239
Rooke D.P. 381, 393

Santiago J.A.F. 117
Schaeidt E. 279
Schmitt H. 13
Skerget P. 3
Stark R.D. 357

Telles J.C.F. 117, 265

Unzueta J. 279

Venturini W.S. 173, 239
von Estorff O. 147

Wood L.A. 161
Wrobel L.C. 135
Wu Y.L. 407

Xie J.F. 407

Zagar I. 3